Basic
Electronics
A Text-Lab Manual

Other Books by Paul B. Zbar

Basic Electricity: A Text-Lab Manual
Basic Television: Theory and Servicing, A Text-Lab Manual (with P. Orne)
Electricity-Electronics Fundamentals: A Text-Lab Manual (with J. Sloop)
Electronic Instruments and Measurements: A Text-Lab Manual
Industrial Electronics: A Text-Lab Manual

Other Books by Albert P. Malvino

Resistive and Reactive Circuits
Transistor Circuit Approximations
Electronic Principles
Electronic Instrumentation Fundamentals
Digital Principles and Applications (with D. Leach)
Digital Computer Electronics: An Introduction to Microcomputers
Experiments for Transistor Circuit Approximations
Experiments for Electronic Principles (with G. Johnson)

Paul B. Zbar
Albert P. Malvino

Basic Electronics
A Text-Lab Manual

Fifth Edition

Gregg Division
McGraw-Hill Book Company

New York   Atlanta   Dallas   St. Louis
San Francisco   Auckland   Bogotá
Guatemala   Hamburg
Lisbon   London   Madrid   Mexico
Montreal   New Delhi   Panama
Paris   San Juan   São Paulo   Singapore
Sydney   Tokyo   Toronto

Sponsoring Editor: Paul Berk
Editing Supervisor: Evelyn Belov
Design and Art Supervisor: Caryl Valerie Spinka
Production Supervisor: Priscilla Taguer

Text Designer: Jorgé Hernandez
Cover Designer: Fine Line, Inc.

Library of Congress Cataloging in Publication Data

Zbar, Paul B.,
  Basic electronics.

  (The Basic electricity-electronics series)
  1. Electronics. I. Malvino, Albert Paul.
II. Title.   II. Series.
TK7816.Z33   1983   621.381′076   82-24900
ISBN 0-07-072803-8

Basic Electronics: A Text-Lab Manual, Fifth Edition

    890 SEM SEM 890987
ISBN 0-07-072803-8

CONTENTS

NOTE ON EXPERIMENT CONTENT

Each of the experiments described below is set up in the following manner:

OBJECTIVES   The objectives are enumerated and clearly stated.

INTRODUCTORY INFORMATION   The theory and basic principles involved in the experiment are clearly stated.

SUMMARY   A summary of the salient points is given.

SELF-TEST   A self-test, based on the material included in Introductory Information, helps the student evaluate his understanding of the principles covered, prior to the experiment proper. The self-test should be taken before the experiment is undertaken. Answers to the self-test questions are given at the end of each experiment.

MATERIALS REQUIRED   All the materials required to do the experiment—including test equipment and components—are enumerated.

PROCEDURE   A detailed step-by-step procedure is given for performing the experiment.

QUESTIONS   The conclusions reached by the student are brought out by a series of pertinent questions.

EXTRA CREDIT   Design problems and questions are included for the more advanced student.

EXPERIMENTS

SERIES PREFACE

Electronics is at the core of a wide variety of specialized technologies which have been developing over several decades. Challenged by a rapidly expanding technology and the need for increasing numbers of technicians, the Consumer Electronics Group Service Committee of the Electronic Industries Association (EIA), in association with the Voorhees Technical Institute, Oklahoma State University, and various publishers, has been active in creating and developing educational materials to meet these challenges.

In recent years, a great many consumer electronic products have been introduced and the traditional radio and television receivers have become more complex. As a result, the pressing need for training programs to permit students of various backgrounds and abilities to enter this growing industry has induced EIA to sponsor the preparation of an expanding range of materials. Three branches of study have been developed in two specific formats. The tables list the books in each category; the paragraphs following them explain these materials and suggest how best to use them to achieve the desired results.

The Basic Electricity-Electronics Series

Title	Author	Publisher
Electricity-Electronics Fundamentals	Zbar/Sloop	Gregg/McGraw-Hill Book Company
Basic Electricity	Zbar	Gregg/McGraw-Hill Book Company
Basic Electronics	Zbar/Malvino	Gregg/McGraw-Hill Book Company

The laboratory text-manuals in the Basic Electricity-Electronics Series provide in-depth, detailed, completely up-to-date technical material by combining a comprehensive discussion of the objectives, theory, and underlying principles with a closely coordinated program of experiments. *Electricity-Electronics Fundamentals* provides an introductory course especially suitable for preparing service technicians; it can also be used for other broad-based courses. *Basic Electricity* and *Basic Electronics* are planned as 270-hour courses, one to follow the other, providing a more thorough background for all levels of technician training. A related instructor's guide is available for each course.

The Radio-Television-Audio Servicing Series

Title	Author	Publisher
Television Symptom Diagnosis— An Entry into TV Servicing	Tinnell	Howard W. Sams & Co., Inc.
Television Symptom Diagnosis (audiovisual materials)	Tinnell	Howard W. Sams & Co., Inc.
Television Servicing with Basic Electronics	Sloop	Howard W. Sams & Co., Inc.
Advanced Color Television Servicing	Sloop	Howard W. Sams & Co., Inc.
Audio Servicing — Theory and Practice	Wells	Gregg/McGraw-Hill Book Company
Audio Servicing—Text-Lab Manual	Wells	Gregg/McGraw-Hill Book Company
Basic Radio: Theory and Servicing	Zbar	Gregg/McGraw-Hill Book Company
Basic Television: Theory and Servicing	Zbar/Orne	Gregg/McGraw-Hill Book Company

The Radio-Television-Audio Servicing Series includes materials in two categories: those designed to prepare apprentice technicians to perform in-home servicing and other apprenticeship functions, and those designed to prepare technicians to perform more sophisticated and complicated servicing such as bench-type servicing in the shop.

The two titles in the apprenticeship servicing course are *Television Symptom Diagnosis—An Entry into TV Servicing* (text, student workbook, instructor's guide) and *Television Symptom Diagnosis*, a series of 33 film loops. The first is a set consisting of a well-illustrated text, a student response manual, and an instructor's guide—all designed for people with no previous electronics training—to provide them with job-entry troubleshooting skills for servicing in the home and shop. The text utilizes the "cue-response" concept of diagnosis, concentrates on identifying abnormal circuit operation and symptom analysis, and develops skills in troubleshooting. In using the response manual, students are exposed to hundreds of television trouble symptoms through color photos and illustrated problems. The instructor's guide is a complete and essential professional course of study that also contains the answers to questions and problems in the text and lab manual.

Television Symptom Diagnosis consists of a series of color-sound motion picture film loops or slides, a student workbook, and an instructor's guide. These audiovisual materials provide integrated learning systems for color-television adjustment and setup procedures, trouble-symptom diagnosis, and the ability to isolate troubles to a given stage in the receiver, concentrating on the requirements for servicing in the customer's home. But these audiovisual materials can also be used to supplement all levels of television courses. This medium is especially suitable for students who may have reading and, in turn, learning difficulties.

Since only a minimum of electronics theory is presented in the two courses described above, it is expected that apprentices completing these programs will be motivated to progress to more comprehensive programs in order to deepen their understanding of electronics, to really know what makes the radio and television receiver work, and to become proficient in servicing all consumer electronics entertainment items. These in-depth studies are provided in the following materials.

The intermediate *Television Servicing with Basic Electronics* (text, student workbook, instructor's guide) goes beyond the basics and expands on the math and use of test equipment introduced in the beginning text. The book continues the diagnostic troubleshooting method.

Advanced bench-type diagnosis servicing techniques are covered in *Advanced Color Television Servicing* (text, student workbook, instructor's guide). Written primarily for color television servicing courses in schools and in industry, this set follows the logical diagnostic troubleshooting approach consistent with the manufacturers' approach to bench servicing.

Audio Servicing (theory and practice, text-lab manual, and instructor's guide) covers each component of a modern home stereo with an easy-to-follow block diagram and a diagnosis approach consistent with the latest industry techniques.

The bench-type service technician courses consist of *Basic Radio: Theory and Servicing* and *Basic Television: Theory and Servicing*. These books provide a series of experiments, with preparatory theory, designed to provide the in-depth, detailed training necessary to produce skilled radio-television service technicians for both home and bench servicing of all types of radio and television. A related instructor's guide for these books is also available.

The Industrial Electronics Series

Title	Author	Publisher
Industrial Electronics	Zbar	Gregg/McGraw-Hill Book Company
Electronic Instruments and Measurements	Zbar	Gregg/McGraw-Hill Book Company

Basic laboratory courses in industrial control and computer circuits and laboratory standard measuring equipment are provided by the Industrial Electronics Series and their related instructor's guides. *Industrial Electronics* is concerned with the fundamental building blocks in industrial electronics technology, giving the student an understanding of the basic circuits and their applications. *Electronic Instruments and Measurements* fills the need for basic training in the complex field of industrial instrumentation. Prerequisites for both courses are *Basic Electricity* and *Basic Electronics*.

The foreword to the first edition of the EIA cosponsored basic series states: "The aim of this basic instructional series is to supply schools with a well-integrated, standardized training program, fashioned to produce a technician tailored to industry's needs." This is still the objective of the varied training program that has been developed through joint industry-educator-publisher cooperation.

Peter McCloskey, President
Electronic Industries Association

PREFACE

The fifth edition of *Basic Electronics: A Text-Lab Manual* is an updated and expanded text-laboratory manual which emphasizes self-learning. It is intended for use in vocational-technical schools, community colleges, and industrial training programs. This manual provides a complete and laboratory-tested program in semiconductor and integrated circuits. As in previous editions, emphasis is on the scientific method, analysis, and logical deduction.

The 55 experiments provide students with intensive experience and procedures which they will find in industry. Many of the topics covered in the fourth edition appear in this fifth edition; these topics include diode circuits, power-supply circuits, transistor biasing, transistor amplifiers, troubleshooting, JFETs, MOSFETs, SCRs, UJTs, and basic digital circuits.

In addition, you will find new experiments covering optoelectronic devices, negative feedback, op-amp circuits, active filters, voltage regulators, IC oscillators, 555 timers, phase-locked loops, mixers, modulators, demodulators, and flip-flops.

The authors wish to thank the members of the Service Education Subcommittee of the Consumer Electronics Group of the Electronic Industries Association for their guidance and help in their review of the manuscript: Jack Berquist (N.A.P.), Greg Carey (Sencore), William Dugger (VPI), Frank Hadrick (Zenith), Charles Howard (Quasar), Gene Jadwin (Sharp), Tom Matterness (Motorola), Paul Neilsen (Dynascan), Irv Rebeschini (Simpson), Dom Sabatini (Panasonic), and Steve Zell (RCA).

Acknowledgment is also made for permission to use equipment photographs and component data sheets to: B & K Company; CBS; Fairchild Semiconductor Co.; General Electric Co.; GTE Sylvania; Hickok Teaching Systems, Inc.; Litronix Co.; Minneapolis-Honeywell Regulator Co.; Motorola Semiconductor Products, Inc.; Radio Corporation of America; Symphonic Radio Electronic Corp.; Tektronix Co.

Paul B. Zbar
Albert P. Malvino

SAFETY

Electronics technicians work with electricity, electronic devices, motors, and other rotating machinery. They are often required to use hand and power tools in constructing prototypes of new devices or in setting up experiments. They use test instruments to measure the electrical characteristics of components, devices, and electronic systems. They are involved in any of a dozen different tasks.

These tasks are interesting and challenging, but they may also involve certain hazards if technicians are careless in their work habits. It is therefore essential that the student technicians learn the principles of safety at the very start of their career and that they practice these principles.

Safe work requires a careful and deliberate approach to each task. Before undertaking a job, technicians must understand what to do and how to do it. They must plan the job, setting out on the workbench in a neat and orderly fashion, tools, equipment, and instruments. Extraneous items should be removed, and cables should be securely fastened.

When working on or near rotating machinery, loose clothing should be anchored, ties firmly tucked away.

Line (power) voltages should be isolated from ground by means of an isolation transformer. Powerline voltages can kill, so these should *not* come in contact with the hands or body. Line cords should be checked before use. If the insulation on line cords is brittle or cracked, these cords must *not* be used. TO THE STUDENT: Avoid direct contact with any voltage source. Measure voltages with one hand in your pocket. Wear rubbersoled shoes or stand on a rubber mat when working at your experiment bench. Be certain that your hands are dry and that you are not standing on a wet floor when making tests and measurements in a live circuit. Shut off power before connecting test instruments in a live circuit.

Be certain that line cords of power tools and nonisolated equipment use safety plugs (polarized 3-post plugs). Do not defeat the safety feature of these plugs by using ungrounded adapters. Do not defeat any safety device, such as fuse or circuit breaker, by shorting across it or by using a higher amperage fuse than that specified by the manufacturer. Safety devices are intended to protect you and your equipment.

Handle tools properly and with care. Don't indulge in horseplay or play practical jokes in the laboratory. When using power tools, secure your work in a vise or jig. Wear gloves and goggles when required.

Exercise good judgment and common sense and your life in the laboratory will be safe, interesting, and rewarding.

FIRST AID

If an accident should occur, shut off the power immediately. Report the accident at once to your instructor. It may be necessary for you to give emergency care before a physician can come, so you should know the principles of first aid. You can learn the basics by taking a Red Cross first-aid course.

Some first-aid suggestions are set forth here as a simple guide.

Keep the injured person lying down until medical help arrives. Keep the person warm to prevent shock. Do not attempt to give water or other liquids to an unconscious person. Be sure nothing is done to cause further injury. Keep the injured one comfortable until medical help arrives.

ARTIFICIAL RESPIRATION

Severe electric shock may cause someone to stop breathing. Be prepared to start artificial respiration at once if breathing has stopped. The two recommended techniques are:

1. Mouth-to-mouth breathing, considered more effective
2. Schaeffer method

These techniques are described in first-aid books. You should master one or the other so that if the need arises you will be able to save a life by applying artificial respiration.

These safety instructions should not frighten you but should make you aware that there are hazards in the work of an electronics technician—as there are hazards in every job. Therefore you must exercise common sense and good judgment, and maintain safe work habits in this, as in every other job.

LETTER SYMBOLS

As noted in the Author's Preface, primary emphasis in this manual has been placed on semiconductor (solid-state) devices and circuits. However, vacuum tubes and their associated circuits are also treated, making it desirable to use letter symbols that have the same meaning throughout the text for both solid-state and vacuum-tube circuits. Accordingly, the IEEE (Institute of Electrical and Electronics Engineers) Letter Symbols for Semiconductor Devices (IEEE Standard #255) were used, with modifications for vacuum tubes.

The following summary of symbols for electrical quantities is intended to clarify their use throughout the text.

Quantity Symbols

1. Instantaneous values of current, voltage, and power, that vary with time, are represented by the lowercase letter of the proper symbol.
   Examples: i, v, p
2. Maximum (peak), average (direct current), and root-mean-square values of current, voltage, and power are represented by the uppercase letter of the appropriate symbol.
   Examples: I, V, P

Subscripts for Quantity Symbols

1. Direct-current values and instantaneous total values are indicated by uppercase subscripts.
   Examples: i_C, I_C, v_{EB}, V_{EB}, p_C, P_C

2. Alternating-component values are indicated by lowercase subscripts.
   Examples: i_c, I_c, v_{eb}, V_{eb}, p_c, P_c
3. Symbols to be used as subscripts:
   E, e emitter terminal
   B, b base terminal
   C, c collector terminal
   A, a anode terminal
   K, k cathode terminal
   G, g grid terminal
   P, p plate terminal
   M, m maximum value
   Min, min   minimum value

   Examples:
   I_E emitter direct-current (no alternating current component)
   I_e rms value of alternating component of emitter current
   i_e instantaneous value of alternating component of emitter current
4. Supply voltages may be indicated by repeating the terminal subscript.
   Examples: V_{EE}, V_{CC}, V_{BB}, V_{PP}, V_{GG}
   The one exception to this system is the occasional use of $V+$ for the plate supply voltage of a tube. Note that $V+$ replaces the more usual $B+$.
5. The first subscript designates the terminal at which current or voltage is measured with respect to the reference terminal, which is designated by the second subscript.

EXPERIMENT

JUNCTION-DIODE CHARACTERISTICS

OBJECTIVES

1. To measure the effects of forward and reverse bias on current in a junction diode
2. To determine experimentally and graph the voltampere characteristic of a junction diode
3. To test a junction diode with an ohmmeter

INTRODUCTORY INFORMATION

Semiconductors

Semiconductors are solids whose resistivity lies between those of electrical conductors and insulators. Transistors, junction diodes, zener diodes, tunnel diodes, integrated circuits, and metallic rectifiers are examples. Semiconductors are used in computers, in radio and TV receivers, and in other electronic products.

Semiconductor devices perform many control functions. They may be used as rectifiers, amplifiers, detectors, oscillators, and switching elements. Some characteristics which make the semiconductor such an attractive member of the electronics family are as follows.

1. Semiconductors are small and light in weight, which permits miniaturization of electronic equipment. Figure 1-1 shows the size of an early transistor.
2. Recent developments have led to microminiaturization. Figure 1-2 shows an integrated circuit (IC) "chip" (microminiature module) containing circuits comprising transistors, resistors, capacitors, wiring, and contacts. This module can do the work previously performed by a vacuum-tube device whose dimensions were about 100 times as large.

Advances in the technology of integrated circuits have brought ICs far beyond the device shown in Fig. 1-2. There are now large-scale integrated circuits (LSICs) which incorporate in one small chip hundreds of transistors, diodes, and resistors. And the end of microminiaturization is not yet in sight.

3. Semiconductors are solids. There is therefore little chance that elements will vibrate. Element vibration in vacuum tubes was the cause of troublesome microphonics.
4. Semiconductors require little power and radiate less heat than tubes. They do not need warmup time and operate as soon as power is applied.
5. Semiconductors are rugged and may be made impervious to external environmental conditions.
6. Semiconductors do not undergo the chemical deterioration which occurs in tube cathodes. The deterioration of tube cathodes eventually results in unacceptable tube performance.

Semiconductor Materials and Impurities

Silicon is the material of which most semiconductor devices are presently constructed. Germanium was used initially in the manufacture of transistors and junction diodes. Today, however, germanium devices are but a small percentage of semiconductor output. Silicon, which is less heat-sensitive, predominates.

Germanium and silicon must be highly purified before they can be processed into effective semiconductor mate-

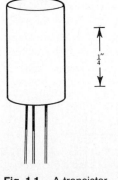

Fig. 1-1. A transistor.

Fig. 1-2. A microminiature integrated circuit (IC) module.

rials. In their pure state these semiconductors have a very low conductivity; that is, their resistivity is high. The conductivity of germanium and silicon may be increased by adding very minute amounts of certain "impurities." The addition of controlled quantities and types of impurities, called *doping*, alters the electron-bond structure within the atoms of these elements and provides them with current carriers, increasing their conductivity.

Current Carriers in a Semiconductor

In a vacuum tube, negatively charged electrons are considered the current "carriers." This concept of negative charge carriers must be modified by the addition of positive charge carriers to explain current flow in semiconductor diodes and transistors. Positive charge carriers are called *holes*. These holes are considered to have mass, mobility, and velocity. Current flow in semiconductors is carried on by the flow of negative charges (free electrons) and positive charges (holes). It is beyond the scope of this material to consider the physics of holes. We shall, however, be concerned with the current resulting from the movement of free electrons and holes and with the control of this current.

Impurities such as arsenic and antimony increase the conductivity of silicon by increasing the number of negative (N) charge carriers (free electrons). For this reason, silicon which has been doped with arsenic or antimony is called *N type*. Some holes exist in N-type silicon, but these are in the minority and hence are called *minority carriers*. Current flow in N-type silicon may be considered as carried on by free electrons, the *majority carriers*.

Impurities such as indium and gallium increase the conductivity of silicon by increasing the number of positive (P) charge carriers (holes). Silicon which has been doped with indium or gallium is designated *P type*. Some free electrons exist in P-type silicon, but these are minority carriers. Current flow in P-type silicon may be considered as carried on by holes, the majority carriers.

Holes have an attraction for free electrons. When a free electron and a hole "meet," the free electron "fills" the hole, neutralizing its charge. The free electron is said to combine with the hole. In this process, both the hole and the free electron are lost as current carriers. While this happens, however, new current carriers are being formed at other points in the semiconductor.

The movement of current carriers may be controlled by applying an external battery voltage V_{AA} across the semiconductor (Fig. 1-3). Holes in the P-type silicon are repelled by the positive terminal of V_{AA} and move toward the negative terminal. Free electrons enter the silicon from the negative terminal of V_{AA} and move toward the holes. Combinations of free electrons and holes take place. While these combinations are being formed, additional mobile electrons and holes are liberated in the silicon from an electron-hole pair. The liberated electrons move toward the positive battery terminal and the holes toward the negative battery terminal. Recombinations and liberations continue to take place; thus a constant current flow in the *external* circuit is maintained.

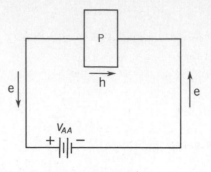

Fig. 1-3. Movement of free electrons and holes is controlled by connecting a battery to a doped silicon crystal.

Operation of a Semiconductor Junction Diode

When P- and N-type silicon are joined as in Fig. 1-4, a junction diode is created. This two-element device has a unique characteristic: the ability to pass current readily in one direction but not in the other.

Let us apply the theory of negative and positive carriers to this diode in an attempt to explain this characteristic. Consider first the effect of connecting a battery V_{AA} across this diode with the polarity shown in Fig. 1-5. Free electrons enter the N-type silicon at the negative terminal of V_{AA}. These in turn repel the free electrons in the N-type silicon, and these free electrons move toward the PN junction. The holes in the P-type silicon are repelled by the positive terminal of the battery and also move toward the PN junction, where combination of free electrons and holes takes place. The current carriers lost in these combinations are replaced by new current carriers resulting from separation of electron-hole pairs. The free electrons created in the P-type silicon are attracted to the positive terminal and flow in the external circuit, as shown. The process is continuous, and current flow is maintained. Moreover, if V_{AA} is increased, current flow in the diode increases.

The manner of connecting the negative battery terminal to the N-type and the positive battery terminal to the P-type silicon results in current flow and is called *forward bias*. Because current flows in this connection, the diode is said to have a low forward resistance.

The *reverse-bias* connection is shown in Fig. 1-6. The positive terminal of the battery attracts free electrons in the N-type silicon away from the PN junction. The negative terminal of the battery attracts the holes in the P type away from the PN junction. Hence there are no combinations of free electrons and holes. Thus the majority current carriers in the diode do not support current flow. In this reverse-bias connection, there is a minute current in the diode. This

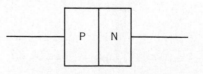

Fig. 1-4. Junction diode.

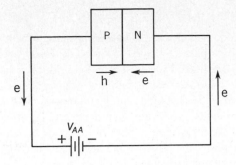

Fig. 1-5. Current flow in a junction diode, forward bias.

Fig. 1-7. Circuit symbol for a semiconductor diode.

current is due to the minority carriers, that is, the holes in the N type and free electrons in the P type. For the minority carriers, battery polarity is correct to support current flow. Only a few microamperes of current flow as a result of the minority carriers. This is shown by the dotted arrows in Fig. 1-6. The reverse-bias connection results in a high reverse resistance in the diode.

There is a limit not only to the forward bias but also to the reverse-bias voltage which may be placed across the diode. If the forward or reverse bias is increased beyond its limiting value, there is a sharp increase in forward or reverse current, respectively. This increase may permanently damage the diode.

Figure 1-7 is the circuit symbol for a semiconductor diode. The terminal marked "Anode" (identified by the arrow head) is connected to P-type material, while that marked "Cathode" is connected to N-type material. Reference to Fig. 1-5 shows that to support current flow in this diode, the positive terminal of a battery must be brought to the anode, the negative terminal to the cathode, in a forward-bias arrangement.

Forward Voltampere Characteristic

The voltampere characteristic of a diode is a graph which shows how current in that diode varies with the voltage applied across it. Experimentally, this can be determined by measuring the current in the diode for a successive number of higher applied voltages, and plotting a graph of current versus voltage. Students will experimentally determine the forward voltampere characteristic of a silicon junction diode

in this experiment and discover some interesting facts about the diode used. They will note that very little current flows in the diode for low levels of applied voltage. Thus below 0.7-volt (V) forward bias, a *silicon* diode draws little current. For forward-bias voltages equal to or higher than 0.7 V, the diode is turned on and permits current to flow. Also, beyond 0.7 V, very slight increases in forward-bias voltage result in large increases of current in the diode. A typical forward voltampere characteristic for a silicon diode is shown in Fig. 1-8.

The turn-on forward-bias voltage for silicon diodes is typically 0.7 V. For germanium diodes it is 0.3 V.

Current, in a semiconductor diode which is forward-biased, increases with an increase in anode-to-cathode voltage. But there is a limit to the amount of current which can flow safely through a diode. Beyond this limit, the diode will overheat and be destroyed.

When the diode is reverse-biased, the small current due to minority carriers remains relatively constant, that is, independent of the bias voltage, up to a certain voltage. Beyond this safe level of reverse bias, a phenomenon called *avalanche breakdown* takes place when a heavy surge of current occurs, which may also destroy the diode. The diode must therefore be operated within these two safe limits. The limits of safe operation will normally be specified by the manufacturer under the headings maximum forward voltage (V_{FM}) and maximum reverse voltage (V_{RM}). Peak forward current (I_{FM}) may also be specified.

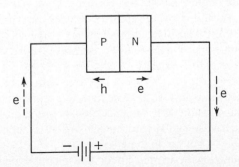

Fig. 1-6. Effect of reverse bias on junction diode.

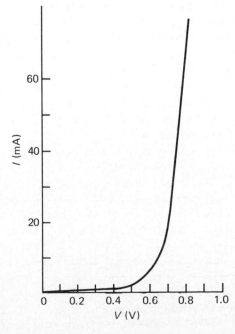

Fig. 1-8. Forward voltampere characteristic of a silicon junction diode.

Junction Diode As a Switch

Once a junction diode is turned on it appears to act like a closed switch. Current in the circuit containing the diode then seems to be limited only by the external resistance in the circuit. A reverse-biased diode does not permit current flow; it acts like an open switch.

The switch analogy is only approximately true. Consider a closed switch. The resistance measured across the contacts of a closed switch is 0, whereas a turned-on diode has a measurable forward resistance R_F. It is true that R_F is small and may be neglected in many applications, but it does exist.

Now consider an open switch. The resistance measured across the contacts of an open switch is infinitely high, and an open switch does not permit current flow through it. A reverse-biased junction diode, however, does permit some current through it. Hence, though its reverse resistance R_R is very high, it is not infinite.

However, as an approximation, it is frequently useful to compare the action of a junction diode to the operation of a switch.

Testing a Semiconductor Diode with an Ohmmeter

A resistance check may be used as a rough test of a semiconductor diode's operation. Recall that the polarity of the terminals of the battery contained within an ohmmeter appears at the leads of the ohmmeter. In Fig. 1-9, lead A is positive, lead B negative. An ohmmeter test of a diode which is operating normally will reveal that the diode has a low forward resistance and a high back (reverse) resistance. Thus, if the positive ohmmeter lead (A in Fig. 1-9) is connected to the anode of a diode, and the negative lead (B) to the cathode, the diode will be forward-biased. Current will flow, and the diode will measure low resistance. On the other hand, if the ohmmeter leads are reversed, the diode will be reverse-biased. Very little current will flow, and the diode will measure a very high resistance. If a semiconductor diode exhibits a very low forward and a low reverse resistance, it is probably damaged (fused). On the other hand, an open diode is indicated if the forward resistance is unusually high or infinite.

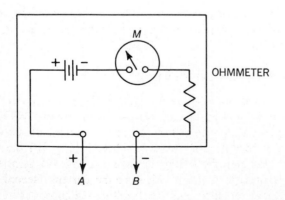

Fig. 1-9. Polarity of ohmmeter leads.

Identifying the Anode and Cathode of a Diode

The cathode end of a diode is usually marked by a circular band or by a plus (+) sign. If the diode is unmarked, it is simple to determine by a resistance check which is the anode, which the cathode. First, the polarity of the ohmmeter leads is determined by checking with a voltmeter across the ohmmeter terminals. Then the ohmmeter-lead position which measures the forward resistance of the diode is determined. In this position, the positive ohmmeter lead is connected to the anode, the negative lead to the cathode.

Low-power-ohms Function of an Ohmmeter

The battery in a nonelectronic ohmmeter, like that in Fig. 1-9, is 1.5 V or higher. Hence it can forward bias a silicon junction diode beyond the 0.7 V required for conduction. Similarly, it can forward bias a germanium junction diode beyond the 0.3 V needed for conduction. That is why it is possible to make the previously discussed ohmmeter checks of semiconductor diodes. In troubleshooting some semiconductor circuits, however, low-power (LP) electronic ohmmeters are used whose lead voltage is lower than 0.7 or 0.3 V. The low-power-ohms (LPΩ) function of this type of ohmmeter *cannot* be used to measure the forward resistance of a diode, nor can it be used to identify the anode and cathode of a diode. Fortunately, the manufacturer provides, in addition to the *low-power-ohms* function, a *normal-ohms* function. Resistance tests of a semiconductor diode may be made using the *normal-ohms* function of the meter.

SUMMARY

1. Semiconductors are used to control current in electronics. They are preferable to vacuum tubes, which have a similar control function, because they are much smaller, consume less power, and allow microminiaturization of electronic devices and circuits.
2. Basic semiconductor materials are silicon and germanium. Silicon is much more widely used than germanium.
3. In their *pure* state silicon and germanium are insulators.
4. When silicon and germanium are *doped* with certain impurities ("dopants"), their resistivity decreases and they become *semi*conductors.
5. After doping, silicon and germanium become either N- (negative current carriers—electrons) type semiconductors, or P- (positive current carriers—holes) type semiconductors. Whether they become N or P depends on the nature of the impurity.
6. When pieces of N-type and P-type silicon are joined together, as in Fig. 1-4, we have a *junction diode*.
7. A junction diode has unidirectional current characteristics; that is, it will permit current to flow through it in one direction but not in the other.
8. A junction diode must be forward-biased to permit current flow through it. To forward-bias a PN junction, the

positive terminal of a power source must be connected to the P-type and the negative terminal to the N-type semiconductor.

9. When the power terminals are reversed—when the positive power terminal is connected to the N-type and the negative power terminal to the P-type semiconductor—the junction diode is reverse-biased and will not permit current flow.

10. Current flow in a junction diode is supported by the movement of electrons and holes.

11. There is a limit to the maximum forward and maximum reverse voltage which may be placed across a junction diode. Operation beyond these limits may destroy the diode by overheating due to excessive current flow.

12. The turn-on point for a silicon diode is 0.7 V; that is, the diode must be forward-biased at least 0.7 V before it will conduct appreciably.

13. The turn-on point for a germanium diode is 0.3 V.

14. Current, in a forward-biased diode, increases with an increase in voltage across it.

15. A junction diode acts like a *closed* switch *after* it is turned on and like an *open* switch *before* it is turned on. However, this analogy is only approximate, because a junction diode does have a measurable forward resistance, and a high but not infinite reverse resistance.

16. A junction diode may be *ohms*-tested by an ohmmeter. When the leads are connected across the diode so that it is forward-biased, the meter shows the low forward resistance (R_F) of the diode. When the leads are reversed, the meter measures the high reverse resistance (R_R).

17. A junction diode *cannot* be ohms-tested on the *low-power-ohms* function of an electronic voltmeter.

SELF-TEST

Check your understanding by answering these questions.

1. The most common semiconductor material is
   _____ .

2. Germanium and silicon in their pure form are
   _____ . (*conductors, insulators*)

3. In silicon which has been doped with impurities such as arsenic, there is an increased number of
   _____ (*negative, positive*) charge carriers and the material is _____ (*N, P*) type.

4. A junction diode may be compared with a resistor because it permits current to flow through it in either direction. _____ (*true, false*)

5. To forward-bias a junction diode, connect the
   _____ (*positive, negative*) lead of a battery to the P-type terminal of the diode and the _____ (*positive, negative*) lead of the battery to the N-type terminal.

6. The forward-bias voltage across a silicon diode must be equal to or greater than _____ V before the diode will conduct appreciably.

7. The voltampere characteristic of a junction diode is a _____ of voltage versus current.

8. After a diode is turned on, an increase in forward voltage across the diode will result in a(n) _____ (*increase, decrease*) of current through the diode.

9. The forward resistance of a silicon diode is
   _____ ; the reverse resistance _____ .

10. The forward resistance of a diode may be checked, approximately, with an _____ .

MATERIALS REQUIRED

- Power supply: Variable regulated low-voltage high-current dc source
- Equipment: Electronic voltmeter; VOM, 20,000-Ω/V; curve tracer
- Resistors: 250-Ω 2-W
- Silicon diode: 1N4154 (other choices: 1N914 or almost any small-signal silicon diode)
- Germanium diode: 1N34A (other choices: 1N4454 or almost any small-signal germanium diode)
- Miscellaneous: SPST (single-pole single-throw) switch

PROCEDURE

Diode Biasing

1. (a) Examine the 1N4154 silicon diode assigned to you and identify the anode and cathode terminals.
   (b) Connect the circuit of Fig. 1-10, with the diode forward-biased. V is an electronic voltmeter set to measure dc volts. M is a VOM set to measure current. Set the output of the variable dc supply so that the voltage V_{AK} *across the diode* measures 0.7 V. Measure the diode current. Record the results in Table 1-1.
   (c) Reverse the diode in the circuit. Readjust the power supply, if necessary, until the voltage V_{AK} across the diode measures 1.5 V. Measure the diode current. Record the results in Table 1-1.

(d) Applying Ohm's law, compute the diode resistance when it is forward- and reverse-biased. Record the results in Table 1-1.

Voltampere Characteristic

2. Reverse the diode in the circuit of Fig. 1-10 so that it is again forward-biased. Set the output of the regulated power supply so that the voltage V_{AK} across the diode is 0 V. Measure and record the current, if any, in Table 1-2.

   Increase the voltage in Table 1-2 in 0.1-V steps to a maximum of 0.8 V. Measure the current. Record the results in Table 1-2. For each condition compute and record the forward resistance of the diode.

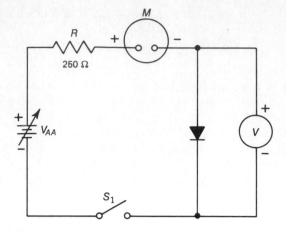

Fig. 1-10. Measuring the effect of forward bias on current flow in a diode.

3. Reverse the diode in the circuit. The diode is now reverse-biased. Measure the current, if any, as the voltage of the supply is varied in 5-V steps from 0 to 40 V and record the results in Table 1-2. For each condition, compute and record the reverse resistance of the diode.
4. Plot a graph of V versus I for both bias conditions.

Voltampere Characteristic Using a Curve Tracer

5. If you have a curve tracer, connect it according to the manufacturer's instructions. Insert the 1N4154 into the diode socket on the tracer and adjust the tracer controls for the same forward and reverse voltages as in Table 1-2.
6. Observe the voltampere characteristic and photograph or draw it on the same graph paper as in step 4. Identify the curve.

Resistance Test

7. Remove the diode from the circuit. Measure the forward and reverse resistance of this diode. Record the results in Table 1-3. Compute the back-to-forward resistance ratio r of this diode. Record the results in Table 1-3.
8. Repeat step 7 for a 1N34A germanium diode.

TABLE 1-1.

Bias	V_{AK}, V	I, mA	R (diode),
Forward	0.7		
Reverse	1.5		

TABLE 1-2.

	Forward			Reverse		
V_{AK}, V	I, mA	R, Ω		V_{AK}, V	I, mA	R, Ω
0				0		
0.1				-5		
0.2				-10		
0.3				-15		
0.4				-20		
0.5				-25		
0.6				-30		
0.7				-35		
0.8				-40		

TABLE 1-3.

Diode	R (Forward), Ω	R (Reverse), Ω	$\dfrac{R\ (Reverse)}{R\ (Forward)} = r$
1N4154			
1N34A			

QUESTIONS

1. Under what conditions will a junction diode turn on? Explain. Refer to your measurements in Table 1-2.
2. What are the limitations, if any, on (*a*) forward bias? (*b*) reverse bias? Were the limitations exceeded in this experiment? Refer to your measurements to substantiate your answer.
3. What portion, if any, of the voltampere-characteristic curve of the forward-biased diode is linear?
4. What can you say about the dc resistance of the diode over this linear portion?
5. How would you identify the anode of an unmarked diode?
6. How can you determine which lead of a nonelectronic ohmmeter is positive, which negative?
7. What is the significance, if any, of the reverse-to-forward-resistance ratio of a diode?

Answers to Self-Test

1. silicon
2. insulators
3. negative; N
4. false
5. positive; negative
6. 0.7
7. graph
8. increase
9. low; high
10. ohmmeter

JUNCTION-DIODE APPROXIMATIONS

OBJECTIVES

1. To analyze circuits with the ideal-diode approximation
2. To improve the analysis with higher approximations

INTRODUCTORY INFORMATION

The dc resistance of a diode is the total diode voltage divided by the total diode current. In the preceding experiment you worked with the dc resistances of forward- and reverse-biased diodes. Dc resistances are useful for ohmmeter checks of a diode.

For diode circuits driven by ac signals, we sometimes use the three approximations discussed in this experiment. These diode approximations are important because they are also used in transistor-circuit analysis.

First Approximation

What does a diode do? It conducts well in the forward direction and poorly in the reverse direction. Boil this down to its essence, and this is what you get: Ideally, a diode acts like a perfect conductor (zero voltage) when forward-biased and like a perfect insulator (zero current) when reverse-biased (see Fig. 2-1a).

In terms of an ideal circuit element, a diode is like a switch that is closed during forward bias (Fig. 2-1b) and open during reverse bias (Fig. 2-1c). This first approximation of a diode is called the *ideal diode*. It gives us a quick and easy way to analyze diode circuits.

For instance, the diode of Fig. 2-2a is forward-biased. To a first approximation, it acts like a closed switch. Therefore, the current through the diode is

$$I = \frac{10 \text{ V}}{2 \text{ k}\Omega} = 5 \text{ mA} \qquad (2\text{-}1)$$

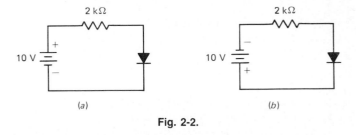

Fig. 2-2.

On the other hand, the diode of Fig. 2-2b is reverse-biased. Ideally it appears as an open switch; therefore, the current through the circuit is 0.

Second Approximation

We need about 0.7 V before a silicon diode really conducts well. When we have a large source voltage, this 0.7 V is too small to matter. But when the source voltage is not large, we may want to take the 0.7 V into account.

Figure 2-3a shows the graph for the second approximation. The graph says no current flows until 0.7 V appears across the diode. At this point the diode turns on. No matter what the forward current, we allow only a 0.7-V drop across a silicon diode. (Use 0.3 V for germanium diodes.) Incidentally, the 0.7 V is known as the *offset* voltage or the *knee* voltage.

Figure 2-3b is the equivalent circuit for the second approximation. Think of the diode as a switch in series with a 0.7-V battery. If the source voltage driving the diode exceeds the offset voltage, then the switch is closed and the diode voltage equals 0.7 V.

As an example, let us use the second approximation for the diode of Fig. 2-2a. The source is large enough to overcome the knee voltage. Therefore, the diode is forward-biased and the current is

$$I = \frac{10 \text{ V} - 0.7 \text{ V}}{2 \text{ k}\Omega} = 4.65 \text{ mA} \qquad (2\text{-}2)$$

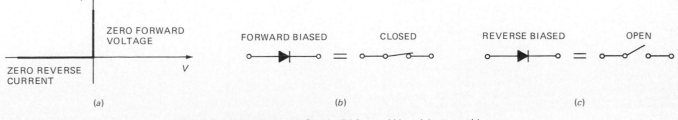

Fig. 2-1. Ideal diode. *(a)* Graph; *(b)* forward bias; *(c)* reverse bias.

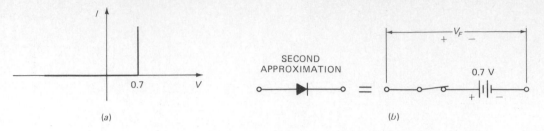

(a)

(b)

Fig. 2-3. Second approximation. *(a)* Graph; *(b)* equivalent circuit for forward bias.

If the diode is reverse-biased as shown in Fig. 2-2b, the second approximation still results in zero current.

Bulk Resistance

Before discussing the third approximation, we need to define the bulk resistance of a diode. Above the knee voltage, diode current increases rapidly; small increases in diode voltage cause large increases in diode current. The reason is this: After overcoming the offset voltage, all that impedes diode current is the resistance of the P and N regions, symbolized by the r_P and r_N in Fig. 2-4a. The sum of these resistances is called the *bulk resistance* of the diode. In symbols,

$$r_B = r_P + r_N \qquad (2\text{-}3)$$

The value of bulk resistance depends on the doping and the size of the P and N regions; typically, r_B is from 1 to 25 ohms (Ω).

Here is how you can calculate the bulk resistance of a silicon diode. A manufacturer's data sheet usually specifies the forward current I_F at 1 V (see Fig. 2-4b). For a silicon diode, the first 0.7 V is required to overcome the offset voltage; the final 0.3 V is dropped across the bulk resistance of the diode. Therefore, we can calculate the bulk resistance by using

$$r_B = \frac{0.3 \text{ V}}{I_F} \qquad (2\text{-}4)$$

where I_F is the forward current at 1 V.

As an example, a 1N456 is a silicon diode with an I_F of 40 mA at 1 V. It has a bulk resistance of

$$r_B = \frac{0.3 \text{ V}}{40 \text{ mA}} = 7.5 \ \Omega \qquad (2\text{-}5)$$

In any circuit using a 1N456, the first 0.7 V is wasted in overcoming the offset voltage. Any additional diode voltage is dropped across the 7.5 Ω of bulk resistance.

Third Approximation

In the third approximation of a diode, we include the bulk resistance r_B. Figure 2-5a shows the effect of r_B. After the silicon diode turns on, the current produces a voltage across r_B. The greater the current, the larger the voltage.

The equivalent circuit for the third approximation is a switch in series with a 0.7-V battery and a resistance of r_B (see Fig. 2-5b). After the external circuit has overcome the offset potential, it forces current through the bulk resistance.

As an example of the third approximation, suppose a 1N456 is used in Fig. 2-2a. Since it has a bulk resistance of 7.5 Ω, Fig. 2-2a may be replaced by Fig. 2-5c. In this circuit, the current is

$$I = \frac{10 \text{ V} - 0.7 \text{ V}}{2 \text{ k}\Omega + 7.5 \ \Omega} = \frac{9.3 \text{ V}}{2007.5 \ \Omega} = 4.63 \text{ mA} \qquad (2\text{-}6)$$

We have analyzed the same circuit (Fig. 2-2a) using the three diode approximations. Here are the results:

$$\begin{array}{lll} I = 5 \text{ mA} & \text{(ideal)} & \\ I = 4.65 \text{ mA} & \text{(second)} & (2\text{-}7) \\ I = 4.63 \text{ mA} & \text{(third)} & \end{array}$$

Which approximation should you use? This depends on the particular circuit being analyzed and the purpose of the analysis.

For preliminary analysis, start with the ideal-diode approximation. This gives you a quick idea of how the circuit works. If 0.7 V is significant compared to the source voltage, then use the second approximation. And when the bulk

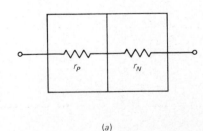

(a)

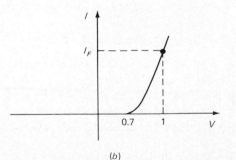

(b)

Fig. 2-4. *(a)* Bulk resistance; *(b)* forward current at 1 V.

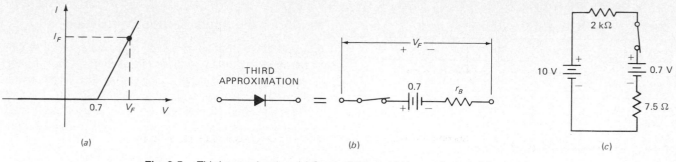

Fig. 2-5. Third approximation. *(a)* Graph; *(b)* forward-bias equivalent; *(c)* example.

resistance is significant compared to the circuit resistance, use the third approximation.

SUMMARY

1. Ideally, a diode acts like a perfect conductor (zero voltage) when forward-biased and like a perfect insulator (zero current) when reverse-biased. This is the first approximation of a diode, also called an *ideal diode*.
2. The circuit equivalent of an ideal diode is a switch: The switch is closed during forward bias, and open during reverse bias.
3. The ideal-diode approximation is used for a preliminary analysis. It gives you a quick idea of what a diode circuit does.
4. If the source voltage driving a silicon diode is not large compared to 0.7 V, then use the second approximation.
5. The circuit equivalent of the second approximation is a switch in series with a 0.7-V battery. When the external circuit turns on the diode, the voltage drop across the diode is 0.7 V no matter what the current.
6. The bulk resistance r_B is the resistance of the P and N regions. Typically, r_B is from 1 to 25 Ω.
7. Bulk resistance produces an *IR* drop across the diode. This *IR* drop is added to 0.7 V to get the total diode voltage.
8. In the third approximation the diode appears as a switch in series with a 0.7-V battery and a bulk resistance r_B.
9. The third approximation is used when bulk resistance is significant compared to the circuit resistance driving the diode.

SELF-TEST

Check your understanding by answering these questions.

1. The first approximation of a diode, also called the _____ diode, is used for preliminary analysis. In this approximation, the diode is either a perfect _____ or a perfect insulator.
2. An ideal diode acts like a _____ switch when forward-biased and like an _____ switch when reverse-biased.
3. The offset voltage of a silicon diode equals _____ .
4. In the second approximation, the external circuit must apply at least _____ before a silicon diode turns on. Then, no matter how much current there is, the diode drop is _____ .
5. The _____ resistance is the resistance of the P and N regions. This is all that impedes current above the offset voltage.
6. In the third approximation of a silicon diode, we visualize a _____ in series with a 0.7-V battery and a _____ resistance.

MATERIALS REQUIRED

- Power supply: Adjustable from at least 1 to 15 V
- Equipment: VOM
- Resistors: Two 220-Ω ½-W, one 470-Ω ½-W
- Diodes: 1N914 (or any small-signal silicon diode)

PROCEDURE

Measuring Two Points on the Forward Curve

1. Connect the circuit of Fig. 2-6*a*. Adjust the source to set up a current of 10 mA through the diode. (You can verify a current of 10 mA by using the VOM as an ammeter or by measuring 2.2 V across the 220-Ω resistor.)
2. Measure the diode voltage *V* and record this value in Table 2-1.
3. Adjust the source to get 50 mA of current through the diode. Measure and record *V* (Table 2-1).

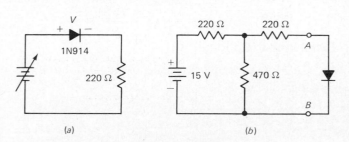

Fig. 2-6. Circuit to measure diode current and voltage.

TABLE 2-1. Two Points on Forward Curve

I, mA	V, V
10	
50	

Calculating the Bulk Resistance

4. In this experiment, we will let the offset voltage be the diode voltage corresponding to a diode current of 10 mA. Record the offset voltage in Table 2-2. (It should be in the vicinity of 0.7 V.)
5. Calculate the bulk resistance using

$$r_B = \frac{\Delta V}{\Delta I} \qquad (2\text{-}8)$$

where ΔV and ΔI are the changes in voltage and current in Table 2-1.

TABLE 2-2. Diode Values

V_{knee} _____

r_B _____

Measuring the Actual Current in a Circuit

6. Connect the circuit of Fig. 2-6b. Get the current I through the diode by either of these two methods: (1) using the VOM as an ammeter or (2) measuring the voltage across the 220-Ω resistor and calculating I. Record I under "Experimental I" in Table 2-3. (Because of the limited choice of current ranges on some VOMs, method 2 often gives a more accurate value of I.)

Calculating the Current with the Three Approximations

7. Calculate the value of I in Fig. 2-6b for an ideal diode. Record this ideal I in Table 2-3.
8. Calculate and record the value of I in Fig. 2-6b using the second approximation. (HINT: Thevenin's theorem.)
9. Repeat the preceding step using the third approximation.

TABLE 2-3. Diode Current in Fig. 2-6b

Experimental I _____

Ideal I _____

Second I _____

Third I _____

QUESTIONS

1. In this experiment, the offset voltage is the diode voltage that: (a) equals 0.3 V (b) equals 0.7 V (c) corresponds to 10 mA (d) corresponds to 50 mA (   )
2. Bulk resistance is not: (a) a ratio (b) a voltage difference divided by a current difference (c) in ohms (d) the same as total diode voltage divided by the total diode current (   )
3. The dc resistance of a silicon diode for a current of 10 mA is closest to: (a) 2.5 Ω (b) 10 Ω (c) 70 Ω (d) 1 kilohm (kΩ) (   )
4. In Fig. 2-6b, the power dissipated by the diode equals the product of voltage and current. This power is closest to: (a) 0 (b) 1.5 milliwatts (mW) (c) 15 mW (d) 150 mW (   )
5. What is the approximate diode current if the source of Fig. 2-6b is changed from 15 to 50 V? (a) 14 mA (b) 43 mA (c) 67 mA (d) 92 mA (   )

Extra Credit (Optional)

6. Use the second approximation to calculate the diode current in Fig. 2-7.

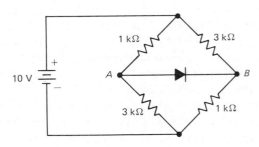

Fig. 2-7. Bridge circuit with diode load.

Answers to Self-Test

1. ideal; conductor
2. closed; open
3. 0.7 V

4. 0.7 V; 0.7 V
5. bulk
6. switch; bulk

ZENER-DIODE CHARACTERISTICS

OBJECTIVES

1. To measure the effects of forward and reverse bias on current in a zener diode
2. To determine and graph the voltampere characteristic of a zener diode
3. To construct a zener voltage regulator and experimentally determine the range over which the zener maintains a constant output voltage

INTRODUCTORY INFORMATION

Zener-Diode Operation

The characteristics of a solid-state diode depend on the semiconductor material from which the diode is constructed, on the nature and extent of "doping" of this material, and on the physical construction and dimensions of the device.

The semiconductor diode you studied in Experiment 1 is operated within its forward-bias current characteristic. There is another class of diodes called *zener diodes* whose unique reverse-bias current and voltage characteristics provide completely different applications from those of the crystal diode. The symbol for a zener diode is shown in Fig. 3-1.

Figure 3-2 is the graph of a typical current-voltage characteristic of a zener diode. When the diode is forward-biased, it acts like a closed switch, and forward current increases with an increase in applied voltage. Forward current is then limited by the parameters of the circuit. When the diode is reverse-biased, a small reverse current I_S, called *saturation current*, flows. I_S remains relatively constant despite an increase in reverse bias, until the zener breakdown region, in the vicinity of the zener voltage V_Z, is reached. In this vicinity reverse current starts rising rapidly because of avalanche effect. Finally, zener breakdown (a sharp increase in current) occurs when the zener voltage V_Z is reached.

In this region small voltage changes result in large current changes. Obviously, there are dramatic changes in effective resistance at the PN junction in this region.

Zener breakdown need not result in the destruction of the diode. As long as current through the diode is limited by the external circuit to a level within its power-handling capabilities, the diode functions normally. Moreover, by reducing reverse bias below the zener voltage, the diode can be brought out of its breakdown level and restored to the saturation-current level.

This process of switching the diode between its zener and nonzener current states can be repeated again and again without damaging the diode. It should be noted, however, that there is a certain time lag, called *recovery time*, in switching the diode from one state to the other.

Ratings

Manufacturers provide a specification sheet for each type of zener diode. Ratings include zener voltage, tolerance range of zener voltage, zener current limits, maximum power dissipation, maximum operating temperature, maximum zener impedance in ohms, thermal derating factor in milliwatts per degree Celsius (°C) (formerly Centigrade), and reverse leakage current. The nature of the material from which the diode is constructed, e.g., silicon, and the intended application of the diode are also indicated.

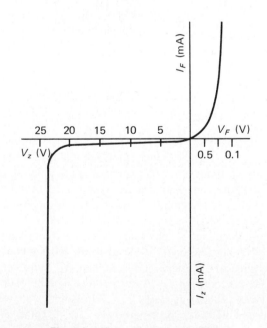

Fig. 3-2. Zener-diode characteristic.

Fig. 3-1. Symbol for a zener diode.

The breakdown voltage in a zener diode depends on the diode material and its construction. Zener diodes have been designed to deliver zener voltages from 1 to several hundred volts. The circuit designer has a wide variety of diodes from which to select the one whose characteristics closely approximate the circuit requirements.

Applications

Zener diodes are used as voltage regulators and as voltage reference standards. Figure 3-3 shows the circuit of a diode used as a shunt regulator. The diode is in parallel with a load resistor R_L. The purpose of the diode is to maintain a constant voltage across the load, within required limits, either as the output of the dc supply changes or as the load resistance, and hence the load current, changes.

NOTE: In an analysis of the circuit of Fig. 3-3, which contains linear (resistors) and nonlinear (zener diode) elements, Ohm's and Kirchhoff's laws apply, as do the network theorems with which you have become familiar.

Consider first the operation of the circuit when the voltage source V_{AA} is constant but the load current I_L changes. Assume a constant output voltage V_{out} is required across the load. The two currents $I_L = V_{out}/R_L$ and $I_Z = V_{out}/R_Z$ combine to form the total current I_T; that is,

$$I_T = I_L + I_Z \qquad (3\text{-}1)$$

The voltage V_R across R is equal to the product of I_T and R.

Thus
$$V_R = I_T \times R \qquad (3\text{-}2)$$

but
$$V_{AA} = V_R + V_{out} \qquad (3\text{-}3)$$

Hence, if V_{AA} remains constant, and it is required that V_{out} remain constant, V_R must remain constant. Therefore the total current I_T must remain fixed despite variations in load current. This can be accomplished only by compensating changes in I_Z. That is, I_Z must change in the manner shown in Eq. (3-4), assuming that I_T is constant and that I_L can vary.

$$I_Z = I_T - I_L \qquad (3\text{-}4)$$

Reflection as to how this result can be achieved leads to the conclusion that the zener diode chosen must be one whose zener voltage $V_Z = V_{out}$. Moreover, it is evident that V_{out} cannot remain absolutely constant. It must vary sufficiently to effect diode-current changes I_Z which will compensate for load-current changes I_L. Therefore a zener diode must be selected whose voltage and current characteristics will fit the requirements of the circuit. In addition, this diode must be operated at the right point in its characteristic.

The diode in Fig. 3-3 can also be used to offset changes in dc supply voltage when the load resistor R_L remains constant, thus ensuring a constant output voltage V_{out} and hence a constant load current I_L. Assume the circuit is operating properly for a dc voltage level V_{AA}. If the supply voltage V_{AA} should now increase, the output voltage V_{out} would tend to rise. As a result, the zener current I_Z would increase, and I_T

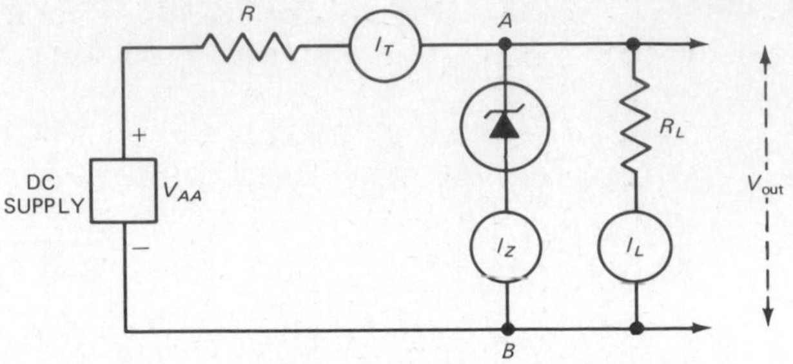

Fig. 3-3. Zener diode used as a shunt voltage regulator.

would increase as would the voltage drop V_R across R. If the regulator circuit has been properly designed, the increased voltage across R, ΔV_R, should equal (approximately) the increased supply voltage ΔV_{AA}, and V_{out} would drop back to its original value. Similarly, a decrease in V_{AA} would result in a decrease in I_Z and hence in I_T. V_R would be reduced, and V_{out} would return to its predetermined level.

The value of R chosen to achieve proper regulation will depend on the characteristics of the diode and on the conditions of variation in V_{AA} and I_L.

In addition to the regulator diode whose operation has just been described, there are voltage reference diodes whose zener voltage is so stable that they can be used as a laboratory standard or as a reference voltage in a more complex voltage-regulator circuit.

Design Considerations

A design value for R and for the zener diode can be calculated from the requirements of the circuit. Assume that a constant 10-V (± 0.7 V) output V_{out} is required for a load whose current I_L may vary from 5 to 20 mA. Power is supplied to the circuit from a constant 20-V dc source. It is required to design a regulating circuit which will achieve this.

Assume a regulating circuit, such as the one in Fig. 3-3, will meet the specifications of the problem. We must select a zener regulator diode whose $V_Z = 10$ V. Assume that such a diode is available which will pass a regulating current I_Z such that the total circuit current I_T remains constant at 30 mA over the range of load-current variation in our problem. By Kirchhoff's voltage law we can write

$$V_{AA} = I_T \times R + V_{out} \qquad (3\text{-}5)$$

and
$$R = \frac{V_{AA} - V_{out}}{I_T} \qquad (3\text{-}6)$$

Substituting in Eq. (3-6) the given values $V_{AA} = 20$, $V_{out} = 10$, and $I_T = 30 \times 10^{-3}$ A, we obtain

$$R = \frac{20 - 10}{30 \times 10^{-3}} = 333 \; \Omega$$

To determine the wattage of R, note that there is a 10-V drop across it. Therefore

$$W = \frac{V^2}{R} = \frac{10^2}{333} = \frac{1}{3} \; W$$

Good engineering practice requires that the resistor be over-rated. Hence a 1-W 330-Ω $\pm$ 5 percent resistor would be used.

The wattage of the diode is determined from the maximum I_Z current required by the circuit. In our problem maximum I_Z = 25 mA (when I_L = 5 mA). Therefore, the minimum wattage W_Z is

$$W_Z = V \times I_Z$$
$$= 10 \times 25 \times 10^{-3} \quad (3\text{-}7)$$
$$= 250 \text{ mW}$$

Again, good engineering practice requires overrating the diode, and a 500-mW diode will suffice.

SUMMARY

1. A zener diode is a semiconductor whose *reverse* volt-ampere characteristic is utilized in some electronic-circuit applications.
2. A zener diode maintains a *constant voltage V_Z* across its output if *reverse-biased* and operated within its rated characteristics.
3. When the diode is operated at its zener voltage V_Z, small changes in voltage across the diode result in relatively large current changes (I_Z) in the diode.
4. Zener diodes are rated for (*a*) zener voltage V_Z, (*b*) tolerance range of V_Z, (*c*) zener current limits, (*d*) maximum power dissipation, and (*e*) maximum operating temperature.
5. There are zener diodes made to deliver voltages from 1 to several hundred volts.
6. Zener diodes are used as voltage regulators and as voltage reference standards.
7. A properly designed shunt voltage regulator (Fig. 3-3) maintains a constant output voltage V_Z across the diode, despite specified variations in input voltage or specified changes in load current.
8. In the shunt regulator of Fig. 3-3, given a constant supply voltage V_{AA}, the zener diode maintains a constant circuit current I_T, despite changes in load current I_L. Thus, if I_L decreases, I_Z increases, and vice versa. In all cases for specified changes in I_L, the zener current I_Z = $I_T - I_L$.
9. In the shunt regulator of Fig. 3-3, if the load current remains constant but the supply voltage V_{AA} increases, then I_Z increases, increasing I_T and the voltage drop

across R, to maintain a constant voltage V_{out} across the zener. The voltages satisfy the equation $V_{out} = V_{AA} - I_T \times R$. Similarly, a decrease in V_{AA} will result in a decrease in I_Z and hence in I_T. The voltage drop across R, which equals $I_T \times R$, will decrease, thus maintaining a constant output voltage across the load.

10. In the shunt regulator of Fig. 3-3, given V_{AA}, I_T, and V_{out}, it is possible to calculate the value of R from the equation

$$R = \frac{V_{AA} - V_{out}}{I_T}$$

SELF-TEST

Check your understanding by answering these questions.

1. When used as a voltage regulator, a zener diode must be _____ (*forward-*, *reverse-*) biased.
2. If a manufacturer specifies that for a specific zener the output voltage is 10 V $\pm$ 10 percent tolerance, V_Z for that diode lies between _____ V and _____ V.
3. Current in a 1-W 10-V zener diode should be limited to a maximum of _____ A.
4. A 20-V 1-W zener diode connected as a voltage regulator in the circuit of Fig. 3-3 supplies an output voltage of _____ V (approx) to the load.
5. In the circuit of question 4, the dc supply V_{AA} is 30 V, and I_T = 0.05 A and is constant over a range of load currents varying from 5 to 35 mA. Within this range the zener diode current varies from _____ to _____ mA.
6. In the circuit of question 5, the value of R which will satisfy the regulator requirements is _____ Ω.

MATERIALS REQUIRED

- Power supply: Variable regulated dc source
- Equipment: EVM; VOM; milliammeter; curve tracer for extra-credit question
- Resistors: 3300-Ω ½-W; 500-Ω 5-W; resistors required for extra-credit procedure
- Semiconductors: 1N3020 (other choice: any 1-W 10-V zener diode)
- Miscellaneous: SPST switch; resistor decade box for extra-credit question

PROCEDURE

Voltampere Characteristic—Reverse Bias

1. Connect the circuit of Fig. 3-4. Switch S is *open*. V_{AA} is a regulated power supply set at 0 V. M is a 20,000-Ω/V VOM set on the *lowest* current range.
2. Close S. Measure the diode current I, if any, with V_{AA} set at 0 V. Record the results in Table 3-1.

3. Set the output of V_{AA} so that the voltage V_{AB} measured across the diode is 2.0 V. Measure the diode current I. Record the results in Table 3-1.

   Repeat step 3 for each value of V_{AB} shown in Table 3-1. Change the range of M as required. Calculate the resistance R_Z of the diode ($R_Z = V_{AB}/I$) and record the results in Table 3-1.

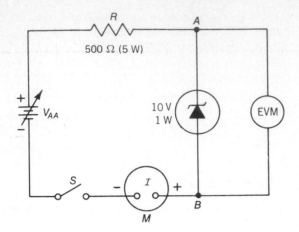

Fig. 3-4. Experimental circuit for observing effect of reverse bias on a zener diode.

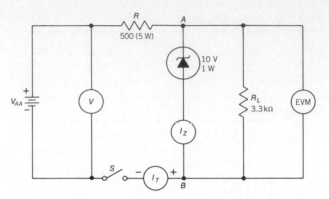

Fig. 3-5. Experiment voltage-regulator circuit.

4. Set the output of V_{AA} so that the diode current I measures 2 mA. Measure the voltage V_{AB} across the diode and record in Table 3-1. Calculate R_Z and record in Table 3-1.

5. Repeat step 4 for every value of current shown and record corresponding values of V_{AB} and R_Z in Table 3-1.

Voltampere Characteristic—Forward Bias

6. Open S disconnecting power from the circuit. Set the output of the power supply at 0 V. Reverse the diode in the circuit.

7. Close S. Measure, and record in Table 3-2, the forward current in the diode at each level of voltage V_{AB} shown in the table. Compute the forward resistance $R_F = V_{AB}/I$. Record the results in Table 3-2.

8. (a) From the data in Tables 3-1 and 3-2, draw a graph of diode current (vertical axis) versus diode voltage.

(b) Draw an expanded graph of diode current versus voltage in the zener region.

(c) Draw separate graphs of diode resistance versus voltage for the reverse- and forward-bias arrangements.

Zener Diode as a Voltage Regulator

9. Connect the circuit of Fig. 3-5. Switch S is open. The output of the power supply V_{AA} is set at 0 V. M is a milliammeter set on the 100-mA range.

10. Close S. Slowly increase the supply voltage V_{AA} until current I_Z in the diode measures 20 mA. Measure the supply voltage V_{AA} and the voltage V_{AB} across the load. Record the results in Table 3-3. Measure the total current I_T. Record the results in Table 3-3.

11. Determine and record the range of variation of V_{AA} over which V_{AB} remains constant within ± 0.1 V of its value in step 8. Measure the variation of I_Z and I_T within this range. Record the results in Table 3-3.

Extra Credit (Optional)

12. (a) Design a regulator circuit from a constant voltage source V_{AA}, using the zener diode whose voltampere characteristic you have just experimentally determined. It is required that the regulator maintain a constant output voltage V_{out} within 0.2 V of the average value of V_{out}, for load currents in the range of 10 to 30 mA. Draw the circuit showing the values of all components and voltages. Explain how you determined these values.

(b) Test the circuit and record your measurements in Table 3-4. Use a resistance decade box as the variable load.

13. With a curve tracer observe the voltampere characteristic of the zener diode. Photograph or draw the curve on the same graph paper as in step 8.

TABLE 3-1. Reverse Bias

V_{AB}, V	I, mA	R_Z, Ω	V_{AB}, V	I, mA	R_Z, Ω
0			5		
2.0			10		
6.0			20		
7.0			30		
8.0			40		
	2		50		

TABLE 3-2. Forward Bias

V_{AB}, V	0	0.1	0.2	0.3	0.4	0.5	0.55	0.6	0.65	0.7
I, mA										
R_F, Ω										

TABLE 3-3. Voltage Regulation

	V_{AB}, V	I_Z, mA	I_T, mA	V_{AA}, V
V_{AB}		20		
$V_{AB} + 0.1$				
$V_{AB} - 0.1$				

TABLE 3-4. Voltage-regulator Design Characteristics

V_{AA}, V	R, Ω	I_L, mA	R_L, Ω	V_{out}, V
		10		
		20		
		30		

QUESTIONS

1. Compare the biasing of a junction diode (Experiment 1) with that of a zener diode in a normal application.
2. Compare the voltampere characteristic of the zener diode graph of procedural step 8*a* in this experiment with the characteristic in Fig. 3-2. Explain any differences.
3. What portion of a zener-diode characteristic is most useful for voltage-regulator applications? Why?
4. (*a*) What is the significance of the graph of procedural step 8*b*? (*b*) How can the graph of procedural step 8*b* be used in the design of a regulator employing a 10-V zener diode?
5. Refer to Table 3-3. Explain how this regulator circuit works.

Extra Credit (Optional)

6. Explain the operation of the regulator circuit you designed in step 12.
7. Will the regulator circuit of Fig. 3-5 compensate for both changes in input voltage V_{AA} and changes in load current I_L? Explain.

Answers to Self-Test

1. reverse-
2. 9; 11
3. 0.1
4. 20
5. 45; 15
6. 200

OPTOELECTRONIC DEVICES

OBJECTIVES

1. To get data for red and green LEDs
2. To display numbers with a seven-segment indicator
3. To transfer a signal through an optocoupler

INTRODUCTORY INFORMATION

We use the sense of sight more than any other. So, with the mention of devices that change light to electricity, the imagination takes off. *Optoelectronics* is the technology that combines optics and electronics. This exciting field includes light-emitting diodes, LED displays, and optocouplers.

LEDs

The *light-emitting diode* (LED) is a solid-state light source. LEDs have replaced incandescent lamps in many applications because they have the following advantages

1. Low voltage
2. Long life [more than 20 years (yr)]
3. Fast on-off switching [nanoseconds (ns)]

In a forward-biased rectifier diode, free electrons and holes recombine at the junction. When a free electron falls into a hole, it drops from a higher energy level to a lower one. As the electron falls, it radiates energy in the form of heat and light. Because silicon is opaque (not transparent), none of the light escapes to the environment.

A LED is different. To begin with, semitransparent materials are used instead of silicon. In a forward-biased LED, heat and light again are radiated when free electrons and holes recombine at the junction. Because the material is semitransparent, some of the light escapes to the surroundings.

By using elements like gallium, arsenic, and phosphorus, a manufacturer can produce LEDs that radiate red, green, yellow, amber, or infrared (invisible) light. LEDs that produce visible radiation are used in instrument displays, calculators, digital clocks, etc. The infrared LED finds application in burglar-alarm systems and other areas requiring invisible radiation.

LEDs have a typical voltage drop from 1.5 to 2.5 V for currents between 10 and 50 mA. The exact voltage drop depends on the color, tolerance, and other factors. For preliminary analysis and design, we will use the second diode approximation with an offset of 2 V. For example, Fig. 4-1a shows a source of +5 V driving a LED through a 100-Ω resistor. The outward arrows on the LED symbolize the radiated light. Allowing 2 V across the forward-biased LED implies a current of

$$I = \frac{5 \text{ V} - 2 \text{ V}}{100 \text{ }\Omega} = 30 \text{ mA}$$

Incidentally, LEDs have low reverse voltage ratings. For instance, the TIL221 (a red LED used in this experiment) has a maximum reverse voltage rating of 3 V. This means accidentally applying a reverse voltage greater than 3 V may destroy or degrade the LED characteristics. One way to protect a LED is by paralleling a rectifier diode as shown in Fig. 4-1b.

LED Arrays

A LED array is a group of LEDs that display numbers, letters, or other symbols. The most common LED array is the *seven-segment* display shown in Fig. 4-2a. The display contains seven rectangular LEDs (A to G). Each LED is called a *segment* because it forms part of the character being displayed.

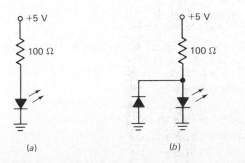

Fig. 4-1. (a) Forward-biased LED; (b) protecting the LED against reverse bias.

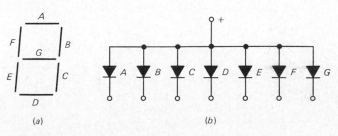

Fig. 4-2. (a) Seven-segment indicator; (b) schematic diagram.

Figure 4-2*b* shows the schematic diagram. A positive voltage drives all anodes. By grounding one or more cathodes, we can form any digit from 0 to 9. For instance, by grounding the cathodes of A, B, and C, we display a 7 in Fig. 4-2*a*. Or by grounding the cathodes of A, B, C, D, and G, we get a 3.

A seven-segment array can also display capital letters A, C, E, and F, plus lower-case letters b and d. (Displaying 0 to 9, plus A, b, C, d, E, and F, is common in microprocessor trainers.)

Photodiodes

A reverse-biased diode has a small current because of its minority carriers. The number of these carriers depends on temperature, but it also depends on the light striking the junction. When the diode package is opaque, no outside light can get through to the junction; therefore, we detect no *photoelectric* effect (light changing an electrical quantity). But when the diode is in a glass package, incoming light does change the amount of reverse current.

A *photodiode* is optimized for its sensitivity to light. In this diode, a glass window lets light pass through the package to the junction. The incoming light produces free electrons and holes. In other words, the light increases the number of minority carriers. The stronger the light, the more minority carriers produced.

Figure 4-3 shows the schematic symbol of a photodiode. The inward arrows represent the incoming light. Also notice the photodiode is reverse-biased. In this way, as the light becomes more intense, the reverse current increases. The reverse current is small, typically in tens of microamperes.

The photodiode is one example of a *photodetector*, a device that can convert incoming light into an electrical quantity. (Other examples of photodetectors are photoresistors, phototransistors, and photo-Darlingtons.)

Optocouplers

An *optocoupler* combines a LED and a photodetector in a single package. Figure 4-4 shows a LED-photodiode coupler. The LED is on the left, the photodiode on the right. The LED supply forces current through the LED. The light from the LED hits the photodiode and sets up a reverse current through resistor R_2. The voltage across the photodiode is given by

$$V_{out} = V_{SS} - IR_2$$

This output voltage depends on how large the reverse current is. If we vary the LED supply, the amount of light changes and this causes the photodiode current to change. As a result,

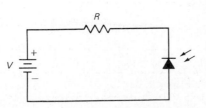

Fig. 4-3. Photodiode is reverse-biased.

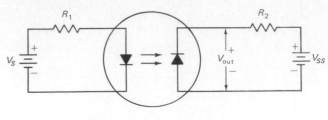

Fig. 4-4. An optocoupler circuit.

V_{out} changes. In fact, if the LED current has an ac variation, V_{out} will have an ac variation.

The key advantage of an optocoupler is the electrical isolation between the LED circuit and the photodiode circuit; typically, the resistance between the input and output circuits is greater than 10^{10} Ω. This is why an optocoupler is also known as an "optoisolator"; the only contact between the input and output circuits is the stream of light.

SUMMARY

1. In a LED, free electrons and holes recombine at the junction to produce heat and light. Because semi-transparent materials are used, some of the light escapes to the surroundings.
2. By using different materials, manufacturers can produce LEDs that emit red, green, yellow, amber, and infrared light.
3. The typical voltage across a LED is from 1.5 to 2.5 V for a current of 10 to 50 mA.
4. The advantages of LEDs are low voltage, long life, and fast on-off switching.
5. The most common LED array is the seven-segment indicator. It can display 0 to 9, as well as some letters of the alphabet.
6. A photodiode is optimized for its sensitivity to light. In this type of diode, light passes through to the junction where it produces free electrons and holes. The stronger the light, the greater the reverse current.
7. An optocoupler combines a LED and a photodetector in a single package. In a LED-photodiode coupler, the light from the LED controls the reverse current in the photodiode. In this way, the input and output circuits are electrically isolated.

SELF-TEST

Check your understanding by answering these questions.

1. In a forward-biased LED, light is radiated when free electrons and holes _____ at the junction.
2. LEDs typically have a voltage drop of 1.5 to 2.5 V for a current of 10 to _____ mA.
3. The most common LED array is the _____ indicator.
4. A photodiode is optimized for its sensitivity to _____ . It should be _____ biased.

5. An optocoupler combines a _____ and a photo-detector.

6. The key advantage of an optocoupler is its electrical _____ .

MATERIALS REQUIRED

- Two power supplies: One at 15 V, another adjustable from at least 1 to 15 V

  $P = VI$

  $.143w$

- Equipment: Electronic voltmeter, VOM
- Resistors: Two 270-Ω 1-W
- Diode: 1N914 (or equivalent small-signal diode)
- Red LED: TIL221 (other choices: Litronix RL-2000 or any red LED that can handle up to 50 mA)
- Green LED: TIL222 (other choices: any green LED that can handle up to 50 mA)
- Seven-segment display: TIL312 (or nearest equivalent)
- Optocoupler: 4N26 (or nearest equivalent)

PROCEDURE

Data for a Red LED

1. Examine the red LED. Notice that one side of the package has a flat edge. This indicates the cathode side. (With many LEDs, the cathode lead is slightly shorter than the anode lead. This shorter lead is another way to identify the cathode.)

2. Connect the circuit of Fig. 4-5 using a red LED. The VOM is connected as an ammeter that measures the current through the LED. The electronic voltmeter measures the voltage across the LED. The 1N914 protects the LED against accidentally applying a reverse voltage.

3. Adjust source voltage V_s to get 10 mA through the LED. Record the corresponding LED voltage in Table 4-1.

4. Adjust the source voltage to set up the remaining currents listed in Table 4-1. Record each LED voltage.

Data for a Green LED

5. Replace the red LED by a green LED in the circuit of Fig. 4-5.

6. Repeat steps 3 and 4 for the green LED.

Using a Seven-segment Display

7. Figure 4-6a shows the pinout for the seven-segment display used in this experiment (top view). It includes a left decimal point (LDP) and a right decimal point (RDP). Connect the circuit of Fig. 4-6b.

8. Figure 4-6c shows the schematic diagram for a TIL312. (If you are using a different device, your instructor will give you a diagram with the correct pin numbers.) Ground pins 1, 10, and 13. If the circuit is working correctly, digit 7 will be displayed.

9. Disconnect the grounds on pins 1, 10, and 13.

10. Refer to Fig. 4-6a and c. Which pins should you ground to display a 0? Ground these pins and if the circuit is working correctly, enter the pin numbers in Table 4-2.

TABLE 4-1. LED Data

I, mA	V_{red}, V	V_{green}, V
10	1.7, 4	2.2, 4.5
20	1.7, 6	2.3, 6.5
30	1.75, 8.3	2.5, 8.9
40	1.8, 10.6	2.6, 11.1

11. Repeat step 10 for the remaining digits, 1 to 9, and the decimal points.

The Transfer Graph of an Optocoupler

12. Connect the circuit of Fig. 4-7a. Adjust the source voltage to 2 V. Measure and record the output voltage (Table 4-3).

13. Repeat step 12 for source voltages shown in Table 4-3.

14. Draw the transfer graph, V_{out} versus V_s, of the optocoupler from the data in Table 4-3.

TABLE 4-2. Seven-segment Indicator

Display	Pins Grounded
0	1 - 13 - 10 - 8 - 7 - 2
1	13 - 10
2	1 - 13 - 11 - 7 - 8
3	1 - 13 - 10 - 8 - 11
4	13 - 10 - 2 - 11
5	1 - 10 - 8 - 2 - 11
6	1 - 10 - 8 - 7 - 2 - 11
7	1 - 10 - 13
8	1 - 13 - 10 - 8 - 7 - 2 - 11
9	1 - 13 - 11 - 8 - 2 - 11
LDP	6
RDP	9

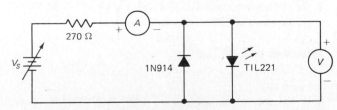

Fig. 4-5. Circuit for LED data.

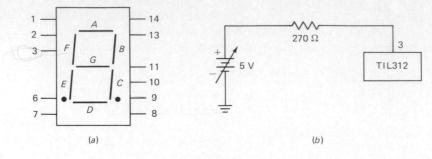

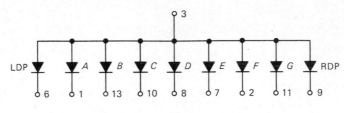

Fig. 4-6. (a) Pinout of TIL312; (b) circuit for TIL312; (c) schematic diagram.

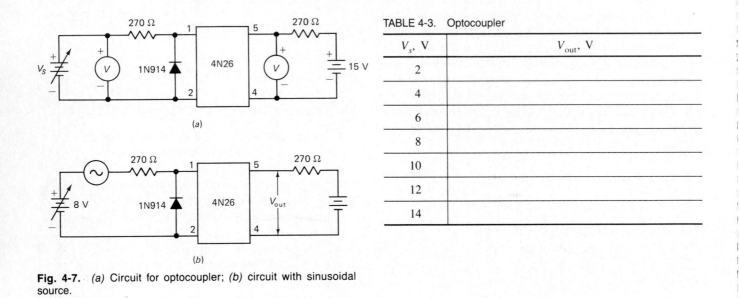

Fig. 4-7. (a) Circuit for optocoupler; (b) circuit with sinusoidal source.

TABLE 4-3. Optocoupler

V_s, V	V_{out}, V
2	
4	
6	
8	
10	
12	
14	

QUESTIONS

1. What is the voltage drop across the red LED when the current is 30 mA? *1.75*

2. Suppose you reverse the source voltage in Fig. 4-5. If $V_s = -15$ V, approximately how much voltage is there across the LED? Explain how the silicon diode protects the LED. *0*

3. What is the voltage drop across the green LED when the current is 30 mA? How does this compare to the voltage across the red LED for the same current? *2.5 .75 HIGHER*

4. When using the seven-segment indicator, which pins did you ground to display a 1? Which pins for an 8? *13-10 1-13-10-8-7-2-11*

5. In Fig. 4-6b, assume an 8 is being displayed. If the voltage drop is 1.6 V between pin 3 and ground, how much current is there through the 270-Ω resistor? *12.6 mA*

6. The seven-segment indicator displayed a 1 more brightly than an 8. Explain why.

7. Why is the 1N914 used in Fig. 4-7a?

8. How much does the output voltage change in Fig. 4-7a when the source changes from 4 to 6 V?

Extra Credit (Optional)

9. Suppose the sinusoidal source in Fig. 4-7b has a peak-to-peak value of 4 V. Describe the output voltage V_{out}.

Answers to Self-Test

1. recombine
2. 50 mA
3. seven-segment
4. light; reverse-
5. LED
6. isolation

THE DIODE LIMITER AND CLAMPER

OBJECTIVES

1. To determine the relationship between the input sine wave and output waveform of a series-connected diode limiter
2. To determine the relationship between the input sine wave and output waveform of a parallel-connected diode limiter
3. To observe the effect on the output waveform of forward- and reverse-biased diode limiters
4. To observe the effect on the output waveform of negative and positive diode clampers

INTRODUCTORY INFORMATION

In electronics it is frequently necessary to square off the extremities of an ac signal voltage or to *limit* an ac voltage to predetermined levels. The electronic device which is used to do this is called a "limiter." Limiters can transform a sine wave into a rectangular wave, can limit either the negative or positive alternation or both alternations of an ac voltage, and can perform other useful waveshaping functions.

Series-Diode Limiters

The unidirectional current characteristics of some semiconductor diodes permit series diodes to serve admirably as limiters.

Consider the circuit of Fig. 5-1a. An ac generator applies a sine-wave voltage v_{in} to a diode in series with a resistor R. In Fig. 5-1b the input voltage v_{in} and output voltage v_{out} are shown in proper time phase. During the positive alternation, the cathode of the diode is positive relative to its anode; that

is, the diode is reverse-biased. Hence no current flows in the circuit, and the output v_{out} across R is 0. During the negative alternation, the diode is forward-biased, acting as a closed switch and permitting current to flow in R. The voltage across R is the negative alternation, with the polarity shown. Actually the diode is not an ideal switch since it does permit some reverse current and has some forward resistance. Part of the input voltage appears as a drop across the diode. Therefore the output across R is somewhat lower than the input negative alternation. This is evident from the level of the voltage v_{in} in Fig. 5-1b and the voltage v_{out}, which is the output voltage across R.

This simple circuit is a positive series limiter, "positive" because the positive alternation has been limited or eliminated from the output. It is called a "series limiter" because the output taken from the load resistor R is in series with the diode.

Figure 5-2a shows that the series-connected diode may also be used as a negative limiter by reversing the polarity of the diode in the circuit. The waveforms of Fig. 5-2b show that during the positive alternation the diode is forward-biased, permitting current to flow in R. The voltage developed across R is positive, following the input alternation. Again, there is a voltage drop across the diode. During the negative alternation the diode is reverse-biased. No current flows. Hence, no output voltage is developed across R. Thus the series-connected diode in Fig. 5-2 acts as a negative limiter.

An additional effect of the diodes in Figs. 5-1 and 5-2 should be noted. Current flow in R is unidirectional (that is, it flows in one direction during the negative or positive alterna-

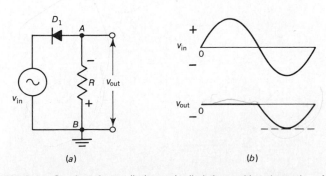

Fig. 5-1. Semiconductor diode used to limit the positive alternation of an ac signal.

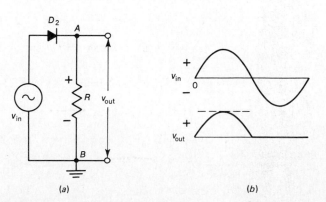

Fig. 5-2. Series-connected diode used to limit the negative alternation of an ac signal.

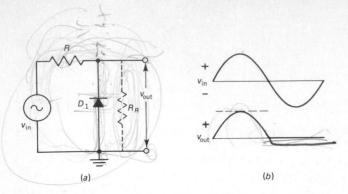

(a)　　　(b)

Fig. 5-3. Parallel-connected diode used to limit the negative alternation.

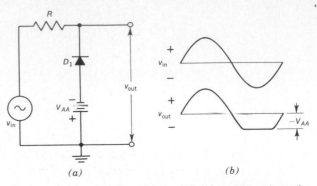

(a)　　　(b)

Fig. 5-5. Biased parallel diode partially limiting negative alternation.

tion of v_{in}, depending on the polarity of the diode). It is therefore a pulsating dc voltage. This effect is called *rectification*. The diode has rectified the ac signal and converted it to a pulsating dc signal, that is, a changing signal voltage of fixed polarity. Since diodes used as rectifiers serve an important function in electronics, they will be covered in other experiments.

Parallel-Diode Limiters

The circuit in Fig. 5-3a is an example of a parallel-connected diode limiter. It is called a parallel limiter because the output is in parallel with the diode.

During the positive alternation, diode D_1 is reverse-biased and exhibits high reverse resistance R_R. R and R_R constitute a voltage divider. If R is very much smaller than R_R, practically the entire positive alternation appears as the output voltage v_{out} across the diode (Fig. 5-3b).

During the negative alternation D_1 is forward-biased. The diode acts as a closed switch. That is, the diode conducting acts ideally like a short circuit. Hence no voltage is developed across the diode, as in Fig. 5-3b.

Since the negative alternation is removed (limited) from the output, Fig. 5-3a is an example of a negative limiter.

By reversing the polarity of the diode, the parallel-diode limiter in Fig. 5-4 is used to remove the positive alternation.

Biased-Parallel Limiters: Partial Limiting

The circuit of Fig. 5-5 accomplishes partial limiting of the negative and positive alternations, respectively, of an input sine wave.

Diode D_1 in Fig. 5-5 is reverse-biased by battery V_{AA} which maintains the anode V_{AA} volts negative relative to its cathode. During the positive alternation of the input voltage v_{in}, the cathode of D_1 is held positive. The diode acts like an open switch, and the positive alternation appears in the output, Fig. 5-5b. During the negative alternation the cathode is driven negative, but the diode will not conduct until v_{in} is more negative than the bias voltage V_{AA} which maintains the anode V_{AA} volts negative. Hence, that part of the negative alternation which is less negative than V_{AA} appears in the output. When the negative alternation of v_{in} reaches the level where it is more negative than V_{AA}, the cathode is driven more negative than the anode and the diode conducts, limiting that portion of the negative alternation between $-V_{AA}$ and $-v_M$ peak.

Biased Double-Diode Limiters

Figure 5-6a shows two biased diode limiters connected in such a manner (in parallel) that the circuit acts as a partial limiter of both the positive and negative alternations. Diode D_1 conducts when the voltage v_{in} reaches a higher negative value than V_{KK1}, thus limiting the negative alternation to the value of V_{KK1}. Diode D_2 conducts when v_{in} reaches a higher positive value than V_{AA2}, limiting the positive alternation to the value V_{AA2}.

It can be seen from the waveform in Fig. 5-6b that the circuit of Fig. 5-6a has converted a sine wave into a waveform which closely approximates a rectangular wave; that is, the extremities of the sine wave have been squared off.

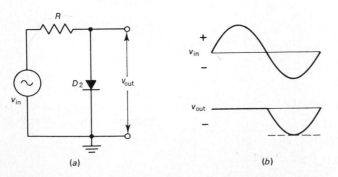

(a)　　　(b)

Fig. 5-4. Parallel-connected diode used to limit the positive alternation.

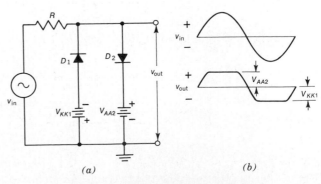

(a)　　　(b)

Fig. 5-6. Biased double-diode limiter.

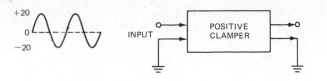
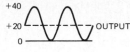

Fig. 5-7. Positive clamper adds a + dc axis to the waveform.

Diode Clamper

The diode limiter or clipper (as it is also known) modifies the input waveform by limiting or "clipping" part of that waveform. Other diode circuits, called *clampers* (or *dc restorers*), do not change the shape of the input waveform; rather, they add a dc level to it. There are positive clampers, negative clampers, and biased clampers.

Positive Clamper

The effect of a positive clamper on an ac waveform with 0 V as its axis is shown in Fig. 5-7. The clamper has added +20 V dc to the 40-V p-p (peak-to-peak) waveform. The result is that the input waveform, which varied from +20 to −20 V, appears in the output of the circuit as a signal varying between 0 and +40 V, with +20 V dc as its axis. The output waveform acts as though a +20-V battery was connected in series with the input.

Figure 5-8 is the circuit diagram of a dc clamper. The circuit operates as follows. On the negative alternation of the 10-V p-p input sine wave the cathode of diode D is driven negative relative to its anode. Therefore D conducts, charging C through the low resistance of the forward-biased diode.

Capacitor C will charge to the peak of the negative alternation, 5 V, with the polarity shown in Fig. 5-8. On the positive alternation D is cut off since its cathode is positive relative to the anode. Capacitor C tries to discharge through R when D is cut off. However, if the time constant RC is large compared with the period of the sine wave, the capacitor will lose very little of its charge and will hold the 5 V across it. As a result, when the negative alternation of the second cycle comes along, the positive voltage on C will cancel the negative input voltage, and diode D will not conduct.

The effect of the circuit is illustrated in Fig. 5-9. Capacitor C has been replaced by a +5-V battery which is in series

with the input signal. If the sinusoidally varying voltage is added, point by point, to the +5 V of the battery, the result is a waveform which varies between 0 and +10 V peak. A high-impedance dc voltmeter connected across the output will measure +5 V (approx).

NOTE: A *long-time constant RC* is defined as one which is equal to or greater than 10 times the period t of the input waveform.

That is

$$RC \geqslant 10t \qquad (5\text{-}1)$$

and

$$t = \frac{1}{F} \qquad (5\text{-}2)$$

where t is measured in seconds (s) and F is measured in hertz (Hz). A numerical example will illustrate this statement. If the frequency of the input waveform is 1000 Hz, then

$$t = \frac{1}{1000} = 1 \times 10^{-3}\,\text{s} \qquad (5\text{-}3)$$

The product RC must therefore be equal to or greater than 10×10^{-3} s.

The foregoing explanation of the circuit must be modified somewhat to reflect the fact that C does lose a small percentage of its charge on every positive alternation. As a result the net voltage on C is not +5 V, but slightly less than 5 V. This loss is restored, however, on the peaks of the negative alternations, when the cathode of D becomes sufficiently negative to turn on diode D, recharging C to its +5-V level.

Negative Clamper

A negative clamper adds a negative dc level to an ac signal. This is accomplished by reversing the polarity of the diode, as in the circuit of Fig. 5-10. In this circuit, C charges on the positive alternation of the input signal. If the input signal

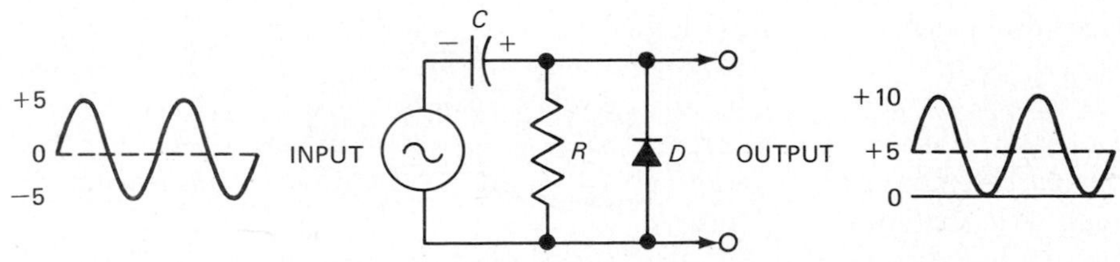

Fig. 5-8. Circuit diagram of a positive clamper.

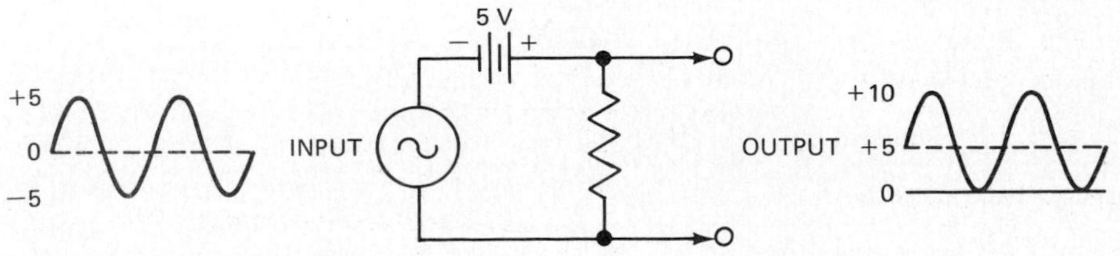

Fig. 5-9. A battery of +5 V replaced the charged capacitor.

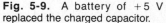

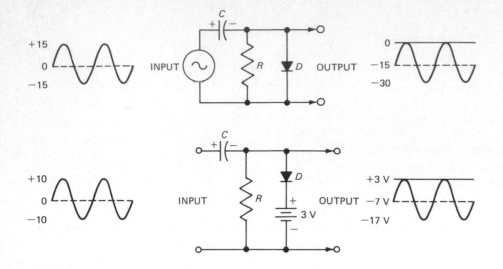

Fig. 5-10. Negative clamper adds a $-$ dc axis to the waveform.

Fig. 5-11. Biased negative clamper.

varies between $+15$ and -15 V, the net effect is to charge C to -15 V. The output waveform now varies from 0 to -30 V, and an EVM, set on dc volts, will measure -15 V in the output.

Biased Clamper

A biased negative clamper is shown in Fig. 5-11. A battery, 3 V in this case, biases the cathode at $+3$ V. Since D cannot conduct until its anode is positive relative to the cathode, the diode must wait until the positive alternation of the input has risen more than $+3$ V on the anode. The diode therefore conducts between the $+3$- and $+10$-V levels of the input signal in Fig. 5-11. As a result, capacitor C charges to -7 V. After this initial charge, the action of the circuit is similar to that of Fig. 5-10. The output waveform is therefore clamped below $+3$ V (Fig. 5-11), and varies between $+3$ and -17 V. Other clamper arrangements, for example positive-biased clampers, are possible. Their operation can be analyzed in the same way as the action of the clamper in Fig. 5-11.

SUMMARY

1. A series-diode limiter is a circuit in which the output is in series with the diode.
2. A parallel-diode limiter is one in which the output is in parallel with the diode.
3. A *positive* limiter eliminates or limits the *positive* alternation of an output waveform.
4. When the *negative* alternation of an output waveform has been limited or eliminated, the circuit which achieves this result is called a *negative limiter*.
5. Diode limiting is possible because of the low forward resistance and high reverse resistance of a diode.
6. The positive parallel limiter presents an infinite resistance during the negative alternation of a waveform.
7. A biased-diode limiter is one in which an external bias source is connected to either the anode or cathode of the diode limiter, as shown in Fig. 5-5.

8. The circuit of Fig. 5-5 is a negative-biased parallel limiter, limiting part of the negative alternation. That part of the alternation which is more negative than $-V_{AA}$, the bias voltage, is excluded from the output.
9. The circuit of Fig. 5-5 operates in the same way as any parallel limiter, except that the limiting action is *delayed* until the bias voltage is reached.
10. A diode clamper does not alter the shape of the input signal, but adds a dc level to an ac waveform.
11. A positive clamper adds a positive voltage level to the signal, while a negative clamper adds a negative voltage level.
12. The polarity of the diode determines whether the circuit is a positive clamper (Fig. 5-8) or a negative clamper (Fig. 5-10).

SELF-TEST

Check your understanding by answering these questions.

1. The sine-wave input to the circuit of Fig. 5-1 is $+9$ V peak positive and -9 V peak negative. The output waveform v_{out} will be _____ 0 _____ to _____ ~9 _____ V (approx).
2. The peak-to-peak voltage of the sine-wave input to Fig. 5-3 is 20 V. That is, it has the peak positive and negative limits, respectively, of $+10$ and -10 V. The output voltage varies between _____ 10 _____ and _____ ~.7 _____ V (approx). This output is in time phase with the _____ +r _____ alternation of the input waveform.
3. In Fig. 5-4, the output waveform is in _____ PAR. _____ with the diode D_2.
4. In Fig. 5-2, diode D_2 acts as a (an) _____ open _____ switch during the negative alternation.
5. In Fig. 5-3, diode D_1 acts as a (an) _____ shot .7 _____ switch during the negative alternation.
6. In the input sine waveform to the circuit of Fig. 5-5 varies between peaks of $+15$ and -15 V, and if $V_{AA} = 7$ V,

v_{out} will vary between the limits _____ and _____ V.

7. In the circuit of Fig. 5-6, the input voltage varies between +9 and −9 V p-p. The output varies between +6 and −5 V. If we consider D_1 and D_2 as ideal switches (zero resistance when conducting), then V_{KK1} = _____ V, and V_{AA2} = _____ V.

8. In the circuit of Fig. 5-8, the input waveform varies between +12 and −12 V. The dc level in the output will measure _____ V. The output waveform will vary between _____ and _____ V.

9. The frequency of the input waveform in Fig. 5-10 is 60 Hz. For the circuit to act as a _____ clamper, the value of $RC \geqq$ _____ s.

MATERIALS REQUIRED

- Power supply: Variable regulated dc source; line-derived-line-isolated 18-V p-p sine-wave source; 6.3-V rms source for tube filaments
- Equipment: Oscilloscope; EVM
- Resistors: 120,000-Ω ½-W
- Semiconductors: Two 1N5625
- Miscellaneous: Two SPST switches; 2500-Ω 2-W potentiometer; components as required for extra-credit step 16

PROCEDURE

Series Limiter

1. If you are using a vertically calibrated laboratory standard oscilloscope, set the vertical gain controls of the oscilloscope for a sensitivity of 5 V per major division. If you are using a service-type oscilloscope with uncalibrated vertical amplifiers, calibrate the vertical amplifiers for voltage measurement at 5 V per major division.

2. Connect the circuit of Fig. 5-1a. D_1 is a 1N5625 diode. R = 120,000 Ω. The input voltage v_{in} is line-isolated, 18 V p-p, 60 Hz.

   Connect the vertical input of your oscilloscope across v_{in} and use v_{in} for external triggering or external synchronization of the oscilloscope, as in Fig. 5-12, or use line triggering/synchronization (sync).

NOTE: If the 60-Hz source is a voltage stepdown transformer operating from the power line, the oscilloscope may be set to "line" triggering or "line" synchronization, instead of "external" triggering or synchronization. Either method of triggering or synchronization will make it possible to observe the phase of the output waveform with reference to the input.

3. Adjust the Time/div. (sweep) and triggering (sync) controls for two or three waveforms. Center these waveforms with respect to the x and y axes as in Fig. 5-13. The waveform designated MN (Fig. 5-13 and Table 5-1) will act as the reference input waveform, v_{in}, for phase measurements.

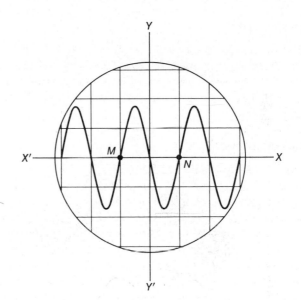

Fig. 5-13. Sine wave centered with respect to X and Y axes.

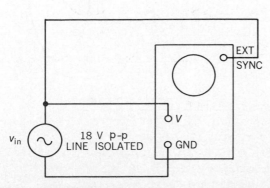

Fig. 5-12. Connecting oscilloscope for observing circuit operation.

TABLE 5-1. Series Limiter

Step	V p-p	Waveform
3	+9 V / 0 / −9	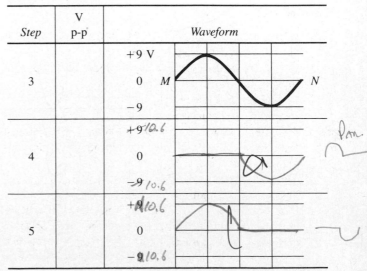
4	+9/0.6 / 0 / −9 /0.6	
5	+9/0.6 / 0 / −9/0.6	

NOTE: If a dual-trace oscilloscope is available, phase measurements can be made directly. Use v_{in} for *external* triggering as in step 2. Apply v_{in}, the reference signal, to the input of vertical channel 1, and v_{out}, the output signal, to vertical channel 2. Adjust the Time/div. control for two or three waveforms, as above.

4. Connect the vertical input of the oscilloscope across R. Observe and measure the amplitude. Record in Table 5-1 the output waveform v_{out} in proper time phase with the input voltage.
5. Reverse the diode so that it is connected as D_2 in Fig. 5-2a. Check to see that the input waveform is still properly centered as in step 1. Observe and measure the output waveform. Record this waveform in Table 5-1.

Parallel Limiting

6. Connect the circuit of Fig. 5-3a. Use the same D_1 and values of R and v_{in} as previously. Observe and measure the output voltage waveform v_{out}, as in steps 1 and 2. Record this waveform in Table 5-2.
7. Reverse D_1 as in Fig. 5-4a. Observe and measure the output voltage waveform v_{out}. Record this waveform in Table 5-2.

Double-Diode Limiter Action

8. Connect the circuit of Fig. 5-14. Switches S_1 and S_2 are open. Set the output of the regulated dc power supply V_{AA} at 10 V, with the variable arm of the potentiometer, point B, set in the center of its range. That is, the arm is set so that the measured dc voltage from F to B (V_{FB}) is equal to the measured dc voltage from B to G (V_{BG}). Then

$$V_{FB} = V_{BG} = 5 \text{ V}$$

CAUTION: *The power supply must not be grounded.*

The voltage V_{AA} must be taken from the positive and negative terminals, with the negative terminal "float-

TABLE 5-2.   Parallel Limiter

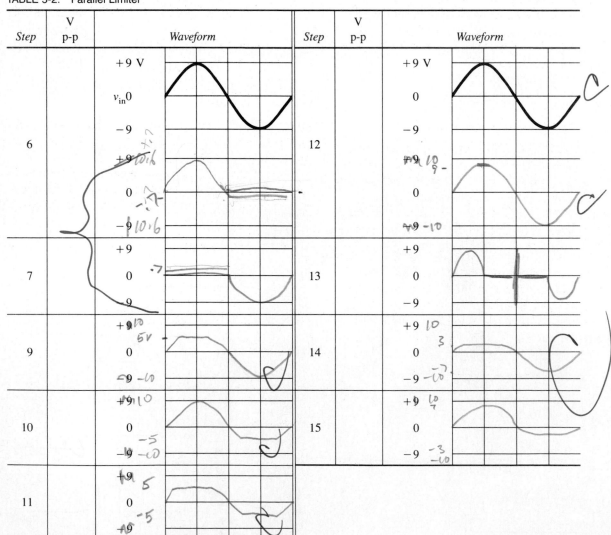

Step	V p-p	Waveform	Step	V p-p	Waveform
6			12		
7			13		
9			14		
10			15		
11					

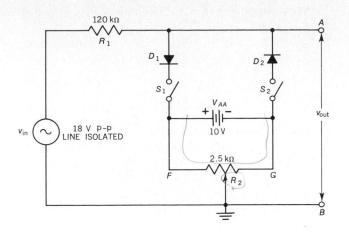

Fig. 5-14. Experimental circuit (biased double-diode limiter).

ing." *Check* v_{in} with the oscilloscope to see that the reference waveform is still properly centered as in Table 5-2. Connect the oscilloscope across the output terminals *AB*.

9. *Close* S_1. Observe and measure the output waveform v_{out}. Record this waveform in Table 5-2. 15v PP

10. *Open* S_1. *Close* S_2. Observe and measure the output waveform v_{out}. Record this waveform in Table 5-2.

11. S_1 and S_2 are both closed. Observe and measure the output waveform v_{out}. Record this waveform in Table 5-2.

12. Gradually increase the voltage V_{AA} until V_{AA} measures 18 V. Observe and measure v_{out}. Record this waveform in Table 5-2.

13. Reduce the voltage V_{AA} to 10 V. With the oscilloscope connected across the ouput, observe the effect on v_{out} of varying R_2 on either side of its center position.

14. Set R_2 so that the measured bias on D_1 is $+3$ V, on D_2, -7 V. Observe and measure v_{out}. Record this waveform in Table 5-2.

15. Set R_2 so that the measured bias on D_1 is $+7$ V, on D_2, -3 V. Observe and measure v_{out}. Record the waveform in Table 5-2.

Clamper (Extra Credit)

16. Experimentally verify the operation of a positive and negative clamper. Explain in detail the circuits you used, including all circuit values, signal source, and the nature of the measurements. Record your results in tabular form. (HINT: Use a *dc* oscilloscope to compare variation of input and output signals and to measure the *dc* level in the output.)

QUESTIONS

1. How does a positive limiter differ from a negative limiter?
2. How does a series limiter differ from a parallel limiter?
3. Is there any basic difference between the output of a vacuum-tube diode and a semiconductor-diode series limiter? Should there be? Explain.
4. Referring to Tables 5-1 and 5-2, compare both the amplitude and waveform outputs of a parallel and series-positive limiter.
5. In a biased limiter, is there any relationship between the amplitude of the output waveform and the bias voltage V_{AA}? Refer to Table 5-2 to substantiate your answer.
6. In a biased, double-diode limiter, what is the relationship, if any, between the amplitude of the output waveform and the bias voltages?

Extra Credit (Optional)

7. Compute the maximum current which can flow through the 2500-Ω potentiometer, R_2 in Fig. 5-14. Show your computations.
8. Which step in the experiment requires the maximum current in the potentiometer R_2? How much of this maximum current is there in R_2? Show your computations.

Answers to Self-Test

1. 0; -9	6. $+15$; -7
2. $+10$; 0	7. -5; $+6$
3. parallel	8. $+12$; 24; 0
4. open	9. negative; 0.167
5. closed	

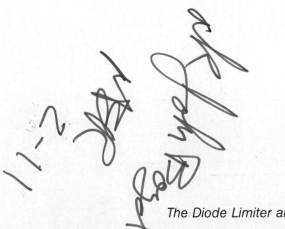

The Diode Limiter and Clamper **29**

HALF-WAVE AND FULL-WAVE RECTIFICATION

OBJECTIVES

1. To observe and measure the output waveforms of a half-wave rectifier
2. To observe and measure the output waveforms of a full-wave rectifier

INTRODUCTORY INFORMATION

DC and ac voltages and currents serve the power requirements of the wide variety of electronic devices. We have noted how alternating current is used to heat the filaments of a vacuum tube. On the other hand, in studying the static voltampere characteristics of solid-state diodes a dc supply served as the power source. Both direct-current and alternating-current power sources then must be available.

Because it is more efficient and economical to transmit, ac power is generally distributed by the power utility companies. This necessitates the rectification (changing) of ac into dc voltages and currents. In this experiment we shall be concerned with electronic means of achieving rectification.

Direct current is current which flows in only one direction. The diode with unidirectional current characteristics is admirably suited to accomplish rectification, since it permits current to flow in only one direction. Either vacuum-tube or solid-state diodes may be used as rectifiers. In this experiment we shall be concerned with solid-state-diode rectification.

Silicon, selenium, germanium, and copper oxide rectifiers are solid-state devices which serve as power rectifiers. We will use a silicon rectifier, the type most widely used in electronics today.

Diffused-Junction Silicon Rectifier

Diffused-junction silicon rectifiers are made by diffusing controlled amounts of "impurities" into thin wafers of silicon. The result is a highly reliable rectifier. The silicon rectifier has several advantages over the selenium, germanium, and copper oxide rectifiers. Silicon can operate at higher temperatures (175°C) than the other solid-state devices. It can be constructed to withstand higher reverse-bias breakdown voltages and to exhibit low reverse-bias current. Silicon rectifiers can handle high forward currents.

An ideal rectifier acts as a zero-resistance closed switch when it is forward-biased and as an infinite-resistance open switch when it is reverse-biased. That is, it is ON when its anode is positive relative to the cathode, OFF when the anode is negative relative to the cathode. Though this ideal is never realized, the silicon rectifier approaches it.

Figure 6-1 shows the voltampere characteristic of a silicon rectifier. When it is forward-biased, the rectifier exhibits extremely low forward resistance R_F. The graph shows that when there is 0.6 V across the diode, it permits 0.2 A of current for a forward resistance $R_F = 0.6/0.2 = 3\ \Omega$. When there is 0.8 V across it, the current is 0.8 A for an $R_F = 0.8/0.8 = 1\ \Omega$. The forward resistance decreases as the current through the diode increases.

The reverse-bias characteristic is equally revealing. Now the current axis is in microamperes and the reverse-bias axis

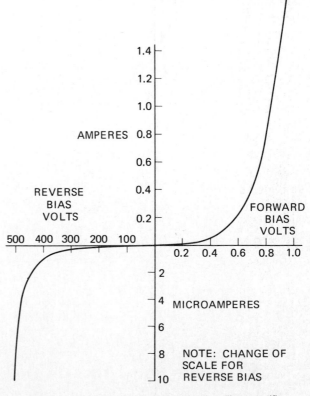

Fig. 6-1. Voltampere characteristic of a silicon rectifier.

is in 100-V divisions. At 300 V, there is approximately 0.4 μA of current, for a reverse resistance

$$R_R = \frac{300}{0.4 \times 10^{-6}} = 750 \text{ M}\Omega$$

At 500 V

$$R_R = \frac{500}{8 \times 10^{-6}} = 62.5 \text{ M}\Omega$$

Though the silicon rectifier is more tolerant of heat than semiconductors made of other materials, it is still heat-sensitive. The graph in Fig. 6-1 is the characteristic of a silicon rectifier at 100°C. If the junction temperature is increased, the forward voltage decreases. If the junction temperature is decreased, the forward voltage increases. If the maximum operating temperature, usually 175°C, is exceeded, the rectifier will fail. To avoid excessive heating of the rectifier junction, heat sinks are used. They dissipate the heat developed and assure trouble-free operation.

Silicon-Rectifier Ratings

Rectifier characteristics usually supplied by the manufacturer include:

1. The *peak inverse voltage* (PIV), which is the maximum reverse bias that can be applied across a rectifier without having the rectifier break down
2. Maximum sine-wave voltage input (rms)
3. Average half-wave rectified forward current with resistive load, at a specified temperature
4. Peak-recurrent forward current at a specified temperature
5. Maximum forward voltage at specified values of current and temperature
6. Maximum reverse current at maximum reverse voltage
7. Operating and storage temperatures
8. A derating factor, to determine the amount of current through a rectifier for a given temperature

The characteristics of a silicon rectifier, such as the maximum current it can safely handle, the maximum reverse bias (PIV), and the maximum input (rms) voltage, are determined by its construction and size. A wide range of silicon rectifiers exist which can deliver load currents from 200 mA to 1000 A, whose PIV ratings vary from 100 to more than 1000 V.

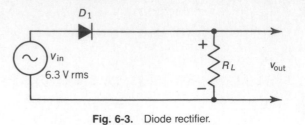

Fig. 6-3. Diode rectifier.

These rectifiers can be connected in parallel for increased load-current requirements, or in series to increase the PIV capabilities of the diode stack.

Silicon rectifiers come in various shapes and sizes including the small flangeless type with two axial leads (like the germanium diodes) (Fig. 6-2a), the top-hat single-ended type (Fig. 6-2b), and the stud-mounted type (Fig. 6-2c). Types *a* and *b* are connected in the circuit like a resistor or capacitor. Type *c* is screwed onto a metal chassis, making the chassis serve as a heat sink for the rectifier.

Half-Wave Rectification

Consider the circuit of Fig. 6-3. A sinusoidal 6.3-V rms voltage is applied across the series-connected diode D_1 and the load resistor R_L. The input voltage v_{in} is an ac voltage which changes in polarity every $\frac{1}{120}$ s. During the positive alternation the anode is positive with respect to the cathode, and current flows. During the negative alternation there is no current, because the anode is negative with respect to the cathode.

It is apparent that current through the diode will result in a voltage drop across R_L, the series-connected load resistor. Moreover, since the variation of current will follow the variation of input voltage, the output voltage v_{out} across R_L should follow the positive alternation which causes current. Figure 6-4 shows the waveforms v_{in} and v_{out}. It should be noted that v_{out} is no longer an ac voltage, but rather a pulsating dc voltage.

The diode may therefore be compared to a valve which opens only when its anode is positive with respect to its cathode. The diode has a certain internal resistance (its forward resistance) which is in series with the line and with the load resistor R_L. Hence the diode can be replaced by an equivalent resistance R_F and the effective line voltage by a

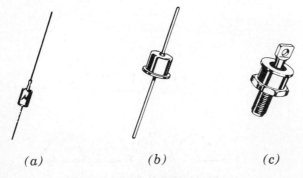

(a)   (b)   (c)

Fig. 6-2. Types of silicon rectifiers.

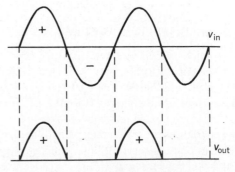

Fig. 6-4. Rectifier waveforms.

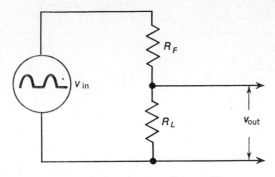

Fig. 6-5. Equivalent circuit for rectifier.

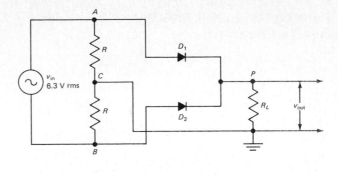

Fig. 6-7. Full-wave rectifier.

generator putting out positive alternations periodically (see Fig. 6-5).

The voltage v_{out} across R_L will therefore be a positive alternation, like the input voltage v_{in}, but smaller than v_{in} (see Fig. 6-6).

The internal forward resistance R_F of the diode rectifier should be small for maximum output v_{out} across R_L. This resistance depends on the type of diode used. The higher the current rating of the solid-state diode, the lower will be the internal resistance (R_F) of the rectifier diode. The voltage drop across the rectifier will normally be limited to about 0.7 V.

The process whereby the diode conducts during one alternation of the input cycle is called *half-wave rectification*.

The student may be struck by the similarity between the half-wave rectifier in Fig. 6-3 and the series-diode-limiter circuit, Fig. 5-2. They are identical electronically, but the function each serves is different. Hence, we shall find that the ratings and characteristics of the components that each circuit uses differ.

Full-Wave Rectification

It is possible to rectify both alternations of the input voltage by using two diodes in the circuit arrangement of Fig. 6-7. Assume 6.3 V rms (18 V p-p) is applied to the circuit. Assume further that two equal-valued series-connected resistors R are placed in parallel with the ac source. The 18 V p-p appears across the two resistors connected between points AC and CB, and point C is the electrical midpoint between A and B. Hence 9 V p-p appears across each resistor. At any moment during a cycle of v_{in}, if point A is positive relative to C, point B is negative relative to C. When A is negative to C, point B is positive relative to C. The effective voltage in proper time phase which each diode "sees" is shown in Fig. 6-8. The voltage applied to the anode of each diode is equal but opposite in polarity at any given instant.

When A is positive relative to C, the anode of D_1 is positive

with respect to its cathode. Hence D_1 will conduct but D_2 will not. During the second alternation, B is positive relative to C. The anode of D_2 is therefore positive with respect to its cathode, and D_2 conducts while D_1 is cut off.

There is conduction then by either D_1 or D_2 during the entire input-voltage cycle.

Since the two diodes have a common-cathode load resistor R_L, the output voltage across R_L will result from the alternate conduction of D_1 and D_2. The output waveform v_{out} across R_L in Fig. 6-8 therefore has no gaps as in the case of the half-wave rectifier.

The output of a full-wave rectifier is also pulsating direct current. In the diagram of Fig. 6-7, the two equal resistors R across the input voltage are necessary to provide a voltage midpoint C for circuit connection and zero reference. Note that the load resistor R_L is connected from the cathodes to this center reference point C.

An interesting fact about the output waveform v_{out} is that its peak amplitude is not 9 V as in the case of the half-wave rectifier using the same power source, but is less than 4½ V. The reason, of course, is that the peak positive voltage of A relative to C is 4½ V, not 9 V, and part of the 4½ V is lost across R.

Though the full-wave rectifier of Fig. 6-7 fills in the conduction gaps, it delivers less than half the peak output voltage that results from half-wave rectification.

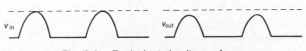

Fig. 6-6. Equivalent circuit waveforms.

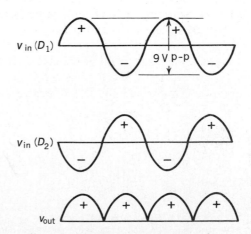

Fig. 6-8. Waveforms of full-wave rectifier.

Transformer-Fed Half-Wave and Full-Wave Rectifiers

Figure 6-7 is not a practical full-wave power-supply rectification circuit because of the use of the two center-tapped resistors R. The voltage drop across R, when its respective diode is conducting, subtracts from the voltage v_{out} and reduces the output voltage. Moreover, the dc voltage and current requirements of electronic circuits vary, depending on the devices used in the circuit, the power consumed, and other factors. The voltage and current requirements of a specific electronic device therefore determine the design of the power supply for that device.

Power transformers are built with a primary winding and one or more insulated secondary windings. The primary winding is line-fed; that is, it operates from the 120-V 60-Hz power line. The secondary windings are either voltage stepup or voltage stepdown windings. Figure 6-9 is a diagram of the transformer T_1 which will be used in this experiment. It contains a 120-V primary and a 26-V center-tapped secondary. The secondary winding is rated at 1 A. This is the maximum current which can be drawn from the secondary. We will see that the center-tap connection on the secondary eliminates the need for the two center-tapped resistors R in Fig. 6-7 and therefore makes possible a practical power rectification circuit.

The circuit of Fig. 6-10 shows how T_1 is connected in a full-wave rectifier circuit. The anodes of rectifier diodes D_1 and D_2 are fed by the secondary voltages AC and BC, respectively. Since C is the center tap, each diode anode receives 13 V rms. The load resistor R_L is connected from the junction of the cathodes of D_1 and D_2, point D, to the center tap on the secondary winding, point C. The output voltage appears across R_L.

When power is applied to the primary of T_1, and switches S_2 and S_3 are closed, D_1 and D_2 operate as a full-wave rectifier. Each diode "sees" only half the voltage appearing across the secondary, and each diode conducts alternately. When switch S_2 is closed but S_3 is open, D_1 acts as a half-wave rectifier. When S_3 is closed but S_2 is open, D_2 acts as a half-wave rectifier. Of course switches S_2 and S_3 are included in this experimental circuit simply to demonstrate half-wave and full-wave rectification. They would not be found in an industrial rectifier circuit.

Power rectifiers D_1 and D_2 are rated for the current they must deliver to a circuit and for the peak forward voltage and the peak inverse voltage they can withstand. Note that the diodes in this experiment are physically larger and heavier

than the small-signal low-current junction diodes used before.

One disadvantage of the circuit of Fig. 6-7 was the voltage loss across R. The need for R in Fig. 6-9 is eliminated by the center tap on the transformer secondary. The dc resistance between the center tap and either end of the secondary winding is very low, and hence the voltage drop across this resistance is negligibly low.

There is one advantage which a transformer power supply has over a transformerless circuit. The output voltage v_{out} of a transformer-fed supply is line-isolated, since there is no direct connection between the primary (line) winding and the secondary.

NOTE: The student may be puzzled by the fact that the output of the full-wave rectifier is pulsating dc rather than constant-level, unvarying dc voltage. The transformation of pulsating dc into constant dc voltage is achieved by filter networks. Power-supply filters will be considered in the next experiment.

Experimental Techniques for Observing Phase Relations in a Circuit

Most oscilloscopes contain "line" sync or "line" triggering facilities. This facility makes it possible to observe phase relations in a circuit receiving its input from a line-derived (60 or 50 Hz) ac source. Of course, dual-trace laboratory-standard oscilloscopes simplify the study of phase relations in an ac circuit over a wide range of input-signal frequencies.

Where neither of these instruments is available, *external sync* or triggering is used to study phase relations. For instance, suppose in Fig. 6-10 it is required to observe the phase of the input waveform AC and the output waveform DC when rectifier D_1 is in and rectifier D_2 is out of the circuit. The procedure is to set the oscilloscope on *external* sync or triggering by connecting transformer points A and C to the *external* sync or triggering jack and ground return, respectively, as in Fig. 6-11. The vertical input of the oscilloscope is connected to point A, the ground to the center tap, point C. After the vertical gain (Volts/division or V/div.) controls are adjusted for the desired waveform amplitude, the oscilloscope sweep (Time/div.) and sync (or triggering) controls are adjusted for a stable presentation of one or two cycles on the screen. The horizontal gain (where available) controls are adjusted so that the waveform is approximately 4 inches (in) wide (if a 5-in oscilloscope is used). The vertical centering controls are adjusted so that the waveform is centered with respect to the X axis. The horizontal centering controls are then set so that the start of the positive alternation of the waveform on the left coincides with a major vertical graticule line, as in Fig. 6-12.

This is the *reference* waveform. The oscilloscope sweep, sync, and horizontal gain and centering controls must not be varied until the required phase relations have been observed.

So that the phase relationship between this input wave and the output across R_L for rectifier D_1 in Fig. 6-10 can be observed, switch S_3 is opened while S_2 is closed. The vertical

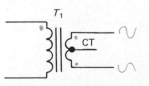

Fig. 6-9. Voltage step-down power transformer with center-tapped secondary.

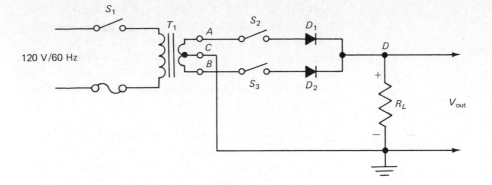

Fig. 6-10. Transformer-fed, full-wave experimental voltage rectifier.

input (hot) lead of the oscilloscope is connected to the top of R_L, point D. The observed output waveform is then in proper time phase with the input reference waveform.

SUMMARY

1. Power companies distribute ac electric power because this method of distribution is more efficient and economical than dc distribution.
2. Because dc voltage and current are required for electronic devices, it is necessary to convert ac into dc by a process called *rectification.*
3. Silicon rectifiers are in greater use in electronics than any other solid-state rectifier.
4. Silicon power rectifiers are rated for their peak inverse voltage (PIV); the peak forward and reverse voltages they can withstand; for the average forward current and the peak recurrent forward current they can provide at a specified temperature; for their operating and storage temperatures; and for the maximum sine-wave voltage input (rms) they can tolerate.
5. A wide range of silicon power rectifiers exists which can meet load current demands in the range of 200 mA to 1000 A.
6. A single rectifier diode, connected as in Fig. 6-3 serves as a half-wave rectifier, in which only *one* alternation of the ac waveform is applied to the load.
7. When two rectifier diodes are used, as in Fig. 6-10, we

have full-wave rectification. Here the two alternations of the input sine wave are processed alternately by diodes D_1 and D_2.

8. The rectified output of a half-wave rectifier appears as unidirectional current pulses (Fig. 6-4). The rectified output of a full-wave rectifier also appears as unidirectional current pulses (Fig. 6-8), but here we have two pulses for every sine wave of input. The rectifiers have transformed the ac waveform into pulsating dc.
9. Each rectifier, when conducting, is not a perfect switch; it exhibits some internal resistance (R_F). Because of this resistance, there is some voltage loss (drop) across each diode.
10. Full-wave power-rectifier circuits normally utilize power transformers with center-tapped secondary windings. Whether the secondary winding of the transformer feeding the rectifiers (Fig. 6-10) is a voltage stepup or stepdown winding depends on the requirements of the electronic device for which the power supply is intended.
11. Either the rectifier output voltage and current can be pulsating dc, as in Fig. 6-10, or the pulses can be filtered, resulting in pure dc.
12. When transformers are used for power-rectifier circuits, the resistance of the secondary windings must be very low to reduce power losses in the ouput.

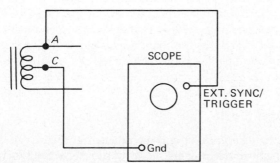

Fig. 6-11. Sync connection for observing phase relations between input and output waveforms in Fig. 6-10. Oscilloscope is set for external sync or external trigger.

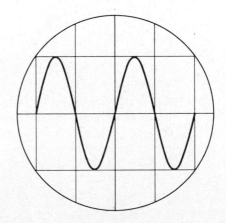

Fig. 6-12. Reference waveform properly centered for observing phase relations in the circuit of Fig. 6-10.

Half-Wave and Full-Wave Rectification **35**

Check your understanding by answering these questions.

1. Because dc power is used mainly in electronics, power companies distribute dc for this need.
_____ (*true, false*)

2. _____ is the process in which ac is converted into dc.

3. A half-wave rectifier, like that in Fig. 6-3, receives a 60-V p-p input. The output waveform across R_L, point D with respect to C, is:
   (*a*) _____ . (*positive, negative*)
   (*b*) In phase with the _____ (*positive, negative*) alternation of the input voltage.
   (*c*) About _____ V p-p.

4. To reverse the polarity of the output waveform in a half-wave rectifier, it is necessary to reverse the _____ in the circuit.

5. In the full-wave rectifier of Fig. 6-10, diode D_1 conducts when _____ is cut off, and vice versa (assume S_2 and S_3 are both closed).

6. If the frequency of the source from which transformer T_1 in Fig. 6-10 receives its power is 60 Hz, the frequency of the output waveform is _____ Hz. (S_2 and S_3 are closed.)

7. In Fig. 6-10 switch S_2 is open, S_3 is closed. The circuit operates like a _____ - _____ rectifier.

8. For the same conditions as in question 7, there is one output pulse, for each alternation of input sine wave. _____ (*true, false*)

9. In Fig. 6-10, when switches S_2 and S_3 are both open, there is no output across R_L. _____ (*true, false*)

10. Phase relations between the input and output waveforms in Fig. 6-10 may be observed by _____ synchronization or triggering of the oscilloscope, or by _____ synchronization or triggering of the oscilloscope.

MATERIALS REQUIRED

- Equipment: Oscilloscope; EVM
- Resistors: 10,000-Ω ½-W
- Solid-state diodes: Two 1N5625
- Miscellaneous: Three SPST switches; power transformer 117-V primary, 26-V center-tapped secondary at 1-A (Triad F40X or equivalent); fused line cord

PROCEDURE

1. Connect the circuit of Fig. 6-10 and set the oscilloscope on *line* sync or triggering. If this facility is not available, set the oscilloscope for *external* sync or triggering, using the circuit of Fig. 6-11. *Have an instructor check your circuit before proceeding.*

2. Connect the vertical input lead of the oscilloscope to the anode of D_1, the ground lead to point C.
   Close switch S_1. **Power on.** *Close* switch S_2, but keep S_3 *open*.
   Calibrate the vertical amplifiers of the oscilloscope for voltage measurement.

3. Adjust the vertical decade attenuator, horizontal gain, sweep, and sync/triggering controls for viewing the reference waveform V_{AC}, as described in the section headed "Experimental Techniques for Observing Phase Relations in a Circuit." The waveform viewed should be identical with the reference waveform in Table 6-1.
   With the oscilloscope, measure the peak-to-peak voltage of V_{AC}. Record the results in Table 6-1. With an EVM, measure the dc voltage, if any, across points AC. Record the results in Table 6-1.

4. *Open* S_2. Connect the vertical input lead of the oscilloscope to the anode of D_2. Close S_3. Draw V_{BC}, in voltage waveform observed, in Table 6-1 in proper time phase with the reference waveform.

   Measure and record the peak-to-peak voltage and the dc voltage, if any, across BC.

5. *Open* S_3. Connect the vertical input lead of the oscilloscope to point D (across R_L).
   Close S_2. Draw the waveform v_{out} observed across R_L, in proper time phase with the reference waveform.
   Measure the peak-to-peak voltage and the dc voltage, if any, across R_L. Record the results in Table 6-1.

6. *Open* S_2. *Close* S_3. Draw the output waveform v_{out} observed across R_L, as above. Measure the peak-to-peak amplitude of the waveform and the dc voltage, if any, across R_L. Record the results in Table 6-1.

7. *Close* S_2. All switches are now closed. Draw the output waveform v_{out}. Measure the peak-to-peak amplitude of the waveform and the dc voltage, if any, across R_L. Record the results in Table 6-1.

Extra Credit

8. Explain in detail an experimental procedure you would use for determining the forward resistance of each rectifier in the circuit of Fig. 6-10.
   Now apply the procedure. Measure (or compute) the resistance of D_1 and D_2. Record the results in Table 6-1.

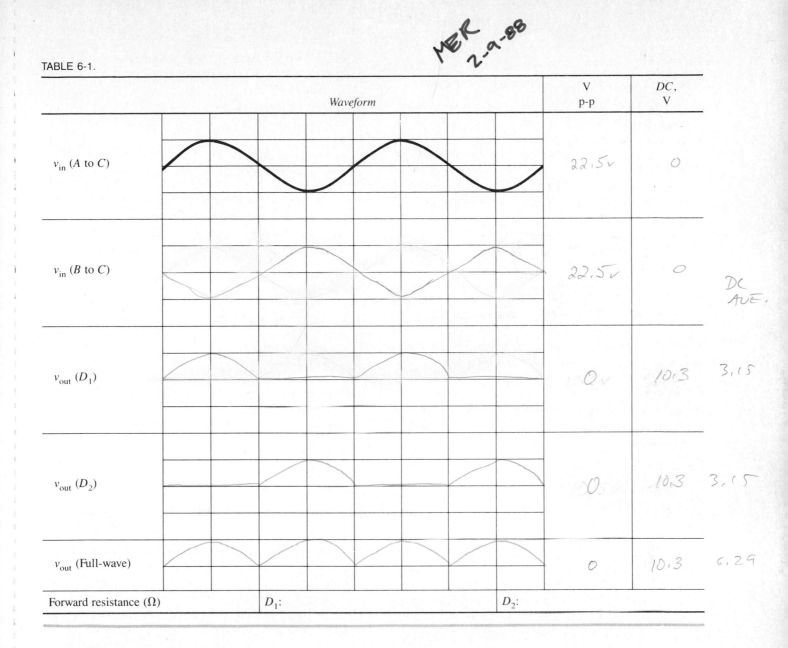

TABLE 6-1.

	Waveform	V p-p	DC, V
v_{in} (A to C)		22.5v	0
v_{in} (B to C)		22.5v	0 DC AVE.
v_{out} (D_1)		0v .10.3	3.15
v_{out} (D_2)		0 .10.3	3.15
v_{out} (Full-wave)		0	10.3 6.29
Forward resistance (Ω)	D_1:	D_2:	

QUESTIONS

1. Explain what is meant by rectification. CHANGING OF AC INTO DC
2. Describe the nature of the output voltage v_{out} of (a) a half-wave rectifier, (b) a full-wave rectifier. ½ OF SIN WAVE   ½'S OF 2 OUT OF PHASE SIN WAVES
3. Compare, with relation to the input frequency, the frequency of the output voltage of (a) a half-wave rectifier, (b) a full-wave rectifier. (Refer to the waveforms in Table 6-1). SAME FREQUENCY
4. From your data, what conclusion can you draw about the relationship between the dc voltage across R_L and the peak input voltage in (a) a half-wave rectifier? (b) a full-wave rectifier? State this relationship mathematically, if possible.
5. In which steps of the procedure was the circuit a half-wave rectifier (a) for D_1? (b) For D_2? 5   6

6. Refer to your data and explain the operation of (a) a half-wave rectifier, (b) a full-wave rectifier.

Extra Credit (Optional)

7. What is meant by forward resistance of a diode?
8. Explain how Kirchhoff's voltage law applies to the circuit of a half-wave rectifier. Illustrate with a diagram, using measured or assumed voltages.

Answers to Self-Test

1. false
2. rectification
3. a, positive; b, positive; c, 30
4. rectifier or diode
5. D_2

6. 120
7. half-wave
8. false
9. true
10. line; external

A STABLE

555

EXPERIMENT

7

TRANSFORMER POWER SUPPLY AND FILTER

OBJECTIVES

1. To measure the effects of filter elements on the dc output voltage and ripple
2. To test and compare the effectiveness of (a) capacitor filter (b) π-type filter
3. To measure and compare the regulation of half- and full-wave transformer-fed power supplies

INTRODUCTORY INFORMATION

Rectification of alternating current to pulsating direct current is achieved by the circuit of Fig. 7-1. Pulses are smoothed by filter networks. The full-wave rectifier was studied in Experiment 6. In this experiment the effect of filtering on the nature of the rectified voltage will be observed.

Capacitors, chokes, and resistors are filter elements. The effectiveness of a capacitor as a filter is related to its capacitance and reactance. Larger capacitances have better filtering action. The filtering action of a choke is related to its inductance.

Capacitor Input Filter

Consider the circuit of Fig. 7-2a. An electrolytic capacitor C_1 replaces the load resistor R in the preceding circuit. C_1 charges alternately through each diode section as the diodes conduct during the alternations when their anodes are positive relative to the common cathode. The polarity of voltage developed across C_1, which charges to the peak of the input voltage, makes the cathode positive relative to ground. There is no path through which the capacitor can discharge, except through its own parallel leakage resistance, which is ordinarily very high. Hence, C_1 maintains a high positive dc

voltage which effectively biases both of the rectifiers to cut off. The rectifier diodes conduct only during the peaks of the positive alternations of input ac voltage, replacing the small charge that C_1 has lost during the discharge interval. An oscilloscope connected across C_1 will show a relatively constant dc voltage with hardly a trace of ripple. An EVM across C_1 will measure a dc voltage equal approximately to the peak of the ac input voltage to each rectifier.

If a load resistor R is connected across C_1 (Fig. 7-2b), the rectified, filtered dc voltage is applied to the load. R draws current from the supply, and the value of R determines how much current is drawn. Since the rectifiers are still cut off during a large portion of the input cycle, the current drawn by the load is actually supplied by C_1, which discharges through R. If the load current is high, that is, if the resistance of R is relatively low, the dc output voltage drops appreciably during the discharge cycle and rises during the interval that C_1 is charging through the rectifiers. The output voltage V_{out} is no longer a steady voltage, but varies between some maximum and minimum value in the manner shown in Fig. 7-2c. This variation in capacitor charge is the ripple that is observed with an oscilloscope across C_1. If C_1 is replaced by a capacitor of higher capacitance, the ripple is decreased. The dc voltage measured with an EVM is lower with load than without.

For low load-current applications, the capacitive filter in Fig. 7-2b may be adequate to maintain a relatively constant dc level. For higher load currents, a more effective filter is required if a ripple-free output voltage is desired.

A more effective filter is shown in Fig. 7-3. A choke coil L and another electrolytic capacitor C_2 have been added to C_1. The output dc voltage V_{out} across C_2, which is designated V_{PG}, is now applied to R. The effect of C_1, L, and C_2 is to improve the filtering action by increasing the charge stored in

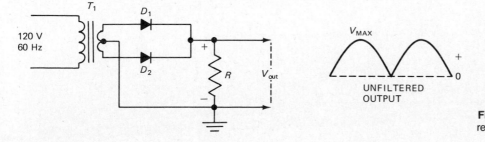

Fig. 7-1. Unfiltered output of full-wave rectifier.

39

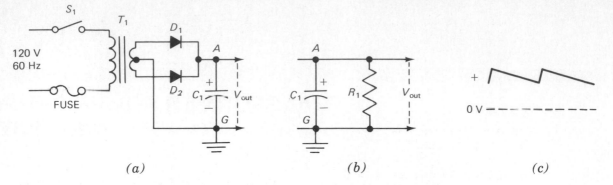

(a)   (b)   (c)

Fig. 7-2. Effect of filter capacitor on output of full-wave rectifier.

these reactive components. A load current drawn from this supply will cause less ripple in the output than an equivalent load will cause in the output of Fig. 7-2b. This arrangement is a pi- or π-type filter, so called because of the Greek letter π which the filter configuration resembles.

Because the first filter element is the capacitor C_1, it is designated a capacitor input filter. A characteristic of this type of filter is that it provides maximum voltage output to the load. Since large capacitors are needed, C_1 and C_2 are electrolytics, connected with the polarity shown. The maximum value of input capacitor which a rectifier can safely handle is usually specified in the manufacturer's manual.

The resistance of the windings of the iron-core choke L in series with the resistor R constitutes a dc voltage divider. The dc voltage V_{PG} between point P and ground is therefore lower than the voltage V_{AG} from A to ground. How much lower it is, is determined by the current I_L through L and the resistance R_L of L, because the dc voltage drop V_{AP} across L equals $I_L \times R_L$. The relationship is

$$V_{AG} - V_{AP} = V_{PG} \qquad (7\text{-}1)$$

For high load currents a large inductance with low internal resistance is required.

The inductance of the iron-core choke is directly related to its effectiveness as a filter element. A characteristic of the choke is to oppose a change in current, while the capacitors oppose a change in voltage. The filter tends to average out the rectified pulses by clipping the peaks and filling in the valleys and thus supplies a relatively constant voltage to the load.

The output dc voltage is termed $V+$. The value of $V+$, then, depends on the ac voltage across the high-voltage secondary, the size of the filter capacitors and choke, and the value of load current. Without any load, the dc output voltage is approximately equal to the peak voltage of each secondary winding of the transformer, that is, from either anode to common.

Frequently a resistor R_C is used to replace the choke as a filter element. Figure 7-4 illustrates this type of filter. It should be noted, however, that a resistor is not as effective a filtering component as a choke. Therefore, if circuit requirements or cost suggest the use of a resistor rather than a choke, larger valued filter capacitors C_1 and C_2 are required to compensate for the loss of the choke.

For solid-state devices requiring low voltages, either regulated supplies (treated later in this book) or power supplies using π-type CRC filters, such as that shown in Fig. 7-4, are employed. For the CRC filtered supplies it is not unusual to find capacitors with values of 500 to 1000 microfarads (μF). CRC filters are not designed for high-voltage supplies because large high-voltage electrolytic capacitors are very bulky and expensive. Low-voltage electrolytic capacitors are cheaper and less bulky.

Voltage Regulation

If a power supply is allowed to operate near peak voltage, it will result in poor voltage regulation. In order to provide better regulation, a certain minimum current must be drawn

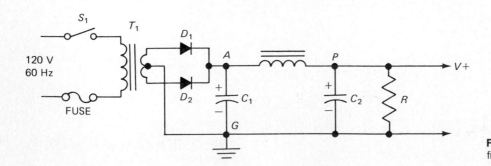

Fig. 7-3. Full-wave rectifier with π-type filter.

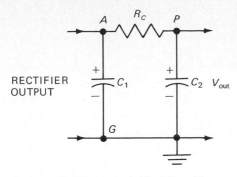

Fig. 7-4. Resistor used in π-type filter.

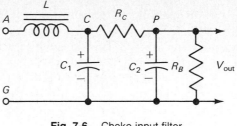

Fig. 7-6. Choke input filter.

from the supply at all times. Any additional variations in load current cause a voltage drop across the rectifier diode and across the resistance of the choke L, or across the filter resistor R. The greater the load current I_L, the greater the voltage drops $I_L R$ across the resistance of the choke or the filter resistor and diode, and hence the lower is the voltage V which can be delivered to the load.

The regulation of a supply is an index which shows how output voltage V varies as it is loaded. The equation for percentage of regulation is

$$\text{Percent regulation} = 100 \times \frac{V_{max} - V_{min}}{V_{min}} \qquad (7\text{-}2)$$

where V_{max} is the no-load voltage and V_{min} is the full-load voltage.

Regulation is improved by the use of a bleeder resistor R_B (Fig. 7-5). This provides bleeder current under all load conditions. R_B also rapidly discharges C_1 and C_2 when power is shut off. Otherwise, these capacitors might become a shock hazard by carrying a charge for long periods of time after power has been removed. For good regulation the bleeder current should be about 15 to 20 percent of total current.

Choke Input Filter

For some high-voltage applications with relatively large variations in load current, better regulation is required than is possible with a capacitor input filter. The use of a choke input filter with a specified minimum bleeder current (Fig. 7-6) provides the improved regulation. However, the output volt-

age V, other conditions being the same as in the circuit of Fig. 7-5, is lower than that previously realized. Of course, the addition of an extra choke has improved the filtering.

The power transformer and the choke coil or coils must be capable of handling the maximum load-current requirement. Thus, if 90-mA load current is to be supplied, the choke must be rated somewhat higher. The same is true of the dc rating for the secondary of the power transformer T.

Where a resistor R_C is used instead of a choke, its wattage must be greater than $I^2 R_C$, where I is the total current drawn, bleed and load.

Regulation of a Half-Wave Rectifier

If only a single rectifier is used in the circuit of Fig. 7-5, or if an anode of either diode is opened, the resulting circuit is that of a half-wave rectifier. For the same load conditions, voltage V drops because now only one of the diodes is supplying the load current instead of two. The filter must therefore supply load current during the alternation when the diode is not conducting. Also, the ripple frequency changes from 120 Hz for full-wave rectification to 60 Hz for half-wave action (based on a line frequency of 60 Hz). At the lower frequency the filter is not as effective, making the ripple voltage much higher, and the dc voltage lower. Full-wave rectifiers are therefore normally used in transformer-type power-supply circuits.

SUMMARY

1. Filters are used to smooth the pulsating dc output of rectifiers. Filter elements are capacitors, chokes, and resistors.

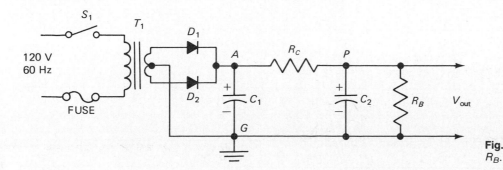

Fig. 7-5. Power supply with bleed resistor R_B.

2. The larger the value of capacitance and inductance, the better the filtering action of conventional π-type filters.
3. For line-derived power-supply filters electrolytic capacitors are used which provide large values of capacitance in relatively small units.
4. The most popular filter arrangement is a π-type unit, using an input capacitor, an inductor or resistor, and an output capacitor, connected as in Figs. 7-3 and 7-4.
5. A capacitor input filter ensures higher output voltage than a choke input filter (Fig. 7-6). A choke input filter provides better regulation.
6. The regulation of a power supply is an indication of how the output voltage varies with load. A small variation between the no-load voltage and the full-load voltage indicates good regulation. Regulation can be improved by adding bleeder resistors as in Fig. 7-5.

5. The resistor R_B in Fig. 7-5 draws about 15 to 20 percent of the total current which the rectifier supplies. This resistor, which is permanently wired into the circuit, is called a _____ resistor.
6. Two functions which the resistor R_B serves in Fig. 7-5 are: It improves _____ , and it _____ capacitors C_1 and C_2 when power is removed from the circuit.
7. Electrolytic capacitors are used as filter elements in line-derived power supplies because of their _____ _____ value in a relatively small unit.
8. A capacitor input filter provides a _____ (higher, lower) voltage output than a choke input filter.
9. A choke input filter provides better _____ than a capacitor input filter.
10. If in Fig. 7-5, diode D_1 were open, the output voltage V would be _____ (higher, lower, the same) than (as) the output voltage if both diodes were in the circuit.

SELF-TEST

Check your understanding by answering these questions.

1. The unfiltered output of the full-wave rectifier in Fig. 7-1 is _____ (positive, negative).
2. The average dc voltage in the output of Fig. 7-1 is approximately _____ percent of V_{max}.
3. With no load (Fig. 7-2a), the pure dc output is approximately equal to _____ .
4. The filter in the output of Fig. 7-3 is a _____ filter.

MATERIALS REQUIRED

- Power supply: 120-V rms 60-Hz source
- Equipment: Oscilloscope; EVM or VOM
- Resistors: 100-Ω, 2700-Ω, ½-W
- Capacitors: Two 100-μF 50-V; 25-μF 50-V
- Solid-state rectifiers: Two 1N5625 or equivalent
- Miscellaneous: Power transformer T_1, 120-V primary, 26.8-V 1-A center-tapped secondary; SPST switch; fused line cord

PROCEDURE

Transformer Supply with Capacitive Filter

1. Connect the circuit of Fig. 7-2a and b. S_1 is off. T_1 is the same power transformer as that used in Experiment 6. C_1 is a 100-μF 50-V capacitor. R is a 2700-Ω ½-W bleeder resistor. Have an instructor check the circuit before proceeding.

2. **Power on.** With a voltmeter measure V_{out}, the dc output voltage across R, and record the result in Table 7-1. With an oscilloscope connected across R, observe, measure, and record in Table 7-1 the ripple waveform and its peak-to-peak amplitude.

3. **Power off.** Discharge capacitor C_1. Connect a 250-Ω 2-W load resistor in parallel with the bleeder resistor. The voltmeter and oscilloscope are still connected across R.

4. Measure V_{out} and the ripple waveform. Record the results in Table 7-1. Compute the load current I_L in the 250-Ω resistor and record the result in Table 7-1. Show your computations.

5. **Power off.** Discharge capacitor C_1.

NOTE: It is always necessary to discharge the filter capacitors after power is turned off when work is to be done in the power supply.

TABLE 7-1. Capacitive Filter

Step	Load, Ω	V_{out}, V	Ripple Waveform	V p-p	I_L (Computed), mA
2	None	9.6	.4 0	.4	X
4	250	8.3	2.7 0	2.7	8.3 / 228.8 = 36 mA
6	250	6.75	6 0	6	6.75 / 228 = 29.5 A

TABLE 7-2. π-Filter Full-wave Rectifier

Step	Point	No Load DC, V	No Load Ripple Waveform	No Load Ripple V p-p	250-Ω 2-W Load DC, V	250-Ω 2-W Load Ripple Waveform	250-Ω 2-W Load Ripple V p-p	Condition
8, 10	A–G	9.65	⟿	.38v	8.65	⟿	.685	π Filter
8, 10	P–G	9.2	⟿	32mv	5.85	⟿	.23	
11	A–G	6.65	⟿	(.5) ok	6.6	⟿	5.2	C₁ Open
11	P–G	4.6	⟿	.81	4.4	⟿	.82	
12	A–G	8.7	⟿	1.9	8.5	⟿	1.9	C₂ Open
12	P–G	(6)	⟿	(6.3)	4.85	⟿	(10.~) 1.6	

6. Replace C_1 with a 25-μF 50-V capacitor. **Power on.** Measure V_{out} and the ripple waveform. Record your results in Table 7-1. Compute and record the load current I_L.

Capacitor Input π Filter

7. Power off. Remove capacitor C_1, the bleeder, and load resistors and connect the circuit of Fig. 7-5. C_1 and C_2 are 100-μF 50-V capacitors; R_C is a 100-Ω ½-W resistor; R_B is a 2200-Ω ½-W resistor.

8. Power on. Measure and record in Table 7-2 the dc no-load voltage across C_1 (point A to G) and V_{out} (point P to G). Observe and measure with an oscilloscope and record the ripple waveform and its peak-to-peak voltage A to G and P to G.

9. Power off. Connect the 250-Ω 2-W load resistor in parallel with R_B.

10. Power on. With load, repeat your measurements in step 8 and record in Table 7-2.

11. Power off. Remove C_1 from the circuit. Leave the other components in the circuit. Repeat steps 8 and 10.

12. Power off. Replace C_1 in the circuit. Remove C_2 from the circuit. Leave the other components. Repeat steps 8 and 10.

Regulation Half-Wave Rectifier

13. Power off. Replace C_2 in the circuit. This restores the π filter in Fig. 7-5. The bleeder resistor is still connected. Remove the 250-Ω load resistor. Remove D_2 from the circuit. We now have a half-wave rectifier and π filter, without load.

14. Power on. Measure and record in Table 7-3 the dc no-load voltage across points A to G and P to G. Observe, measure, and record the ripple waveform and its peak-to-peak voltage at points A to G; points P to G.

15. Replace the 250-Ω load resistor in the circuit. Repeat the measurements in step 14.

TABLE 7-3. Half-Wave Rectifier Measurements with and without Load

Point	No Load DC, V	No Load Ripple Waveform	No Load Ripple V p-p	250-Ω 2-W Load DC, V	250-Ω 2-W Load Ripple Waveform	250-Ω 2-W Load Ripple V p-p
A–G	9.6	⟿	.87	8.(?)	⟿	1.9
P–G	9.6	⟿	.26	5.8	⟿	10.3

QUESTIONS

Refer specifically to the data in Tables 7-1, 7-2, and 7-3 in answering these questions. Identify the table and data on which your answer is based.

1. Compare the full-wave and half-wave supplies with π filter under 250-Ω load as to (*a*) dc voltage at input and output of filter, (*b*) ripple voltage at input and output of filter.

2. Explain the difference, if any, in output voltage between the full-wave and half-wave rectifiers under load.

3. Explain the difference, if any, in ripple voltage between the outputs of the full-wave and half-wave rectifiers under load.

4. Calculate the regulation of the full-wave and half-wave rectifiers, using a 250-Ω load. Show your calculations. Which has the better regulation?

5. How does a capacitive filter (Fig. 7-2) compare with a π filter, for a full-wave rectifier under load, as to (a) dc output, (b) ripple? Explain why.

6. (a) Which value capacitor is more effective as a filter element, the larger or smaller? (b) Why? (c) Where in your experiment did you determine this fact?

7. At which point in the π filter, input or output, is the ripple voltage higher? Why?

8. In the experimental full-wave rectifier (Fig. 7-5), which open capacitor reduced the dc output voltage more, C_1 or C_2? Why?

9. In the experimental full-wave rectifier (Fig. 7-5), which open capacitor increased the ripple voltage in the output more, C_1 or C_2? Why?

10. What is the objection to having a high ripple in the output voltage V at P (Fig. 7-5)?

11. At which point in the π filter, input or output, is the dc voltage higher? Why?

12. A full-wave power supply and filter have been designed to deliver 250 mA at 40 V, with ½-V ripple. What would be the effect of increasing the load current appreciably (assuming the components can handle the increased current without damage) on (a) ripple? (b) dc voltage? Why?

Answers to Self-Test

1. positive
2. 63.6 percent
3. V_{max}
4. π
5. bleeder
6. regulation; discharges
7. large capacitance
8. higher
9. regulation
10. lower

POWER-SUPPLY TROUBLESHOOTING

OBJECTIVES

1. To construct and measure the output of a supply which provides both positive and negative output with respect to a common return
2. To troubleshoot a power supply

INTRODUCTORY INFORMATION

Positive and Negative Supply

It is frequently necessary to provide a source of $V-$ and $V+$ voltage, measured with respect to a common return. A simple modification of the $V+$ supply in Fig. 8-1 will achieve this (see Fig. 8-2). The circuit in Fig. 8-2 is the same as that in Fig. 8-1, except that R_B, the bleeder resistor, has been replaced by two resistors, R_{B1} and R_{B2}, and filter capacitors C_3 and C_4 have been added.

The junction of R_{B1} and R_{B2}, point G, is chosen as the common reference or "ground" point. Point P_1 is positive and P_2 is negative with respect to G. $V+$ is therefore taken from P_1 to G and $V-$ from P_2 to G. The resistance ratio of the divider R_{B1} and R_{B2}, in conjunction with the respective load resistances of $V+$ and $V-$, determines the amplitude of $V+$ and $V-$ voltage. Note that the total value of voltages $V+$ to $V-$ for similar loading conditions is the same as $V+$ to ground in the circuit of Fig. 8-1. The circuit of Fig. 8-2 has not increased the total dc voltage output of the power supply. It has reduced the available *positive* voltage $V+$ (P_1 to G) by the amount of *negative* voltage $V-$ (P_2 to G).

C_1, C_2, C_3, and C_4 are electrolytic capacitors, and it is therefore necessary to observe proper polarity in connecting them in the circuit. Note that the polarity of C_4 has been properly indicated because point G is positive with respect to P_2.

Voltage, Ripple, and Resistance Analysis of a Power Supply

In troubleshooting a defective power supply it is necessary for the technician to know the *normal values* of dc and ac voltage, ripple, and resistance expected at test points in the circuit. The discussion that follows proposes guidelines for determining these normal values. Where the manufacturer specifies the usual test point readings, they will serve as the standard of reference for troubleshooting measurements. DC and ac voltages and resistances are of course measured with a volt-ohm-milliammeter (VOM) or an electronic volt-ohm-milliammeter (EVM). Ripple measurements are made with an oscilloscope.

DC Voltage and Ripple

Consider the circuit of Fig. 8-1. The dc voltage measured from point A to G is *approximately* equal to one-half the peak-to-peak ac voltage across the secondary winding of T_1. The output dc voltage ($V+$), point P with respect to G, is lower than the voltage from A to G. However, $V+$ is much better filtered (it has a lower ripple) than the voltage from A to G. The ripple voltage at A and P increases with load. $V+$, however, should have a very low ripple voltage, even at

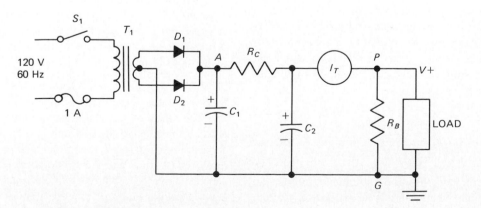

Fig. 8-1. Power supply connected to furnish $V+$ with respect to common return G.

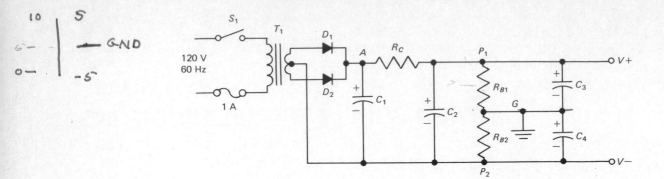

Fig. 8-2. Power supply connected to furnish $V+$ and $V-$ with respect to common return G.

full load. The size of the electrolytic filter capacitors therefore depends on the extent of the load current and on the level of ripple voltage which the circuit can tolerate.

The voltage $V+$ in Fig. 8-1 can be found if the dc voltage V_{AG}, the total dc current I_T drawn from the supply, and the filter resistor R_C are known, since

$$V+ = V_{AG} - I_T \times R_c \qquad (8\text{-}1)$$

Example. Determine $V+$ in the circuit of Fig. 8-1, if the peak-to-peak voltage across the secondary of T_1 is 50 V, $R_C = 100\ \Omega$, and $I_T = 0.1$ A.

Solution. $V_{AG} = \frac{1}{2} \times 50$ (approx) $= 25$ V. Therefore

$$V+ = 25 - 0.1\ (100) = 15\ \text{V}$$

A leaky capacitor C_1 or C_2 reduces the output voltage $V+$, and the ripple voltage increases. If capacitor C_1 or C_2 is open, this reduces $V+$ appreciably, and causes a large increase in ripple.

AC Voltage

Under normal operation the line voltage (120 V/60 Hz in Fig. 8-1) is present across the primary of transformer T_1. The voltage specified for the secondary of the transformer should be measured across the secondary winding. Line voltage and voltages across the one or more secondary windings of a transformer are given in rms values. Voltages measured from the anodes of D_1 and D_2, respectively, to the center tap on the secondary should be approximately equal to one-half the measured value across the secondary winding. AC voltage in a transformer power supply such as that in Figs. 8-1 and 8-2 is limited to the voltage across the primary winding (line voltage), to the voltages across each of the secondary windings, and to the voltages from each end of a secondary winding to any tap on that winding.

Resistance

NOTE: *Before measuring resistance, turn power* **off** *and discharge all electrolytic capacitors.*

Resistance measurements in Fig. 8-1 are made from $V+$ to ground to determine whether C_1 or C_2 is short-circuited. Other resistance measurements may be made to determine

continuity of the transformer windings and the resistance of R_C or the choke, the bleeder resistor, the on-off switch, and the rectifier diodes. Where the circuit arrangement makes it impossible to measure the resistance of a component without measuring the combination resistance of another component in parallel with it, it may be necessary to disconnect *one end* of the component from the circuit and then measure.

Specific resistance values are related to the parameters of the circuit. The following are suggested as guidelines in the circuit of Fig. 8-1, where C_1 and C_2 are 100-μF 50-V capacitors, $R_C = 100\ \Omega$, $R_B = 5000\ \Omega$, the load is disconnected, and T_1 is a 4:1 stepdown transformer whose secondary winding is rated at 1 A.

- P to ground, 5000 Ω
- A to ground, 5000 Ω
- R_C *(A to P)*, 100 Ω
- Forward resistance (R_F) of D_1 or $D_2 \lessgtr 500\ \Omega$
- Back resistance (R_R) of D_1 or $D_2 > 1\ \text{M}\Omega$
- The ratio $R_R/R_F \gtrless 2000$
- Resistance of secondary winding of $T_1 = 0.2\ \Omega$
- Resistance of primary winding of $T_1 = 1.2\ \Omega$

When the bleeder and load are both disconnected from the circuit, capacitors C_1 and C_2 give a charging indication when an ohmmeter is first connected from point P or point A to G. The meter will first read close to zero resistance. Then if the meter leads are left connected, the capacitors will slowly charge toward the supply voltage in the ohmmeter. The charging time constant is *long* because of the large values of C_1 and C_2. Therefore the measured resistance will *gradually* increase. The resistance may finally measure 1 MΩ or higher. A brief capacitor charging indication is evidence that C_1 and C_2 are not shorted.

NOTE 1: An ohmmeter check is not a conclusive test of a leaky capacitor because the ohmmeter voltage is relatively low. The capacitor may charge on an ohmmeter test but break down when rated voltage is applied. A dynamic test of a capacitor is to test for *dc* voltage at P and at A. If the voltage is lower than normal and the ripple is excessive, one or both capacitors may be defective. Unhook one lead of a suspected leaky capacitor and replace with a good one. If the circuit functions properly, the defect is in the original capacitor.

NOTE 2: The forward and back resistances of a rectifier diode are measured by placing the ohmmeter leads across the diode and reading the resistance, then reversing the leads across the diode and again reading the resistance. The forward resistance (R_F) should be relatively low and the back resistance (R_R) very high. If R_F and R_R are close in value, the diode is defective.

Troubleshooting a Power Supply

In troubleshooting an electronic device defects may sometimes be traced to the power supply. Thus, in Fig. 8-1, if the dc output voltage V_{PG} is lower than normal, or if the ripple voltage at P or A is higher than normal, the trouble may be in the power-supply circuit, *or* in the load, for a large increase in load current can give these indications.

A first step in the troubleshooting process is to isolate the trouble to the load or to the supply by disconnecting the load from the supply. If the measured dc voltage at P is now higher than the rated $V+$ voltage under load, and if the ripple voltage at P is appreciably lower than the rated ripple under load, the trouble is in the load circuit. However, if the voltage V_{PG} without load is still low, and/or the ripple is still higher than normal, the trouble is in the power supply.

Troubleshooting the supply requires measuring both the dc voltages and the ripple voltages at P and A with respect to ground. The results of these measurements may give a clue as to the trouble.

No $V+$ Voltage

If $V_{PG} = 0$ V, the trouble may be due to defects in any of the following components: (*a*) open line cord or defective plug, (*b*) open fuse, (*c*) open switch, (*d*) open or shorted transformer winding, (*e*) defective D_1 and D_2, (*f*) shorted C_1, (*g*) shorted C_2, (*h*) open R_C, (*i*) shorted R_B (very unlikely). The problem then is to find the defective component and replace it with a known good one. This may involve voltage measurements, resistance measurements, and parts substitution.

If V_{PG} measures 0 V, the next check is to measure the *dc* voltage V_{AG}. If there is voltage, the trouble is a shorted C_2, an open R_C, or a very unlikely shorted R_B. A resistance check of these components will determine the trouble.

If V_{AG} also measures 0 V dc, the trouble may be a shorted C_1. However, before resistance-checking this component, another voltage check is indicated. Measure the ac voltage across the secondary of T_1. It is possible to eliminate four of the components with this one check. If the ac voltage across the secondary is normal, then the line cord, fuse, switch, and power transformer are okay.

If there is no ac voltage across the secondary, remove the power plug from the ac outlet and measure the ac voltage at the outlet. If that is normal, proceed as follows. Connect an ohmmeter across the two *hot* prongs of the power plug. Close the switch. If the meter indicates continuity in he circuit (about 1 to 2 Ω), then the power plug, line cord, switch, fuse, and primary of the transformer are okay. If the meter shows

infinite resistance, then it is necessary to make a continuity check of the plug and each wire in the line cord, fuse, switch, and primary of T_1. One of these components will be open and should be replaced.

If the fuse is open, it may be because of a temporary overload or trouble in another circuit component. The simplest test is to replace the fuse and apply power to the circuit. If the fuse blows again, the trouble is elsewhere.

If all components from the line cord through the primary of T_1 are okay, the trouble is an open secondary in T_1. Check for continuity in the secondary to confirm this conclusion.

NOTE: A shorted transformer winding would also give no or low ac voltage across the secondary, but in that case there would be such an increase in ac current that the fuse would also blow.

Now assume that there is ac voltage across the secondary winding but that V_{PG} and V_{AG} both measure 0 V dc. The trouble then may be a shorted C_1 or open rectifiers D_1 *and* D_2. (Note that if only one rectifier were open, there would still be dc voltage at A and P, though the voltage would be lower than normal.) *One* other possibility is an open center tap on the secondary of the transformer or open wiring from the center tap to point G. The defective component or the open lead from the center tap to point G may be found by resistance measurements.

As a final check the *connections and wiring* between components should be checked for continuity.

Low $V+$, High Ripple

An increase in load current or leaky electrolytic capacitors are the usual reasons for low output V and high ripple. Of course, as we noted previously, an open D_1 or D_2 can also cause this problem.

An oscilloscope check across PG (Fig. 8-1) will indicate not only the amplitude of ripple voltage but also the frequency of the ripple. The frequency should be 120 Hz for a full-wave rectifier. If it is 60 Hz, then either D_1 or D_2 is defective. These can be checked by determining their forward resistances. When the defective rectifier is replaced, the dc voltage and ripple levels should return to normal.

The usual cause for low $V+$ *and* high ripple is an open or leaky C_1 or C_2. The simplest procedure is to replace these capacitors, one at a time, with good capacitors until both dc voltage and ripple levels are restored to normal.

SUMMARY

1. The dc test points in a power supply with a π-type filter are the input to the filter (point A in Fig. 8-1) and the output from the filter (point P). At each of these points, the dc voltage is measured with respect to common return, point G.
2. In a filtered dc supply the dc input to the filter (V_{AG} in Fig. 8-1) is higher than the dc output from the filter, V_{PG}. The relationship among V_{AG}, V_{PG}, and the volt-

age drop V_{AP} across resistor R_C (or filter choke) is $V_{AG} = V_{PG} + V_{AP}$.

3. The ripple voltage at the output of a power-supply filter (V_{PG}) is always lower than the ripple voltage at the input to a π-type filter (V_{AG}).

4. The ripple voltage on $V+$ under load is higher than the ripple voltage on $V+$ without load. A power supply must still furnish a relatively low ripple voltage on $V+$ under normal load.

5. AC test points in a transformer-fed power supply are (a) across the primary winding and (b) across the secondary winding. The ac voltage measured across the primary should be the line voltage. The ac voltage across the secondary is as specified by the manufacturer.

6. The dc voltage at the input to a π-type filter in a full-wave rectifier power supply is approximately one-half the peak-to-peak voltage across the secondary of the transformer.

7. In troubleshooting a power supply it may be necessary to measure the input and output dc voltages of the filter; ripple voltage at the same points; ac voltage across the primary and secondary of the transformer; resistance (with power off) at the output of and input to the filter; resistance and continuity of all the remaining components and wiring in the circuit.

8. In troubleshooting it is always desirable to isolate the trouble to part of a circuit, thus eliminating as "good" those components in that part of the circuit which tests out. For example, in the circuit of Fig. 8-1, if dc voltage and ripple checks across points AG are normal, then all the components to the left of C_1, including C_1, are good. Similarly, if an ac voltage check across the secondary of T_1 checks normal, then the ac input circuit components are good, with the possible exception of the tap on the secondary of the transformer.

SELF-TEST

Check your understanding by answering these questions.

1. In the circuit of Fig. 8-2, if $R_{B1} = R_{B2}$, $C_3 = C_4$, and $V+$ voltage (measured from P_1 to G) = $+10$ V, $V-$ voltage (measured from P_2 to G) = _____ V.

2. In the circuit of Fig. 8-1, the dc voltage $V_{AG} = 57$ V. If $R_C = 150$ Ω and $I_T = 0.15$A, then $V_{PG} =$ _____ V.

3. If the filter capacitor C_2 opens in Fig. 8-1, the most noticeable effect will be a(n) _____ (increase, decrease) in ripple across the load.

4. In Fig. 8-1, $V+$ voltage under load has decreased, the ripple voltage across the load has increased, and the ripple frequency is 60 Hz. The trouble is a defective _____ .

5. The dc test points in the circuit of Fig. 8-1 are at _____ and _____ .

6. The dc test points in the circuit of Fig. 8-2 are at _____ , _____ , and _____ .

7. The ripple test points in the circuit of Fig. 8-2 are at _____ , _____ , and _____ .

8. With the load open, the resistance measured across PG (with power off) in Fig. 8-1 is 0 Ω. The most likely cause of trouble is a _____ _____ .

9. Zero voltage ac is measured across the primary of T_1. Assume the meter is okay. The trouble can be the result of an open _____ _____ , _____ , _____ , _____ , or connecting _____ , or a dead _____ _____ .

10. Trouble in a power supply has been traced to one of two silicon rectifiers. An ohmmeter check of one shows $R_F = 100$ kΩ, $R_R = 1$ MΩ. This rectifier is _____ (good, bad).

MATERIALS REQUIRED

- Power supply: 120-V rms 60-Hz source
- Equipment: Oscilloscope; EVM or VOM
- Resistors: 100-Ω, two 1200-Ω, 2700-Ω ½-W; 250-Ω 2-W
- Capacitors: Two 100-μF 50-V; two 25-μF 50-V
- Solid-state rectifiers: Two 1N5625 or equivalent
- Miscellaneous: Transformer T_1, 120-V primary, 26.8-V 1-A CT secondary; SPST switch; components for troubleshooting; fused line cord

PROCEDURE

Full-wave Supply for $V+$ and $V-$

1. Connect the circuit of Fig. 8-2. $C_1 = C_2 = 100$ μF 50 V; $C_3 = C_4 = 25$ μF 50 V; D_1 and D_2 are 1N5625 silicon rectifiers; $R_C = 100$ Ω/2 W; $R_{B1} = R_{B2} = 1200$ Ω ½ W; T_1 is the transformer used in Experiment 7.

2. **Power on.** Measure and record in Table 8-1 the no-load dc voltage from $V+$ to ground, $V-$ to ground, $V+$ to $V-$, and A to $V-$. Observe and measure with an oscilloscope the no-load ripple waveform, its frequency, and the peak-to-peak voltage of the ripple from $V+$ to ground, $V-$ to

ground, $V+$ to $V-$, and A to $V-$. Record the results in Table 8-1.

3. Connect a 250-Ω load from $V+$ to $V-$ and repeat the measurements in step 2. Record your results in Table 8-1.

4. **Power off.** Remove the lead connecting the anode of D_1, to the secondary winding of T_1. The result is a half-wave rectifier circuit which supplies $V+$ and $V-$. Disconnect the 250-Ω load.

5. **Power on.** Repeat steps 2 and 3 and record the results in Table 8-2.

$$\%VR \quad \frac{9.6-8.6}{8.6} \times 100 = 11.63\%$$

TABLE 8-1. Full-wave V+ and V− Supply

Point	Load, Ω	DC, V	Ripple Waveform	Ripple Frequency, Hz	Ripple V p-p
V+ to G	No	4.65		121.9	14 mV
V+ to G	250	2.9		121.9	86 mV
V− to G	No	−4.6		121.9	14 mV
V− to G	250	−2.9		121.9	96 mV
V+ to V−	No	9.2		121.9	27 mV
V+ to V−	250	5.8		121.9	190 mV
A to V−	No	9.6		121.9	260 mV
A to V−	250	8.6		121.9	1.8 mV

TABLE 8-2. Half-wave V+ and V− Supply

Point	Load, Ω	DC, V	Ripple Waveform	Ripple Frequency, Hz	Ripple V p-p
V+ to G	No	4.5		3.2 ×5mS / X 62.5	50 mV
V+ to G	250	2.6		62.5	.31
V− to G	No	−4.5		62.5	50 mV
V− to G	250	−2.6		62.5	.32 v
V+ to V−	No	9		62.5	100 mV
V+ to V−	250	5.1		62.5	.62 v
A to V−	No	9.4		62.5	.55 v
A to V−	250	7.6		62.5	3.3 v

TROUBLESHOOTING REPORT

Student's Name(s) _____ Experiment # _____

Class _____ Date _____

1. Describe trouble. _____

_____

2. Preliminary inspection: describe troubles found, if any. _____

_____

3. Describe procedure (list steps in order performed).

Test Point	Voltage, AC or DC	Resistance, Ω	Waveforms and V p-p	Other

4. Describe trouble found and list parts used for repair. _____

_____

5. Was circuit operation normal after repair? _____ 6. Instructor's signature _____

Fig. 8-3. Standard troubleshooting report.

Troubleshooting

6. Your instructor will insert trouble into the power supply of Figs. 8-1 or 8-2. Troubleshoot the circuit, keeping a step-by-step record of each check that you make in the order performed. Record these checks in the standard trouble-shooting report (Fig. 8-3) which will be used for all servicing procedures. When you have found the trouble, correct it and notify your instructor, who will put another trouble into the circuit. Service this circuit, using a separate troubleshooting report to record your results. Service as many troubles as time will permit.

QUESTIONS

1. In the circuit of Fig. 8-2 what is the relationship between $V+$, $V-$, and the voltage measured from $V+$ to $V-$?
2. Is the relationship in question 1 confirmed by your measurements? Refer specifically to the table and measurements.
3. In the circuit of Fig. 8-2 compare the no-load ripple on $V+$ and $V-$ to ground with the full-load ripple. If there is any difference, explain why it exists.
4. Compare the no-load $V+$ and $V-$ outputs and ripple of the full-wave and half-wave rectifiers in Fig. 8-2 and explain the difference, if any.
5. Compare the full-load $V+$ and $V-$ outputs and ripple of the full-wave and half-wave rectifiers in Fig. 8-2 and explain any differences.
6. Trouble in an electronic device has been isolated to the dc power supply. Assume it is the supply of Fig. 8-1. $V+$ voltage is lower than normal, and the ripple on $V+$ is much higher than normal. What are the troubles which could give these effects?
7. Explain the procedure you would follow to find the defective component for the conditions in question 5.
8. In Fig. 8-2 $V+$ voltage to ground measures $+30$ V; $V-$ to ground measures 0 V. What is the most likely cause of trouble?
9. In Fig. 8-1 the ac voltage measured from the anode of D_1 to the anode of D_2 is 60 V rms. The ac voltage measured from either anode to the center-tap connection on the secondary is 0. The output of the dc supply ($V+$) is 0. What is the trouble in the circuit?

Answers to Self-Test

1. -10
2. 34.5
3. increase
4. rectifier
5. A; P
6. A; P_1; P_2
7. A; P_1; P_2
8. shorted C_2
9. line cord; primary; switch; fuse; wiring; power outlet
10. bad

THE VOLTAGE DOUBLER

OBJECTIVES

1. To observe the input and output waveforms of a full-wave voltage-doubler supply and to measure the dc voltages at key points
2. To observe the input and output waveforms of a cascade voltage-doubler supply and to measure the dc voltages at key points
3. To measure the effects of defective capacitors on dc voltage and ripple

INTRODUCTORY INFORMATION

Line-Operated Half-Wave Power Supply

Figure 9-1 is the schematic diagram of a typical silicon half-wave supply which derives its input voltage directly from the line. Since no transformer is used, there is *no isolation* from the line, and the ground return G of the supply is directly connected to the line.

Refer to Fig. 9-1. D_1, the rectifier, is connected to deliver a positive dc voltage to the load, points P to G. C_1, R_1, and C_2 constitute a π-type RC filter. R_2, a surge-current-limiting resistor, protects the rectifier against initial overload caused by the high charging currents of electrolytics. Thermistors which have a high cold resistance and very low hot resistance are sometimes used as protective resistors. About 100 V are delivered by this type of supply under normal load.

The rectifier conducts when the instantaneous value of the positive alternation exceeds the voltage across C_1. This neglects the small voltage drop across the diode and the series resistance of the supply. The resultant 60-Hz current is filtered and appears as a relatively ripple-free dc voltage ($V+$) at point P. R_3 is a bleeder resistor.

Full-Wave Voltage Doubler

An advantage of the silicon half-wave rectifier supply is that no power transformer is required. Hence the cost of the supply is appreciably reduced. A disadvantage, however, is that the dc voltage is limited to about 100 V.

The voltage-doubler power supply overcomes this disadvantage, since it is capable of delivering approximately twice the dc voltage of the half-wave rectifier operating off a power-line input.

The circuit of Fig. 9-2 is a full-wave voltage doubler. During the positive alternation of the line voltage, when the anode of D_1 is positive with respect to its cathode, D_1 conducts and charges C_1 to the peak of the line voltage, with the polarity shown. During this first alternation, D_2 does not conduct because the cathode is positive with respect to the anode.

During the negative alternation of the line voltage, D_2 conducts when its anode is positive relative to its cathode and charges capacitor C_2 to the peak line voltage with the polarity shown. The voltages of C_1 and C_2 add to give twice the peak line voltage (without load) across AG. The dc output V_{out} from this supply must be filtered to be useful.

C_1, C_2, and C_3 are electrolytic capacitors. C_3 must have a higher voltage rating than either C_1 or C_2 because the voltage across it is twice as high (approximately) as that across either C_1 or C_2. R_B is a bleeder and R_1 is a surge-current-limiting resistor to protect the rectifier from damage when power is first applied.

Load current requirements determine the size of the filter components. Low-current filters can use a resistor instead of a choke L.

The ripple frequency of this supply is 120 Hz.

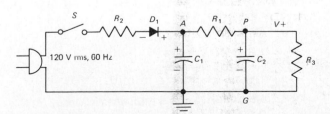

Fig. 9-1. Line-derived half-wave power supply.

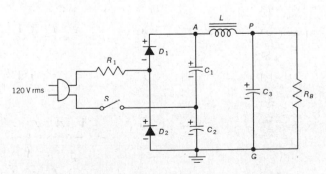

Fig. 9-2. Full-wave voltage doubler using silicon rectifier.

This circuit has several disadvantages. The first is that capacitors C_1 and C_2 must be fairly well matched to avoid excessive 60-Hz ripple in the output. The second limitation is that the 120-V line is above the ground return of the dc output voltage, and the vacuum-tube filaments of series-connected filament circuits using this supply are therefore above ground. This is generally undesirable. Hence, this circuit is avoided.

Transformer-Fed Full-Wave Voltage Doubler

The full-wave doubler can be used advantageously with a power transformer, as in Fig. 9-3. If T is a 1:2 voltage stepup transformer, the circuit of Fig. 9-3 ensures an output dc voltage greater than 500 V. Note that transformer T does not need a center tap, as does the conventional full-wave rectifier studied in Experiment 7. The arrangement of Fig. 9-3 therefore permits the full-transformer secondary voltage to be applied to each rectifier. Thus with a 240-V (approx) secondary, the full-wave *doubler* circuit develops more than 500 V dc. If a 500-V supply is required from a conventional full-wave rectifier, a 1:8 (approx) stepup transformer is needed. Its cost is higher than that of transformer T in Fig. 9-3.

Cascade Voltage Doubler

A more practical, transformerless voltage doubler used with vacuum-tube series filament circuits is that shown in Fig. 9-4. This is a cascade (half-wave) voltage doubler and filter that employs silicon rectifiers and derives its ac input directly from the line. C_1, C_2, and C_3 are electrolytic capacitors. D_1 and D_2 are silicon rectifiers, R_1 a surge-current-limiting resistor, and R_B a bleeder resistor.

To understand how this circuit operates, assume that when power is applied, D_1 is not in the circuit. On the negative alternation of the line voltage at R_1, D_2 conducts and charges C_1 to the peak line voltage (169 V) with the polarity shown.

Now consider that D_1 is also in the circuit. C_1, just charged, now acts like a battery in series with the power source. On the positive alternation, therefore, D_1 "sees" a positive voltage equal to twice the line-voltage peak (see Fig. 9-5). D_1 now conducts and charges C_2 with the polarity shown. The voltage across C_2 is therefore equal to approx-

imately twice the peak line voltage, or 338 V. The function of D_2, therefore, is to charge C_1 so that rectifier D_1 can receive twice the peak positive line voltage. The dc voltage from A to G results from conduction of D_1, and D_1 conducts only on positive alternations. Hence this is a half-wave rectifier, and the ripple frequency is 60 Hz.

If this supply must deliver 300 to 400 mA, 150- to 300-μF capacitors are employed.

The filter is connected in a conventional manner. Note that electrolytic capacitors C_2 and C_3 must have about twice the voltage rating of C_1.

This supply is capable of delivering 250 to 300 V, depending on the size of the filter components and on the load current.

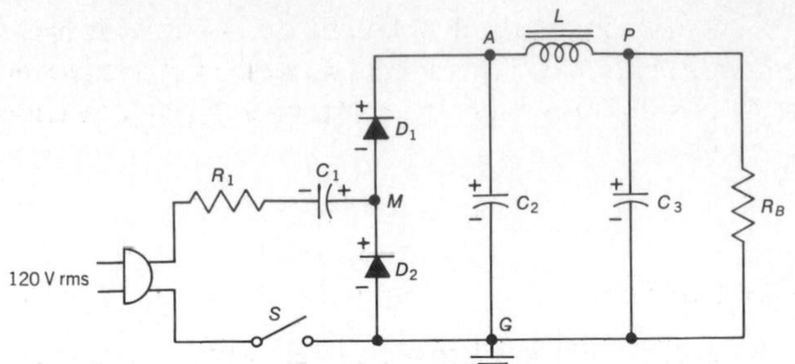

Defective Capacitors
and Their Effect on Operation

In the full-wave voltage-doubler circuit of Fig. 9-2, if either C_1 or C_2 is open, there will be approximately half the dc output with an appreciable increase in filter ripple. An open C_3 will not affect the dc output, but filter ripple will increase.

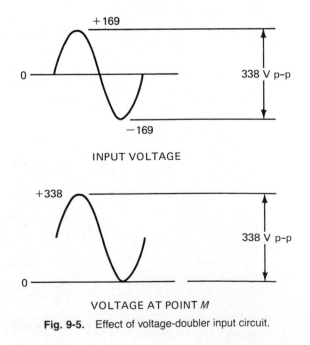

Fig. 9-4. Cascade voltage doubler (half wave).

INPUT VOLTAGE

VOLTAGE AT POINT M

Fig. 9-5. Effect of voltage-doubler input circuit.

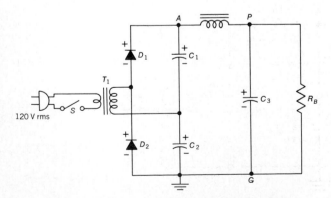

Fig. 9-3. Transformer-fed full-wave voltage doubler.

In the cascade voltage doubler of Fig. 9-4, if C_1 is open, there will be no dc output. An open C_2 will greatly reduce dc output and increase filter ripple. An open C_3 will increase filter ripple.

SUMMARY

1. Transformerless line-derived power supplies are sometimes used.
2. These supplies eliminate the need for heavy, costly power transformers.
3. The disadvantage of these supplies is that they are directly connected to the line.
4. One type of line-derived power supply is the half-wave rectifier circuit in Fig. 9-1. It develops about 100 V dc with 60-Hz ripple.
5. A line-derived doubler using silicon rectifiers with filtered dc output is shown in Fig. 9-2. The output of this full-wave supply is about double that of Fig. 9-1 with 120-Hz ripple.
6. A *transformer-fed* full-wave voltage doubler is shown in Fig. 9-3. This supply has the advantage of line isolation.
7. Another type of doubler is the *cascade* half-wave voltage doubler shown in Fig. 9-4. This supply can operate directly from the line, or it can be transformer-fed.
8. Open and leaky filter electrolytic capacitors affect the dc output of these doubler supplies just as they affect the output of the power supplies studied in Experiment 7. They *reduce* the dc voltage and *increase the ripple*.

SELF-TEST

Check your understanding by answering these questions.

1. The line-derived power supply in Fig. 9-2 _____ (is, is not) isolated from the ac line.
2. For an input of 120 V rms, the output of the half-wave supply of Fig. 9-1 is about _____ V.
3. For a 120-V rms input, the output (a) of the full-wave doubler (Fig. 9-2) is about _____ V; (b) of the half-wave doubler (Fig. 9-4) is about _____ V.
4. The ripple frequency in the output of Fig. 9-2 is _____ Hz.
5. The ripple frequency in the output of Fig. 9-4 is _____ Hz.
6. The transformer-fed full-wave doubler supply of Fig. 9-3 develops _____ as much voltage as the non-doubler full-wave supply using the same transformer.
7. In Fig. 9-4, an open C_3 will result in a _____ (high, low) output ripple voltage.
8. In the circuit of Fig. 9-4, an open D_1 will result in _____ (low, zero, high) dc output voltage (V_{PG}).

MATERIALS REQUIRED

- Power supply: 120-V rms 60-Hz source
- Equipment: Oscilloscope; EVM or VOM
- Resistors: 5600-Ω ½-W; 100-Ω 1-W; 500-Ω 5-W
- Capacitors: Three 100-μF 50-V
- Solid-state rectifiers: Two 1N5625 or the equivalent
- Miscellaneous: Transformer T_1, 120-V pri, 26.8-V/1-A sec; SPST switch; fused line cord

PROCEDURE

Full-Wave Voltage Doubler

NOTE: Only one-half of the secondary winding of the power transformer is used in this experiment.

1. Connect the circuit of Fig. 9-6. R_B is a 5600-Ω bleeder resistor. D_1 and D_2 are 1N5625 silicon rectifiers.
2. **Power off.** Discharge all capacitors. Measure resistance from points P to G. Ohmmeter-lead polarity must correspond to capacitor polarity. Do *not* apply power if

resistance is appreciably less than 5600 Ω, but check the circuit to determine why resistance is low.

3. **Power on.** Measure and record in Table 9-1 the dc no-load voltage from points A and P to ground G, respectively.

4. Observe the waveforms, and measure with an oscilloscope the no-load ripple voltage from points A and P to ground, respectively. Record the results in Table 9-1.

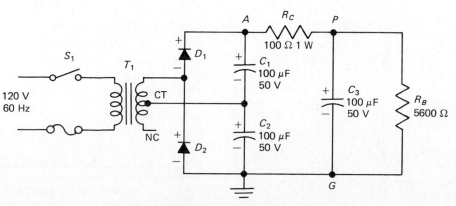

Fig. 9-6. Experimental full-wave voltage doubler.

TABLE 9-1.

| | No Load | | | With Load (500 Ω) | | |
| | DC, V | Ripple | | DC, V | Ripple | |
Point		Waveform	V p-p		Waveform	V p-p
A–G						
P–G						

5. **Power off.** Connect the 500-Ω 5-W load resistor in parallel with R_B.

6. Repeat steps 3 and 4 with this load, and record your measurements in Table 9-1.

7. **Power off.** CAUTION: *Discharge all capacitors before proceeding.* Disconnect C_1.

8. **Power on.** Measure and record in Table 9-2 the dc voltages and ac waveforms at points *A* and *P* to ground.

9. **Power off.** Repeat steps 7 and 8 for each of the components listed in Table 9-2.

Cascade Voltage Doublers (Half-Wave)

10. **Power off.** *Discharge all capacitors before proceeding.* Connect the circuit of Fig. 9-7.

11. Repeat steps 2 to 9, recording your results in Tables 9-3 and 9-4.

TABLE 9-2.

| | Point A (to Ground) | | | Point P (to Ground) | | |
Open Component	DC, V	Waveform	V p-p	DC, V	Waveform	V p-p
C_1						
C_2						
C_3						

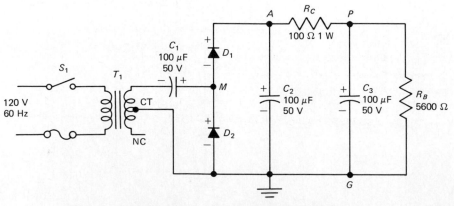

Fig. 9-7. Experimental cascade half-wave voltage doubler.

TABLE 9-3.

Point	No Load DC, V	No Load Ripple Waveform	No Load Ripple V p-p	With Load (500 Ω) DC, V	With Load (500 Ω) Ripple Waveform	With Load (500 Ω) Ripple V p-p
A–G						
P–G						
M–G						

TABLE 9-4.

Open Component	Point A (to Ground) DC, V	Point A (to Ground) Waveform	Point A (to Ground) V p-p	Point P (to Ground) DC, V	Point P (to Ground) Waveform	Point P (to Ground) V p-p
C_1						
C_2						
C_3						

QUESTIONS

1. Which experimental circuit, if any, has a lower ripple in the output under load? Explain why.
2. Which experimental circuit, if any, has a higher dc output voltage? Explain why.
3. (a) In the circuit of Fig. 9-6 which capacitors most affect the level of dc output? (b) Which most affect the output ripple?
4. Explain the result on dc output of an open capacitor C_1 in Fig. 9-7.
5. Why does the dc output voltage V_{PG} drop when a load is connected to the power supply in Fig. 9-6?
6. Because of the bleeder and load resistors, direct current flows through R_C. Compute this current when a 500-Ω load resistor is connected across (a) R_B in Fig. 9-6, (b) R_B in Fig. 9-7.
7. How could the filtering of the supply voltages in Figs. 9-6 and 9-7 be improved? Be specific, giving values of all suggested component changes.
8. Explain the effect on dc output of a short-circuited capacitor C_1 in Fig. 9-7.
9. Compare the dc output voltage of a full-wave rectifier in a nondoubler circuit (Experiment 7) with that in the doubler supply of this experiment.
10. Concerning the circuits in question 9, what advantages, if any, does the full-wave doubler have over the non-doubler supply? Disadvantages?

Answers to Self-Test

1. is not
2. 100
3. 200; 200
4. 120
5. 60
6. 4 times
7. high
8. zero

OBJECTIVES

1. To verify that conduction in a bridge rectifier results from the conduction, alternately, of two series-connected rectifiers
2. To observe and measure the input and output waveforms
3. To measure the effects of a filter network on the dc voltage output and ripple

INTRODUCTORY INFORMATION

Theory and Operation

The bridge rectifier employing silicon diodes has become increasingly popular with designers.

Figure 10-1 is the circuit diagram of a transformer-fed bridge rectifier. The high-voltage secondary winding of transformer T supplies four silicon rectifiers, D_1 through D_4. Operation of the circuit is as follows: Assume that during the positive alternation (alternation 1) of the input sine wave, point C is positive with respect to D (the voltages at the opposite ends of a transformer winding are $180°$ out of phase). This makes the anode of D_1 positive with respect to its cathode, and D_1 is therefore forward-biased. Similarly the cathode of D_3, connected to point D, is negative relative to its anode. Hence, D_3 is forward-biased. It is evident also that D_2 and D_4 are reverse-biased during alternation 1. Thus, in a circuit D_1 and D_3 will conduct during alternation 1 while D_2 and D_4 will be cut off.

Figure 10-2a shows that during the positive alternation there *is* a complete path for current for rectifiers D_1 and D_3, which are connected in series with the load resistor R_L.

Current flows through R_L, through D_1, through winding CD, and through D_3, with the polarity shown.

Figure 10-2b shows the positive-voltage waveform developed during alternation 1 across R_L. During the negative alternation (alternation 2), D_1 and D_3 are reverse-biased and are cut off. If D_2 and D_4 were not in the circuit, D_1 and D_3 would act as a half-wave rectifier.

Figure 10-2c shows that during the negative alternation (alternation 2), that is, when point C is negative relative to point D, the anode of D_2 is positive with respect to its cathode, and the cathode of D_4 is negative with respect to its anode. Hence, rectifiers D_2 and D_4 are forward-biased, while D_1 and D_3 are reverse-biased. Now D_2 and D_4 conduct, permitting current through R_L. The polarity across R_L is the same as in Fig. 10-2d.

Thus D_1 in series with D_3 rectifies during the positive alternation of the input, while D_2 in series with D_4 rectifies during the negative alternation. A bridge rectifier is therefore a full-wave rectifier. The center tap (CT) of the secondary is not connected in the bridge rectifier. In a conventional circuit rectifier, the CT acts as the common return, and the voltage across each diode is one-half the voltage across the transformer. Hence, if the same transformer is used, the output voltage of a conventional full-wave rectifier (Fig. 10-3) is only one-half that of a bridge circuit.

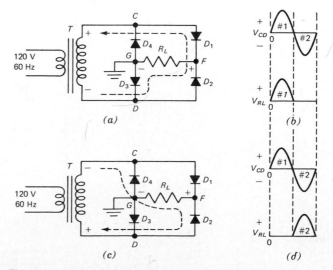

(a)

(b)

(c)

(d)

Fig. 10-2. *(a)* and *(b)* Action of bridge rectifier on positive alternation; *(c)* and *(d)* on negative alternation.

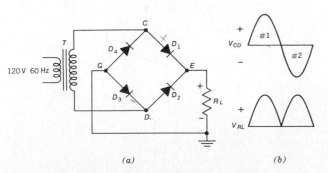

(a)

(b)

Fig. 10-1. *(a)* Bridge rectifier; *(b)* waveforms.

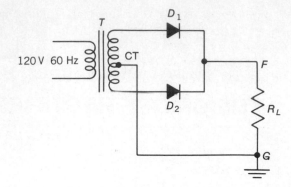

Fig. 10-3. Transformer-fed duo-diode full-wave rectifier.

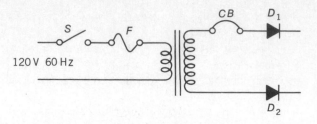

Fig. 10-5. A fuse (F) protects the primary; a thermal circuit breaker (CB) protects the secondary of the transformer.

Because two rectifiers are always operating in series in a bridge rectifier, the peak inverse voltage is divided across the rectifiers. Hence, the PIV for each rectifier is the transformer peak, whereas in the conventional duo-diode full-wave rectifier the PIV is about twice the transformer voltage.

One disadvantage of a bridge rectifier is that on each alternation, the direct current in the circuit must flow through two series-connected diodes. The forward dc voltage drop (loss) across the two rectifiers is therefore greater than the drop across a single rectifier. However, the small drop across silicon diodes can normally be tolerated.

Filter Circuit

The same type of filter arrangement can be used with a bridge rectifier as with any other rectifier circuit. Observe that a π filter is used in Fig. 10-4. For a 300-mA load (approx), capacitors rated at 80 to 100 μF are conventionally employed. The choke L varies from 1 to 8 henrys (H), depending on how much ripple can be tolerated.

A filter resistor may be used to replace the choke. In that case larger capacitors are used.

For the bridge rectifier the voltage rating of the filter capacitors must be at least *twice* for the full-wave rectifier using the same transformer.

Overload Protection

Fuses and thermal circuit breakers are used to protect the transformer and the other circuit components against overload. The circuit of Fig. 10-5 shows both devices. A fuse is connected in the primary, and a resettable circuit breaker is used in the secondary. The fuse, in this arrangement, acts as protection against an ac overload. For instance, if the secondary winding should be short-circuited accidentally, the circuit breaker or the fuse in the primary will blow. However, a dc overload in the load or in the rectifier circuit will cause the circuit breaker to open, thus protecting the output circuit.

Slow-blow fuses are being increasingly used in electronic devices to withstand initial or temporary current surges. Circuit breakers also are designed to withstand sudden current surges.

SUMMARY

1. The full-wave bridge rectifier (Fig. 10-1) employs four rectifier diodes. Of these, opposite pairs (D_1 and D_3 or D_2 and D_4) in series with the load conduct alternately.
2. The bridge-rectifier output consists of pulsating dc (Fig. 10-1b) just like the output of the conventional two-diode full-wave rectifier in Fig. 10-3. However, if the same transformer T is used in both circuits, the amplitude of the voltage of the bridge rectifier is twice the amplitude of the

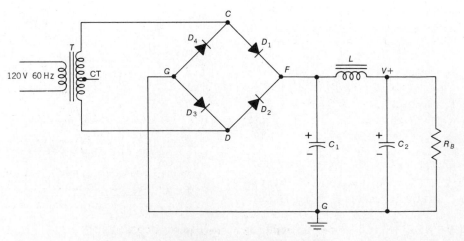

Fig. 10-4. A transformer-fed bridge rectifier with filter.

conventional full-wave rectifier. The bridge rectifier does *not* use a CT secondary.

3. A π-type filter, either CLC or CRC, is used to smooth the output of a bridge rectifier. The capacitors must be rated to withstand the higher voltages developed in the bridge rectifier.

4. Power supplies in electronics are protected for overload by fuses or circuit breakers. In Fig. 10-5, a fuse protects the primary; a circuit breaker protects the secondary.

5. Slow-blow fuses are frequently used to protect the circuits against instantaneous overloads.

SELF-TEST

Check your understanding by answering these questions.

1. In Fig. 10-1, an open rectifier D_1 will result in _____ (*full, half*)-wave rectification.

2. In the circuit of Fig. 10-1, an open D_2 will still produce voltage across R_L, but only when C is positive relative to D. _____ (*true, false*)

3. If the rms voltage across the secondary of T is 50 V, the peak of the positive alternation across R_L is _____ V (approx).

4. In Fig. 10-4, the rms voltage across the secondary of T is 50 V. The dc voltage across FG will be approximately _____ V.

5. In Fig. 10-5, the input is protected against overload by _____ ; the output by _____ .

MATERIALS REQUIRED

- Equipment: Oscilloscope, EVM, 0–100 mA dc milliammeter
- Resistors: *2700-Ω, 5600-Ω ½-W; 100-Ω 1-W; 250-Ω *2-W; 500-Ω 5-W
- Capacitors: Two 100-μF 50-V
- Solid-state rectifiers: Four 1N5625 or equiv.
- Miscellaneous: Power transformer T_1, 120 primary, 26.8-V/1-A secondary; five SPST switches; fused line cord

NOTE: Starred (*) resistors are for extra-credit procedure.

PROCEDURE

Operation of a Bridge Rectifier

1. Connect the circuit of Fig. 10-6. The silicon rectifiers are D_1 to D_4. T is the same transformer as that used in the preceding rectifier experiments. Resistor R_L is 5600 Ω.

NOTE: Be certain that the silicon rectifiers are connected with the proper polarity in the circuit. Switches S_1 through S_5 are open (power OFF).

2. Calibrate the vertical amplifiers of your oscilloscope at 15 V/div. Set the oscilloscope on *Line-Trigger/Sync*. Connect the oscilloscope across CD, hot lead to C, and ground lead to D. Close S_5. **Power on.**

3. Adjust the time/base controls until there are two waveforms on the screen. Position these vertically so that the peak positive and peak negative alternations are equally centered with respect to the x axis on the graticule of the oscilloscope. Adjust the horizontal centering control until one waveform is symmetrical with respect to the y axis, with positive alternation on the left. The waveform should appear as in Table 10-1. Measure, and record in Table 10-1, the peak positive and peak negative amplitude of the waveform. They should be the same. If they are not, recenter the waveform vertically until they are equal. This is the reference waveform. Do not readjust any of the oscilloscope controls until you are finished with steps 4 to 11. All waveforms will be time-related to the reference waveform.

4. Connect the vertical leads of the oscilloscope across R_L, the hot lead to F, and the ground lead to G. Switches S_1 through S_4 are still open. Observe and measure the waveform, if any, across R_L in time phase with the reference wave. Draw the waveform in Table 10-1.

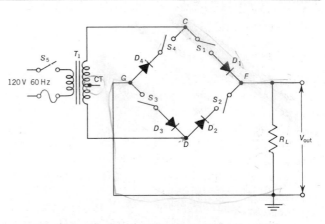

Fig. 10-6. Experimental bridge rectifier without filter.

5. *Close S_1.* (S_2, S_3, and S_4 are still open.) Observe and measure the waveform, if any. Draw the waveform in Table 10-1 in proper time phase with the reference.

6-11. Set the switches as indicated in Table 10-1. For each step, observe and measure the waveform, if any, across R_L. Draw the waveform in time phase with the reference in Table 10-1.

Filtering the Output of a Bridge Rectifier

12. **Power off.** Remove R_L from the circuit. Connect a π-type filter between F and ground, terminated by a 5600-Ω bleeder resistor R_B, as in Fig. 10-7. A 100-mA dc milliammeter M measures the total direct current in the circuit. S_1 through S_4 are all closed as in step 11.

CAUTION: *Be certain the filter capacitors and M are connected with the proper polarity.*

TABLE 10-1.   Bridge Rectifier Operation—Unfiltered Output

Step	Switch Position S_1	S_2	S_3	S_4	Waveform	Peak Voltage
					+20 / 0 / −20 Reference	20
4	Open	Open	Open	Open	0	0
5	Closed	Open	Open	Open	0	0
6	Closed	Closed	Open	Open	0	0
7	Closed	Open	Open	Closed	0	0
8	Closed	Open	Closed	Open	0	20
9	Open	Closed	Open	Open	0	0
10	Open	Closed	Open	Closed	0	20
11	Closed	Closed	Closed	Closed	0	20

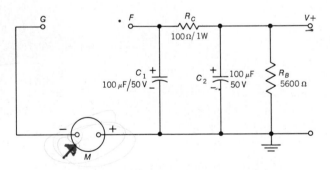

Fig. 10-7.   Adding a filter to the bridge rectifier.

13. **Power on.** With an EVM, measure the rms voltage across *CD*. Record the voltage in Table 10-2. Measure also, and record in the "No Load" column, the dc voltage point *F* to ground, $V+$ to ground, and the dc current. With an oscilloscope, observe and measure the ripple voltage, *F* to ground and $V+$ to ground. Draw the ripple-voltage wave in Table 10-2.

14. Repeat the measurements in step 13 with a 500-Ω 5-W load resistor connected across R_B. Record the results in the "With Load" column.

TABLE 10-2.   Bridge Rectifier—Filtered Output

RMS Volts across CD	Point	No Load DC, mA	DC, V	Ripple Waveform	V p-p	With Load (500 Ω) DC, mA	DC, V	Ripple Waveform	V p-p
14	F	3.4mA	19.2		.27	2.6 mA	18		2.2
14	V+	X	19		60m	25 mA	14.5		.3

Extra Credit (Optional)

15. Determine the no-load and load characteristics (dc voltage, dc current, and ripple) of a conventional two-diode full-wave silicon rectifier using the same filter components as for the bridge rectifier. The bleeder resistor R_B is 2700-Ω. Use a 250-Ω 2-W load resistor. Draw a circuit diagram, prepare a table, and enter the data. Give the step-by-step procedure followed.

QUESTIONS

1. Describe the operation of a bridge rectifier.
2. What is the ripple frequency of the bridge rectifier? Refer to your data to substantiate your answer.
3. Determine the regulation of the bridge rectifier in this experiment. Show your computations.
4. How does the rms voltage across the secondary winding compare with the no-load $V+$ voltage of a (a) bridge rectifier? (b) two-diode full-wave rectifier (using tubes or silicon diodes)?
5. How does the regulation of a bridge rectifier compare with that of a voltage doubler for the same load? Use experimental data to substantiate your answer.

Extra Credit

6. Refer to the data in Table 10-2.
   (a) What other data, if any, are required to compute the dc current, assuming we had no milliammeter?

(b) Compute the dc current with and without load. Show your computations.
(c) Compare the computed and measured values. Explain any discrepancy.
7. Explain how a thermal circuit breaker works and where it is used.
8. What is an advantage in using (a) a thermal circuit breaker? (b) a fuse?

Answers to Self-Test

1. half
2. false
3. 70
4. 65–70
5. F; CB

TRANSISTOR FAMILIARIZATION

OBJECTIVES

1. To become familiar with transistor basing
2. To measure the effects on emitter-base current of forward (normal) and reverse bias in the emitter-base circuit
3. To measure the effects on *collector* current of forward and reverse bias in the emitter-base circuit
4. To measure I_{CBO}

INTRODUCTORY INFORMATION

The Transistor: A Three-element Device

Our concern until now has been with two-element electronic devices. We found a variety of diodes each with unique characteristics, serving different purposes. In our experiments we employed diodes as rectifiers of alternating current, as voltage and current regulators, and as limiters. In this experiment we shall become familiar with a three-element electronic device, the *transistor*, whose characteristics dramatically extend the range and scope of electronic applications.

Scientific interest in semiconductors led to the development of the transistor. One advantage of the transistor is that it is small and light, permitting miniaturization of electronic equipment. The transistor operates with low supply voltages and uses little power. It does not require any warmup period and operates as soon as power is applied. A transistor has fewer circuit connections. A disadvantage of transistors is their sensitivity to heat, but an advantage is that they do not generate much heat.

There are many types of transistors. These may be classified according to the basic material from which they are formed. In this category we find germanium and silicon transistors. Most transistors are now made of silicon. Transis-

tors may be classified also according to the process by which they are constructed. Here we find various types of junction transistors such as grown-junction, alloy-junction, drift-field, mesa, epitaxial mesa and planar transistors, and point-contact transistors. Transistors may also be classified according to the number of elements. Thus there are the triode, or three-element transistor, and the tetrode, or four-element transistor. They may also be classified according to their ability to dissipate power. Here we find a wide range from the low-power (less than 50 mW) to the high-power (2 W and higher) type.

Transistors come in different shapes and sizes (see Fig. 11-1). There are variations also in the basing arrangements and in the manner of mounting the transistor in the circuit. Some are socket-mounted. The sockets conform to the physical basing of the transistor. Some transistors have flexible leads for soldering directly into the circuit.

Figure 11-2 is a cutaway view of a very early transistor called a *point-contact* type. This figure shows a small germanium crystal and two thin wires (cat whiskers) making contact with the crystal. These elements are encased in a moistureproof housing. Terminal leads from these elements are brought out through the housing for circuit connections. Point-contact transistors are now obsolete.

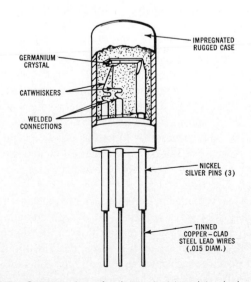

Fig. 11-2. Cutaway view of point-contact transistor (enlarged four times). *(CBS)*

Fig. 11-1. Some transistor shapes. *(Sylvania Electric Products, Inc.)*

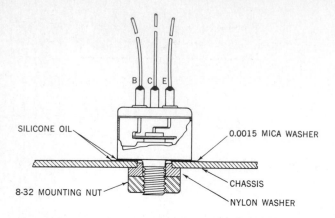

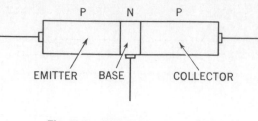

Fig. 11-4. PNP junction transistor.

Fig. 11-3. Cutaway view of junction transistor. *(Minneapolis-Honeywell Regulator Co.)*

Figure 11-3, a cutaway view of an alloy-junction transistor, shows a thin semiconductor-crystal wafer. Indium pellets are alloyed onto each face of the base. The two pellets and the semiconductor crystal constitute the elements of this transistor. The elements are enclosed in a hermetically sealed housing, and terminal leads are brought out of the housing for connection to the circuit.

Junction Transistors

Transistors are an extension of the semiconductor diode. The PNP transistor illustrated in Fig. 11-4 is an example. This is a junction transistor formed by sandwiching a very thin strip of N-type silicon between two "wide" strips of P-type silicon. Three leads are brought out from individual metallic plates which make contact with the respective semiconductor crystals. The entire assembly is encased in a moistureproof housing. The geometry of junction transistors will differ from that in Fig. 11-4, depending on the method of constructing the junction. Moreover, the characteristics of the various types of junction transistors depend on the method of fabrication.

Refer to Fig. 11-4. The P wafer on the left is designated the "emitter," the N wafer in the middle, the "base," and the P wafer on the right, the "collector." The base is about 1 mil (0.001 in) thick. For purposes of biasing, this transistor may be considered as two diodes. The emitter-base constitutes one diode, the collector-base the other diode. When used as an amplifier, the transistor is biased as shown in Fig. 11-5. The emitter-base is biased in the forward or low-resistance direction by V_{EE}, the collector-base in the reverse or high-resistance direction by V_{CC}.

Holes are the majority-current carriers in the emitter-base diode, and they emanate from the P-type emitter. If the collector were not present, current flow in the emitter-base section would take place in the manner described in Experiment 1. However, the presence of the P-type collector and its connection to the negative terminal of collector battery V_{CC} radically alter the path of hole-current flow. Only a small

percentage of the holes emitted by the emitter combine with the free electrons in the base. The other holes (about 95 percent) pass through the very thin base crystal and are attracted to the negative battery terminal on the collector. The emitter-collector circuit is externally completed through the two batteries V_{EE} and V_{CC} connected in series-aiding. From this description of current flow in a transistor, it is apparent (Fig. 11-5) that the emitter is the source of current carriers, that the emitter-base current is very small, and that the emitter-collector current is high. It may be seen also that changes in emitter-base bias will result in changes in emitter current. Thus an increase in forward bias will result in an increase in emitter current and hence in collector current. Base current will increase or decrease very little when emitter current increases or decreases. It is evident, therefore, that collector current may be controlled readily by changes in emitter-base bias. The designations "emitter" and "collector" may now be readily associated with their functions.

A junction transistor may also be made with an NPN configuration (Fig. 11-6). As in the preceding case, biasing of the emitter-base must be in the forward direction and of the collector-base in the reverse direction. Because of the use here of an N-type crystal as emitter and collector and of a P-type as the base, battery polarities must be reversed, as compared with the biasing of a PNP transistor. In the NPN transistor, electrons are the majority-current carriers. In the PNP transistor, holes are the majority-current carriers.

A simple scheme to remember battery polarity connections to a transistor is the following: The battery polarities to the emitter and base are the same as the letter of the emitter and base impurity designations. Thus, in the PNP type, the emitter receives the P (positive) (+) terminal, and the base receives the N (negative) (−) terminal. Similarly, in the

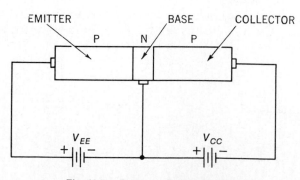

Fig. 11-5. Biasing a PNP transistor.

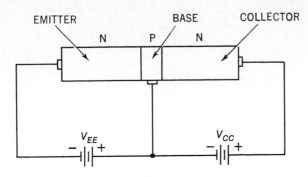

Fig. 11-6. Biasing an NPN transistor.

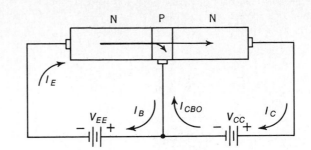

Fig. 11-8. Electron-current flow in external circuit of NPN transistor.

NPN type, the emitter receives the N (negative) (−) terminal, the base the P (positive) (+) terminal. The collector receives a battery polarity opposite to its impurity-type designation. Thus the P-type collector receives the N (negative) (−) terminal of the battery, and the N-type collector receives the P (positive) (+) terminal.

Figure 11-7 is a simplified diagram showing the direction of electron-current flow in the *external* circuit of a PNP transistor. Notice that it is *electron current* flowing in the external circuit and hole flow *within* the P-type crystal. The current noted as I_{CBO} is very small and will not be discussed now. Note that the current flowing through the emitter circuit is "total" current and that this current divides into base current and collector current. It is important to emphasize that the emitter-base current (determined by base-emitter bias voltage) is actually the cause of the emitter-collector current. As we shall see later, the emitter-base current also controls the emitter-collector current.

Figure 11-8 is a simplified diagram showing the direction of electron-current flow in the external circuit of an NPN transistor. Comparing with Fig. 11-7, we note that the direction of current flow in the external circuit of an NPN transistor is opposite to that of a PNP type.

I_{CBO} and Thermal Runaway

An important characteristic of the collector-base circuit should be noted. This is I_{CO} (also called I_{CBO}), the collector current flowing with the collector-base junction biased in the reverse direction and the emitter-base open-circuited. This leakage current (I_{CBO}) is due to minority carriers in the collector and base. I_{CBO} is in the range of a few microamperes (μA) for germanium and a few nanoamperes (nA) for silicon, and it increases with an increase in temperature.

An important factor which affects the operation of a transistor is its operating temperature. A transistor is very sensitive to temperature changes. Increased temperature results in increased current. This in turn leads to added heat and more current. If this chain reaction, which is called *runaway*, is uninterrupted, it may result in complete destruction of the transistor because of excessive heat. In the design of transistor circuits, thermal runaway is prevented by negative-feedback arrangements with which you will become familiar. The normal range of temperature within which a transistor may be operated safely is specified by the manufacturer. Silicon transistors are more tolerant of heat than are germanium transistors, and their temperature operating range is therefore much wider than germanium.

Transistor Symbols, Basing, and Mounting

The schematic symbols for a PNP and an NPN transistor are shown in Fig. 11-9a and b, where the element with the arrow is the emitter and its symmetrical counterpart is the collector. The PNP transistor is characterized by the fact that the emitter arrow points to the base, whereas the arrow points away from the base in the NPN type. Note that electron-current flow inside the transistor is *opposite* to the direction of the arrow.

Two methods are generally used for bringing transistor connections out of the envelope. The first employs long, flexible leads. An example of this is the transistor shown in Fig. 11-10a. The three leads are all in line and may be identified as emitter, base, and collector in the order shown.

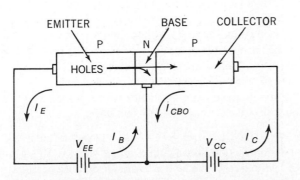

Fig. 11-7. Electron-current flow in external circuit of PNP transistor.

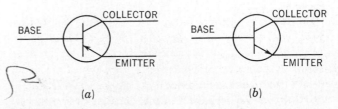

Fig. 11-9. Schematic symbol for *(a)* PNP and *(b)* NPN transistors.

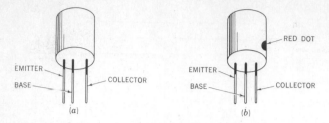

Fig. 11-10. Methods of identifying transistor leads.

The collector spacing is greater than that of the other two elements. A variation of this system also uses three in-line leads. The collector is identified by a red dot which is painted close to it on the transistor case, as in Fig. 11-10b. The usual method for connecting the transistors of Fig. 11-10a and b into the circuit is to solder the flexible leads to the proper terminals in the circuit.

The second method uses rigid pins for bringing transistor connections out of the envelope, as in Fig. 11-11a. This is a plug-in type of transistor and requires a matching socket to hold it. Another basing arrangement permits the selection of either the first or second mounting method. In this case, the transistor is supplied with flexible leads welded to rigid base pins (see Fig. 11-2). The flexible leads are used if the transistor is to be soldered into the circuit. If a socket is to be used, the flexible leads are easily removed with a pair of cutters. A triangular type of socket and basing arrangement used on other socket-mounted transistors (Fig. 11-11b and c) is also used. The connection layout duplicates the accepted transistor-diagram configuration.

Other typical arrangements of transistor basing and socket arrangements are shown in Fig. 11-12.

Transistors are rated according to their ability to dissipate power. Thus a transistor used as a low-level audio amplifier may have a power rating which is low, say 50 mW. A transistor used as an output amplifier must have a higher wattage rating. The casings of power transistors are especially designed to permit ready cooling. For example, some power transistors use radial fins for conducting the heat away (see Fig. 11-1). Other types use a metal shell which mounts onto the metal chassis of the equipment where it is employed (see Fig. 11-3). This transistor (Fig. 11-3) has the collector connected to the transistor housing. The chassis then conducts the heat away. This type of transistor has a

higher power rating when physically clamped onto the metal chassis as described and a lower power rating when mounted off the chassis. Good design technique requires that transistors be mounted in the coolest part of the chassis. Fans are sometimes installed inside equipment to cool the transistors and other components. The following list gives the symbols used to denote transistor parameters.

V_{BB}	Supply voltage to the base
I_{CB}	Collector-to-base current (the second subscript may be used in any of the above to avoid confusion)
V_{KJ}	Circuit voltage between elements, for example, between elements K and J
V_{CB}	Voltage between collector and base

Rules for Working with Transistor Circuits and for Making Measurements in Them

Special Tools

Special tools are required for use in transistor circuits because of the small size of the transistors and their associated components. Moreover, transistorized equipment is usually miniaturized, and ordinary tools will be inapplicable. A listing of special tools includes:

- Miniature cutters
- Miniature needle-nose pliers
- An assortment of service-type tweezers
- A 25-W pencil-type soldering iron
- A soldering aid
- A jeweler's eyepiece (loupe) or a magnifying lens (stand-mounted)

Precautions against Damaging Transistors

Though they are sturdy devices, transistors may be easily damaged if improperly handled. Thus you should exercise great care in working with low-power transistors employing flexible leads, since these leads are fragile and easily broken. The same precaution applies to the handling of transistor components. These components have been miniaturized, and leads may be easily broken if not handled carefully. For maximum life, the transistors and associated components used in this and subsequent experiments should be permanently mounted on bread-boarding devices.

Transistors can be destroyed almost instantly, in contrast to vacuum tubes, which can tolerate moderate overloads for extended periods of time. A short circuit from base to collector, in an operating circuit, will almost always destroy a transistor. A meter probe or a tool short-circuiting a pair of terminals may thus destroy a transistor.

You will avoid many problems with solid-state circuits if you follow these guidelines as you work:

Inserting Transistors into a Circuit. Do not install a transistor or remove one from a circuit with power ON. This practice may permanently damage a transistor because of high transient currents which may develop. Be certain that the power is

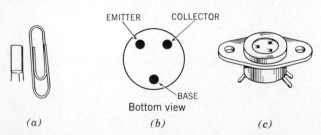

Fig. 11-11. *(a)* Plug-in transistor. Triangular basing identification; *(b)* basing diagram (bottom view); *(c)* socket. *(CBS)*

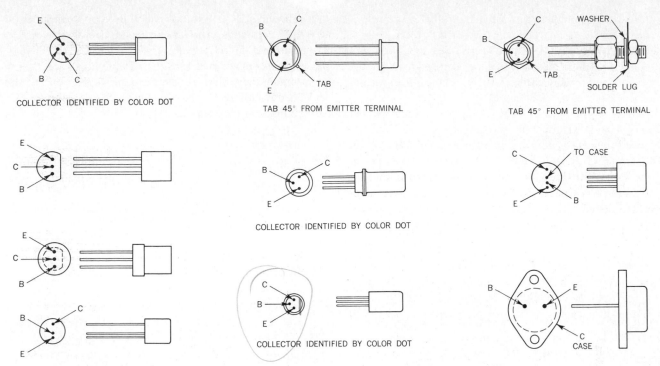

COLLECTOR IDENTIFIED BY COLOR DOT

TAB 45° FROM EMITTER TERMINAL

TAB 45° FROM EMITTER TERMINAL

COLLECTOR IDENTIFIED BY COLOR DOT

COLLECTOR IDENTIFIED BY COLOR DOT

Fig. 11-12. Typical transistor basing and socket arrangements. *(Symphonic Radio Electronic Corp.)*

off before inserting or removing a transistor or other component from a circuit.

Collector Bias and Voltage Values. Be certain that the polarity of bias voltage on the collector is correct before applying power. The collector must be reverse-biased. Moreover, the voltage on the collector and emitter must not exceed the values specified. You should therefore measure these voltages and adjust them for the proper value before applying power to the circuit.

Check of Circuit Connections. All connections should be checked against the circuit diagram before power is applied. Transistors should not be connected to a voltage source without some limiting resistance in the circuit.

Transistor Soldering. Soldering of transistors where necessary should be accomplished quickly. Low-wattage irons (25 W) are recommended. Transistor pigtails should be kept as long as possible consistent with circuit design considerations to reduce heat transfer. The same type of heat sink should be used when soldering transistor leads in a circuit as when soldering germanium diodes. An effective heat sink is created when long-nose pliers are used for grasping the transistor pigtail between the transistor body and the point of heat application, as in Fig. 11-13.

Voltage Measurements. All test equipment should be isolated from the power line. If the equipment is not isolated, an isolation transformer should be used. In making voltage measurements in transistor circuits, care should be taken to minimize the possibility of accidental short circuits between closely spaced terminals. Accidental short circuits may apply improper or excessive voltage to the transistor elements and may destroy the transistor.

Resistance Measurements. Transistors may be damaged during resistance measurements. Thus, if a shunt-type ohmmeter is used on the low-resistance ranges, it may supply excessive current to the transistor and destroy it. You must therefore be cautious in making ohmmeter checks in transistor circuits. A good rule to follow is to check the ohmmeter leads' potential and polarity before checking transistor junctions. Also, if it is necessary to check resistance of components in a transistor circuit, the transistor should be removed first if it is a plug-in type. When resistance measurements

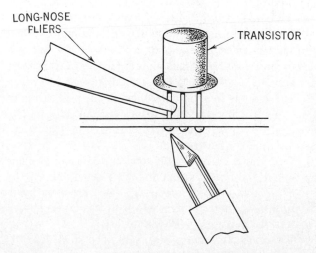

Fig. 11-13. Pliers act as heat sink when transistors are soldered in circuit.

with the transistor in-circuit are made, allowance must be made for conduction through the transistor. A recommended method is to make two sets of resistance measurements, reversing the ohmmeter leads for the required reading. The higher reading is more correct because on the higher reading the transistor was reverse biased.

Electronic voltmeters (EVMs) frequently contain a *low-power-ohms* (LPΩ) function, specifically designed for making resistance measurements in transistor circuits. On the LPΩ function there is insufficient voltage developed to forward-bias transistor functions. Hence the LPΩ function is used for resistance measurements in circuits containing solid-state devices.

Use of Signal Generators as a Signal Source in Transistor Circuits. Excessive signal-generator output may destroy a transistor. Hence generator output should be set at minimum at the start of signal injection. As an additional safety measure, the generator should not be coupled directly into the circuit. Loose capacitive coupling, wherever possible, is recommended.

SUMMARY

1. Transistors are solid-state devices which extend the range and application of the two-element diode.
2. Silicon is the element from which most transistors and other semiconductors are made today.
3. Junction transistors consist of a very thin element called the *base*, sandwiched between two elements called the *emitter* and *collector* (Fig. 11-4).
4. Transistors are either PNP or NPN type. The first letter designates the emitter material, the middle letter, the base, and the last letter, the collector material. Transistors therefore contain two *junctions*, the emitter-base junction and the collector-base junction.
5. The characteristics of junction transistors depend on the extent of doping of the emitter, base, and collector; on the geometry of the transistor; and on the method of fabrication.
6. For purposes of biasing, the transistor may be considered as consisting of two diodes, the emitter-base diode and the collector-base diode.
7. In most applications the emitter-base diode is forward-biased and the collector-base diode reverse-biased.
8. The emitter *injects* (emits), or is the source of, the current carriers *in* a transistor. The collector receives (collects) most of the current carriers. The base controls the collector current.
9. The emitter current is the total current, and it divides into the collector current and the base current.
10. Current flow inside a transistor is carried on by the majority-current carriers, which are electrons in N-type material and holes in P-type material.
11. There are also *minority* carriers. The small *minority*-carrier current in the collector-base junction with the emitter open is called *leakage current* or I_{CBO}.

12. Though I_{CBO} is very small in silicon transistors, it increases with heat and may destroy a transistor unless checked.
13. Transistors are heat-sensitive. The temperatures within which they may be operated are specified by the manufacturer.
14. Transistors may be solder-in or plug-in type. The plug-in type uses sockets varying in shape and size to match the transistor.
15. Transistors are rated according to ability to dissipate power, varying from the very low (50 mW) to high (2 W or more).
16. High-power transistors employ heat sinks to conduct the heat away from the transistor.
17. In soldering transistors, low-wattage soldering irons are used to prevent overheating.
18. Transistors should be handled carefully to prevent damage to their fragile leads.
19. Always turn power off before inserting or removing a transistor from a circuit.
20. Resistors may be destroyed accidentally when measuring voltages at their terminals. Care must therefore be exercised to prevent short-circuiting terminal leads during measurement in the circuit.
21. Transistors may be damaged during ohmmeter checks of their elements if the ohmmeter supplies excessive current to the transistor. Therefore, the low current ranges of an ohmmeter should be used.
22. In making resistance checks in a circuit containing a transistor or other solid-state device, the *low-power-ohms* function of an EVM should be used. This prevents the transistor elements from being forward-biased and ensures proper resistance readings.

SELF-TEST

Check your understanding by answering these questions.

1. Present-day transistors are usually made of _____ (*germanium, silicon*).
2. Three-element transistors contain _____ (*two, three*) junctions.
3. The emitter-base junction of a transistor is _____-biased; the collector-base junction is _____-biased.
4. In a PNP transistor the emitter is _____ (*positive, negative*) relative to the base, while the collector is _____ (*positive, negative*) relative to the base.
5. Battery polarities in an NPN transistor are _____ (*reverse, same*) as compared with those in a PNP transistor.
6. The arrow head on the emitter symbol of a transistor shows the direction of conventional current flow, which is opposite to electron flow. _____ (*true, false*)

7. Transistors are hardy devices and therefore little care need be exercised in handling them. _____ (*true, false*)

8. A dot on the transistor case, or the spacing between elements, is frequently used to identify transistor elements. _____ (*true, false*)

9. Transistors are sensitive to _____ . Therefore _____ sinks should be used in soldering transistor terminals in a circuit.

10. _____ should be exercised in making voltage measurements at transistor elements in a circuit, to prevent _____ to the transistor.

MATERIALS REQUIRED

- Power supply: 1½- and 6-V dc sources
- Equipment: Two multirange milliammeters (or 20,000-Ω/V VOMs); EVM
- Resistors: 100- and 820-Ω ½-W
- Semiconductors: 2N6004 and 2N6005 transistors appropriately mounted, or equivalent
- Miscellaneous: 2500-Ω 2-W potentiometer; two SPST switches

PROCEDURE

PNP Biasing

1. You will receive one PNP and one NPN junction transistor, properly identified. Examine these transistors. Draw a basing diagram for each, identifying the transistor and each of the elements.

2. Connect the circuit of Fig. 11-14, with power switch S_1 off. Set R_2 for maximum resistance. This will apply minimum emitter bias when power is applied. R_1 is a current- (bias-) limiting resistor in the emitter circuit. R_3 is a limiting resistor in the collector circuit. Maintain

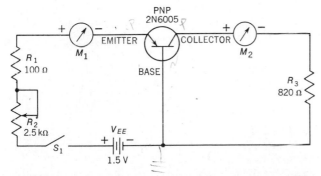

Fig. 11-14. Current-voltage measurements in the emitter-base circuits.

proper battery and meter polarity. Set meters M_1 and M_2 on their high ranges. After power is applied, decrease the meter range until it is suitable for reading the current.

Have instructor approve circuit connections before applying power.

3. **Power on.** Observe and measure current in the emitter and collector circuits. Record the data in Table 11-1. With an EVM, measure emitter-base and collector-base voltage. Record the data in Table 11-1. Show polarity of voltage.

4. **Power off.** Revise circuit Fig. 11-14 by adding V_{CC} and S_2, as in Fig. 11-15. This second battery biases the collector-base circuit.

NOTE: In the circuit of Fig. 11-15, and in those circuits containing *two* power switches, the instruction power OFF or power ON will mean opening or closing, respectively, *both* switches.

5. **Power on.** Maintain R_2 at maximum resistance. Observe and measure emitter and collector currents. Record the data in Table 11-1.

Measure and record emitter-base voltage and collector-base voltage. Indicate polarity of voltage.

TABLE 11-1. Transistor Biasing

	PNP					NPN			
Transistor:					Transistor:				
Step	Emitter-base		Collector-base		Step	Emitter-base		Collector-base	
	Current	Voltage	Current	Voltage		Current	Voltage	Current	Voltage
3	.19 m	.64	.18 m	.2	X	X	X	X	X
5	.18 m	.64	.18 m	≈6	12	.2 m	.5	.19 ma	5.8
6	6.7 m	.75	6.6 m	.64	13	2.1 m	.72	6.9 m	.69
8	.1	1.5	0	.64	15	0	1.58	0	6
10	X	X	0	6.4	17	0	X	0	6

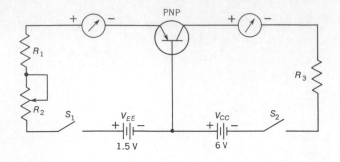

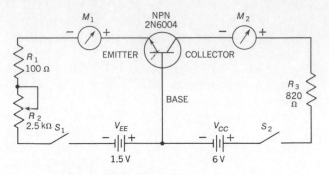

Fig. 11-15. Current-voltage measurements in the emitter and collector circuits of a PNP transistor.

Fig. 11-17. Current-voltage measurements in the emitter and collector circuits of an NPN transistor.

6. Set R_2 for minimum resistance, hence for maximum emitter bias. Switch meter ranges as required. Observe and measure emitter and collector current, and emitter-base and collector-base voltages. Record the data in Table 11-1. Indicate polarity of voltage.

7. **Power off.** Reverse emitter battery (V_{EE}) polarity and meter M_1 connections (see Fig. 11-16). Set R_2 for minimum resistance.

8. **Power on.** Observe and measure current in emitter and collector circuits. Record the data in Table 11-1. Measure, and record in Table 11-1, emitter-base voltage and collector-base voltage. Indicate polarity of voltage.

9. *Open S_2.* Open the emitter-base circuit by opening S_1.

10. *Close S_2.* Observe and measure collector current. Record the data in Table 11-1. This is the value of I_{CBO} for the circuit conditions. Measure, and record in Table 11-1, collector-base voltage. Indicate polarity of voltage.

13. Set R_2 for minimum resistance. Switch meter ranges as required. Observe and measure emitter and collector current and emitter-base and collector-base voltage. Record the data in Table 11-1. Indicate polarity of voltage.

14. **Power off.** Reverse emitter battery (V_{EE}) polarity and meter M_1 connections (see Fig. 11-18). Set R_2 for minimum resistance.

15. **Power on.** Observe and measure current in emitter and collector circuits. Record the data in Table 11-1. Measure, and record in Table 11-1, emitter-base voltage and collector-base voltage. Indicate polarity of voltage.

16. *Open S_2.* Open the emitter-base circuit by opening switch S_1.

17. *Close S_2.* Observe and measure emitter and collector currents. Record the data in Table 11-1. Measure, and record in Table 11-1, collector-base voltage. **Power off.**

NPN Biasing

11. **Power off.** Remove PNP transistor from the circuit and substitute NPN transistor, as in Fig. 11-17. Reverse polarity of V_{CC}, as in Fig. 11-17.

12. *Close S_2.* Set R_2 for maximum resistance. *Close S_1.* Observe and measure emitter and collector currents. Record the data in Table 11-1. Measure, and record in Table 11-1, emitter-base and collector-base voltage. Indicate polarity.

Extra Credit

18. The relationship between base, emitter, and collector current is

$$I_B = I_E - I_C$$

Verify this relationship experimentally for the PNP and NPN transistors. Draw the schematic diagram of the circuit you used, and give the detailed procedure you followed. Enter your measurements in a table especially prepared for them.

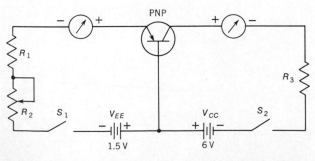

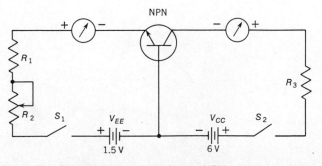

Fig. 11-16. Effect of reverse emitter bias on PNP transistor operation.

Fig. 11-18. Effect of reverse emitter bias on NPN transistor operation.

QUESTIONS

1. Which figure, 11-14 to 11-18, shows proper biasing of the emitter and collector circuit for (*a*) a PNP transistor? (*b*) an NPN type? Explain why. (Refer to the results in Table 11-1 for explanation.)
2. What is the effect on collector current of increasing emitter bias? Refer to your measured data to substantiate your answer.
3. What are the effects on collector current of reverse bias on the emitter-base circuit? Refer to your data to substantiate your answer.
4. What is I_{CBO}? Explain how I_{CBO} was measured.
5. What are holes? In which type of semiconductor material are holes the majority-current carriers? Why?
6. What precautions must be observed in working with, and making measurements in, transistor circuits? List as many as you know.

Answers to Self-Test

1. silicon
2. two
3. forward; reverse
4. positive; negative
5. reversed
6. true
7. false
8. true
9. heat; heat
10. Care; damage

CURRENT GAIN (β) IN A COMMON-EMITTER CONFIGURATION

OBJECTIVES

1. To measure the effects on I_C of varying I_B
2. To determine beta (β)

INTRODUCTORY INFORMATION

The usefulness of a transistor lies in its ability to control current. Associated with the property of control is the ability to amplify current, voltage, and power. The related characteristics of amplification and control make it possible to use this semiconductor in a wide range of applications from communications to automation.

Transistor Circuit Configurations

A three-element transistor may be connected in a circuit in any one of three different arrangements. These are (1) the grounded, or common, emitter, (2) the grounded, or common, base, and (3) the grounded, or common, collector. The circuits are shown in Fig. 12-1*a, b,* and *c,* respectively. Each of these circuits has unique properties with which you will become familiar. The grounded emitter is the configuration most frequently used because it provides voltage, current, and power gain. In each circuit a signal voltage is applied to the input and the processed signal is taken from the output. In the case of the common emitter, the input signal is applied between base and emitter, the output taken from the collector to emitter. The name *common emitter* derives from the fact that the emitter is common to both the input and output circuits. In the grounded, or common, base, the base is common to both the input and output. In the common-

collector arrangement, the collector is at some dc potential V_{CC} required for reverse biasing of the collector circuit. However, assuming V_{CC} is an ideal battery with no internal resistance, the impedance in the collector circuit is 0. No signal voltage is developed in the collector, which then is at signal 0 level. The collector now serves as the common reference for both input and output signal. Since the collector is common to both the input and output, the circuit is called a *common* or *grounded collector.*

NOTE: Figure 12-1 shows a PNP transistor connected in each of three configurations. An NPN transistor may be similarly connected, however, with a change in battery polarities.

Current Gain in Transistors

Alpha (α)

In the common-base circuit, control of collector current is effected by variations in emitter current. This fact was established in Experiment 11, where measurement revealed that an increase or decrease in emitter current resulted in an increase or decrease, respectively, in collector current. This is considered in evaluating one important control characteristic (alpha) of a transistor in the common-base arrangement. Alpha is defined as the ratio of a change in collector current ΔI_C (read delta I_C), to the change in emitter current ΔI_E, with the collector voltage held constant. Thus

$$\alpha = \frac{\Delta I_C}{\Delta I_E} \quad (V_{CB} \text{ const}) \tag{12-1}$$

Alpha is called the *forward-current-transfer ratio* and is also designated by the letters h_{fb}.

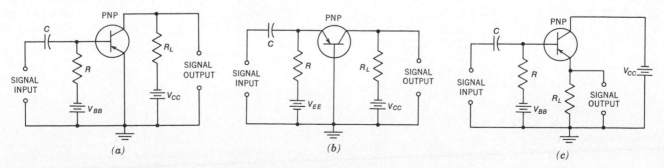

Fig. 12-1. Transistor circuit configurations: *(a)* common emitter; *(b)* common base; *(c)* common collector.

In Experiment 11 it was determined that in a junction transistor in the common-base circuit, the collector current was slightly less than the emitter current. It can therefore be expected that ΔI_C will be slightly less than ΔI_E. Therefore the ratio $\Delta I_C/\Delta I_E$, hence α, will be slightly less than 1 since the numerator is less than the denominator. For the junction transistor in the common-base configuration there is no current gain between input and output. There is actually a slight current loss. However, a common-base transistor amplifier exhibits voltage and power gain. Current "gain" is considered in transistor discussions because the transistor is a current-driven device.

Beta (β)

In the grounded-emitter configuration, the input signal is applied to the base. Current gain is now designated by β (the Greek letter beta) and is defined as follows:

$$\beta = \frac{\Delta I_C}{\Delta I_B} \quad (V_{CE} \text{ const}) \tag{12-2}$$

Equation (12-2) states that β is the ratio of the change in collector current (ΔI_C) effected by a change in base current (ΔI_B) with collector voltage (V_{CE}) maintained at a constant value. β, then, is the current amplification factor in a grounded-emitter amplifier. Another designation for β is h_{fe}.

In a junction transistor, values of β are always greater than 1. The reason becomes clear when we examine the base circuit of a PNP transistor connected as a grounded-emitter amplifier (Fig. 12-2). The direction of emitter current and collector current are opposite in the base circuit as shown here. Base current I_B is the difference between I_E and I_C. Emitter and collector current are of comparable magnitudes. Hence base current is small. Small changes in base current must therefore produce large changes in collector current.

Determining Beta Experimentally. A test setup for measuring β in an NPN transistor is shown in Fig. 12-3. Base current is varied by adjustment of R_2. M_1 is used for measuring base current. Collector current is measured by M_2. R_4 is used for maintaining a constant collector voltage V_{CE} measured by EVM, M_3.

To determine β, R_2 is first adjusted for some current reading I_{B1}. R_4 is adjusted to maintain a given value of collector voltage V_{CE}. Collector current I_{C1} is measured by

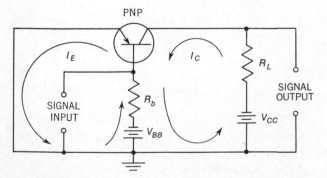

Fig. 12-2. Current in grounded-emitter amplifier.

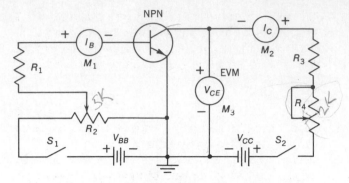

Fig. 12-3. Test circuit for measuring β in an NPN transistor.

M_2. R_2 is then varied for a slightly higher (or lower) value of base current I_{B2}. R_4 is readjusted to maintain the same value V_{CE} as previously measured. Collector current I_{C2} is now read. β may be computed as

$$\beta = \frac{\Delta I_C}{\Delta I_B} = \frac{I_{C2} - I_{C1}}{I_{B2} - I_{B1}} \tag{12-3}$$

Proper polarity must be maintained in connecting the batteries and meters in the circuit. If a PNP transistor is used, reversal of meter and battery polarities is required.

There is an important relationship between α and β in a junction transistor which can easily be demonstrated. The relationship between current in the elements of a junction transistor is given by

$$I_E = I_C + I_B \tag{12-4}$$

For a small change in emitter current ΔI_E, there is a corresponding change in collector current ΔI_C, and in base current ΔI_B. Therefore

$$\Delta I_E = \Delta I_C + \Delta I_B \tag{12-5}$$

and

$$\Delta I_B = \Delta I_E - \Delta I_C \tag{12-6}$$

The ratio

$$\frac{\Delta I_C}{\Delta I_B} = \frac{\Delta I_C}{\Delta I_E - \Delta I_C} \tag{12-7}$$

Dividing numerator and denominator of the right-hand member of Eq. (12-7) by ΔI_E, we find that

$$\frac{\Delta I_C}{\Delta I_B} = \frac{\Delta I_C/\Delta I_E}{1 - \Delta I_C/\Delta I_E} \tag{12-8}$$

When V_{CE} is constant, the ratio $\Delta I_C/\Delta I_B$ equals β. With V_{CB} constant, $\Delta I_C/\Delta I_E$ equals α. Therefore for a constant V_{CE} and a constant V_{CB}, Eq. (12-8) may be written as

$$\beta = \frac{\alpha}{1 - \alpha} \tag{12-9}$$

From Eq. (12-9) follows the relationship

$$\alpha = \frac{\beta}{1 + \beta} \tag{12-10}$$

It is evident that as α approaches the value 1, β becomes increasingly larger. Thus a transistor whose α is 0.98 has a β

.00iUA   *.001 m*

whose value is 49, whereas an α of 0.99 is associated with a β of 99. Since such a small change in α in the region of 1 gives rise to such a large change in β, measurement of α must be very accurate if formula errors in β are to be avoided.

SUMMARY

1. In a grounded-emitter (CE) configuration (Fig. 12-2) the input signal is applied to the base.
2. In this configuration current gain is designated as beta (β).
3. Beta is defined as the ratio of the change in collector current ΔI_C, effected by a change in base current ΔI_B, with collector-emitter voltage V_{CE} held constant. Thus

$$\beta = \frac{\Delta I_C}{\Delta I_B} \quad (V_{CE} \text{ const}) = h_{fe}$$

$$\beta = \frac{I_{C2} - I_{C1}}{I_{B2} - I_{B1}}$$

4. When V_{CE} and V_{CB} are both held constant, there is an important relationship between α and β which is given by either of these two formulas:

$$\beta = \frac{\alpha}{1 - \alpha}$$

or

$$\alpha = \frac{\beta}{1 + \beta}$$

If we have measured either α or β, we may find the other by substituting in the appropriate formula.

SELF-TEST

Check your understanding by answering these questions.

1. Current gain in a CE configuration is called
   _____ .

2. Current gain in a CE configuration is always _____ than 1.
3. In a CE configuration the transistor element which is common to both the input and output signals is the _____ .
4. In a CE circuit the input signal is applied to the _____ .
5. In determining β experimentally, the voltage V_{CE} must be _____ _____ .
6. The formula which defines β in terms of collector and base current is:

$$\beta = \text{_____} \quad (\quad)$$

7. In the circuit of Fig. 12-3 the following measurements were made with $V_{CE} = 5$ V: $I_{B1} = 50$ μA, $I_{C1} = 4.5$ mA; $I_{B2} = 75$ μA; $I_{C2} = 9.5$ mA. Beta must therefore equal _____ .
8. The value of α corresponding to the β determined from the measurements in question 7 is _____ .

MATERIALS REQUIRED

- Power supply: Two low-voltage dc sources
- Equipment: Two multirange micromilliammeters (or 20,000-Ω/V VOMs); EVM
- Resistors: 100- and 4700-Ω ½-W
- Semiconductors: 2N6004 with appropriate socket; 2N6005 for extra-credit question
- Miscellaneous: 2500- and 5000-Ω 2-W potentiometers; two on-off switches; other parts as required for extra-credit question

PROCEDURE

Measuring Beta (h_{fe})

1. Connect the circuit of Fig. 12-3 using the following component values: a 2N6004 transistor, $R_1 = 4700$-Ω ½-W resistor, $R_2 = 5000$-Ω 2-W potentiometer, $R_3 = 100$-Ω ½-W resistor, $R_4 = 2500$-Ω 2-W potentiometer, $V_{EE} = 1.5$ V, and $V_{CC} = 9$ V.

   M_1 and M_2 are multirange micromilliammeters or equivalent ranges on 20,000-Ω/V VOMs. R_4 *must be set for maximum resistance before power is applied.*
2. Check circuit for proper connections.
3. **Power on.** Vary R_2 from maximum to minimum resistance, thus increasing base current I_B from its minimum to maximum value in this circuit. Observe the effect on I_C as I_B is increased. Record the results in Table 12-1.
4. Adjust R_2 for 10-μA base current. Adjust R_4 for 6 V at the collector. Measure and record I_C.

5. Adjust R_2 for maximum base current. Adjust R_4 to maintain 6 V. Measure and record I_C.
6. Adjust R_2 for 30 μA of I_B. Adjust R_4 for 6 V. Measure and record I_C.
7. Adjust R_2 for 40 μA of I_B. Adjust R_4 to maintain 6 V. Measure and record I_C.
8. **Power off.** Compute and record β. Show your computations.

Extra Credit

9. Explain in detail how you would measure β of a PNP transistor. Draw the circuit diagrams of the test setup.
10. Measure and compute β of a 2N6005 transistor. Record the measurements in a specially designed table. Show your computations.

TABLE 12-1. Test Data for Measuring β

Step	I_B, μA	I_C, mA	Effect on I_C of Increasing I_B	
4	10	1.8mA	Step	INCREASED
5	Maximum: 244uA	51.5mA	3	

Step	I_B, μA	I_C, mA	Collector Voltage	
6	30	5.5mA	IE = 5.53	6
7	40	7.4mA		6
8		$\beta = \dfrac{\Delta I_C}{\Delta I_B} =$ _____ = 182.6		

QUESTIONS

1. Define β.
2. Explain in detail a procedure for measuring β.
3. Using the value of β determined in this experiment, find α. Show formula and all work.
4. Given a transistor with $\beta = 25$ and $I_C = 3$ mA when $I_B = 100$ μA. Assume linear operation. (*a*) What is the value of I_C when $I_B = 125$ μA? (*b*) What is the value of I_B when $I_C = 2$ mA?

Extra Credit

5. What are the differences, if any, in determining the current gain of NPN and PNP transistors?

1   180 $\dfrac{1.8m}{10u}$

2

3   .183 $\dfrac{5.5m}{30u}$

4   .185 $\dfrac{7.4m}{40u}$

$I_E = I_B + I_C$

Answers to Self-Test

1. beta (β)
2. greater
3. emitter
4. base
5. held constant

6. $\dfrac{\Delta I_C}{\Delta I_B}$ (V_{CE} const)
7. 200
8. 0.995025

TRANSISTOR DATA AND THE COLLECTOR CHARACTERISTIC CURVE FOR THE CE CONNECTION

OBJECTIVES

1. To become familiar with the nature of the data found in transistor manuals
2. To determine experimentally and plot the family of collector (V_{CE} versus I_C) characteristic curves for the CE configuration
3. To use a transistor curve tracer to display the family of average collector characteristics, CE configuration

INTRODUCTORY INFORMATION

Transistor Data

Transistors are designed with unique characteristics to meet certain application requirements. The manufacturer provides data sheets in which these characteristics are given. Data are furnished both in *tabular* and in *graphic* form. It is important to understand these data charts and graphs.

Compilation of data sheets is found in transistor manuals supplied by transistor manufacturers or in transistor reference handbooks published commercially. The nature of the data depends on the source and on the intended use of the transistor. The following specifications are generally available.

1. *A brief description of the transistor and suggested applications.* As an example the 2N4074 is listed in the RCA manual as a small-signal 0.5-W, Si NPN epitaxial-planar type used in high-voltage high-current audio and

video amplifier service in commercial and industrial equipment. [JEDEC (Joint Electron Device Engineering Council) *TO-104,* Outline No. 32.]

2. *Mechanical data, including dimensions, basing, and mounting.* Figure 13-1a is an outline drawing *(TO-104)* of the 2N4074 transistor and its dimensions. The basing diagram (Fig. 13-1b) locates the positions of pins 1, 2, and 3. These numbers, referenced to the circuit symbol for this transistor (Fig. 13-1c), identify the emitter (1), base (2), and collector (3).

3. *Maximum ratings.* These ratings are usually based on the *absolute maximum system* defined by JEDEC and standardized by EIA (Electronic Industries Association) and NEMA (National Electrical Manufacturers Association). The definition states that "absolute-maximum ratings are limiting values of operating and environmental conditions which should not be exceeded . . . under any condition of operation." For example, for the 2N4074, maximum ratings are given as follows.

It should also be noted that dissipation ratings (the power dissipated in the form of heat) are usually specified for ambient, case, or mounting-flange temperatures up to 25°C. If the transistor or solid-state device is to be operated at a higher temperature, the dissipation value must be derated, either by use of derating curves or by the formula specified in the characteristic sheet of the device.

4. *Characteristics and other engineering data.* Figure 13-2 gives some characteristics for the 2N4074. Other characteristics are omitted here.

Not all transistor manuals list their characteristics in

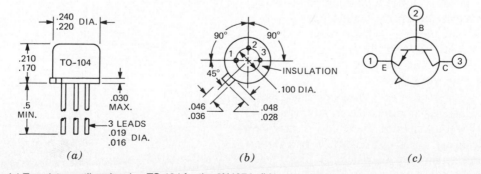

Fig. 13-1. *(a)* Transistor outline drawing TO-104 for the 2N4074; *(b)* basing diagram, bottom view; *(c)* circuit symbol. *(RCA)*

77

Collector-emitter voltage

$V_{BE} = -1$ V	V_{CEV}	40 V
Base open	V_{CEO}	40 V
Emitter-to-base voltage	V_{EBO}	8 V
Collector current	I_C	300 mA
Emitter current	I_E	-300 mA
Transistor dissipation		
T_C up to 75°C	P_T	2 W
T_A up to 25°C	P_T	0.5 W
Temperature range		
Operating (junction)	$T_J(\text{opr})$	-65 to 175°C
Storage	T_{STG}	-65 to 175°C
Lead-soldering temperature	T_L	225°C
(10 s max)		

NOTE: The symbols used above have the following meanings.

V_{CEV}	Collector-to-emitter voltage, with specified voltage between base and emitter (-1 V, in the case above)
V_{CEO}	Collector-to-emitter voltage, base open
P_T	Total nonreactive power input to all terminals
T_C	Case temperature
T_A	Ambient temperature
T_J	Junction temperature
T_{STG}	Storage temperature

Fig. 13-2. A partial list of characteristics for the 2N4074 at case temperature 25°C.

this manner. Moreover, the symbols used, though fairly standardized, are not always the same. Technicians should be familiar with the transistor manual they are using. Within its covers are data on solid-state devices, their operation, symbols, and applications.

Transistor manuals frequently contain interchangeability charts which list the product numbers of other manufacturers and give suggested replacement numbers. This information is helpful when exact replacements cannot be obtained. However, such listings do not take into account special characteristics required in some applications. These characteristics are controlled by sorting from the normal production spread of the standard type. Random use of the unsorted transistors will result in some which will not work in a particular application. Exact manufacturer's replacement parts should always be used, if possible, to ensure proper circuit operation.

Manuals also contain a listing of obsolete and discontinued numbers, circuits of useful devices, and dimensional outlines of all transistors.

5. *Transistor characteristic curves.* Transistor characteristic curves included in transistor manuals give information, in graphic form, about a transistor. These curves show the effects of variation of transistor parameters and are therefore an important source of information to the circuit designer and technician.

Characteristic curves for specific transistors are usually found in transistor manuals. Where they are not available they can be plotted in any one of several ways which will be described here.

Average Collector Characteristics, CE Configuration

Figure 13-3 shows the V_{CE} versus I_C family of average collector characteristics for the type 2N217 common-emitter connection. Each curve in the family is obtained by plotting I_C as a function of V_{CE} (for various values of V_{CE}), while maintaining a constant value of base current. Note that each curve reflects an increasing collector-current level for an increased input base-current level. It is interesting to note that to the right of the ordinate $V_{CE} = -1$ V, collector current is practically independent of collector voltage and is mainly dependent on base current. Another noteworthy fact is the termination point of each curve. Thus for 0.1-mA base-bias current the curve ends at $V_{CE} = -12$ V. At this point collector current is approximately 12 mA. The product of V_{CE} and I_C at the terminal point on the curve is therefore 144 mW. This value is close to 150 mW, the rated power dissipation of the transistor. The terminal point of each of the other curves is similarly limited to this power range. Transistors should not be operated beyond their rated power dissipation.

The curves in Fig. 13-3 were experimentally determined while the ambient temperature of the transistor was maintained at 25°C. If the temperature had been higher, collector current would have been higher, a factor which would have placed greater restrictions on the permissible dissipation.

From Fig. 13-3 it is possible to compute β.

$$\beta = \frac{\Delta I_C}{\Delta I_B} \qquad (V_{CE} \text{ const}) \qquad (13\text{-}1)$$

On the graph, the value of I_C is read for two consecutive values of I_B for some fixed value of V_{CE}. Those values are

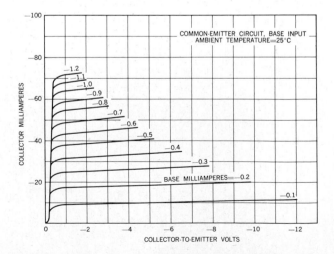

Fig. 13-3. Average collector characteristics for type 2N217 common emitter. *(RCA)*

then substituted in Eq. (13-1) and β computed. For example, when $V_{CE} = -4$ V and $I_B = 0.1$ mA, $I_C = 10$ mA (approx). Also for $V_{CE} = -4$ V and $I_B = 0.2$ mA, $I_C = 18$ mA. Since I_B and I_C are in the same units (milliamperes), it is possible to substitute directly in Eq. (13-1). Thus

$$\beta = \frac{18 - 10}{0.2 - 0.1} = \frac{8}{0.1} = 80$$

for the conditions specified.

Similarly, at $V_{CE} = -4$ V and $I_B = 0.3$ mA, $I_C = 26$ mA. Again, since $I_B = 0.2$ mA and $I_C = 18$ mA at $V_{CE} = -4$ V, the value of β is again 80. Therefore, if the transistor were operated at 0.2-mA base bias, and $V_{CE} = -4$ V, linear changes of base current between 0.1 and 0.3 mA would result in linear changes of collector current. That is, within the specified boundaries, the relationship between collector current and base current is

$$I_C = \beta I_B + k \tag{13-2}$$

For the above conditions, Eq. (13-2) is

$$I_C = 80 I_B + 2 \tag{13-3}$$

where I_C and I_B are given in milliamperes.

Test Circuits to Determine Average Collector Characteristics (V_{CE} versus I_C)

Point-by-Point Method

The test circuit shown in Fig. 13-4 is used for plotting the characteristic curves of an NPN transistor in the CE configuration, employing the point-by-point method. The same test circuit with battery and meter polarities reversed could be used for a PNP type. In this circuit, base current may be set to a specified value by adjusting R_1. The procedure is as follows: R_1 is adjusted to a reference value of I_B, at which value it is desired to plot the curve. M_1 monitors base current, and R_1 is used to maintain a constant level of I_B. M_2 measures collector current, while M_3 measures collector voltage. Predetermined values of collector voltage are selected in turn, and collector current is measured and recorded. The results are then plotted and graphed. To obtain a family of curves,

this procedure is repeated for specified values of base current.

Oscilloscope Display

A more rapid method for plotting NPN transistor characteristic curves of the grounded-emitter connection involves the use of the test circuit in Fig. 13-5. The curve is displayed on the oscilloscope and may be photographed. To obtain a family of curves, R_2 is adjusted in succession to predetermined values of base current, and each curve is in turn photographed in a multiple exposure.

The input circuit, base-to-emitter, is conventional. The output circuit requires explanation.

There is no dc voltage applied to the collector-emitter circuit. Instead, the circuit is energized by the pulsating dc voltage applied to the emitter as a result of the action of the transformer T and the rectifier D. Transformer T is a step-down filament transformer operating from the 120-V ac line, supplying 6.3 V rms to the secondary. For one half-cycle of input voltage, when the polarity across the secondary is as shown in Fig. 13-5, rectifier D conducts, effectively completing the emitter-to-collector circuit. The polarity of the pulsating voltage applied (collector is positive with respect to emitter) is proper for collector current. Resistor R_3, only 100 Ω, does not materially affect the collector circuit. However, current flow in the collector circuit causes a voltage drop across R_3. Thus instantaneous values of voltage developed across R_3 are proportional to the instantaneous values of collector current. The voltage across R_3 is applied to the vertical input of the oscilloscope. Hence, the vertical deflection of the oscilloscope beam is proportional to the instantaneous collector current. The emitter-collector voltage is applied to the horizontal input of the oscilloscope and supplies the sweep for the oscilloscope. Thus horizontal deflection of the oscilloscope's beam is proportional to this varying voltage. The presentation on the oscilloscope is therefore a graph of collector current (vertical) versus collector voltage (horizontal) for a specific value of base current.

During the first alternation, as described, collector current flows. During the second alternation, the rectifier acts as an open circuit. Hence no current flows in the collector-emitter circuit. This safeguard is essential; D did not open the circuit, forward bias applied to the collector would destroy the transistor.

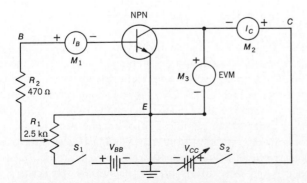

Fig. 13-4. Test circuit for determining V_{CE} versus I_C characteristics for grounded-emitter connection of an NPN transistor.

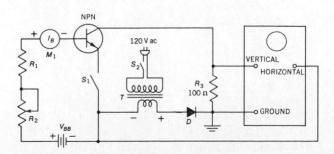

Fig. 13-5. Test circuit for displaying the characteristic curve of an NPN transistor in a grounded-emitter configuration.

A modification of this circuit arrangement may be used for displaying transistor characteristic curves in the common-base connection. For PNP devices, battery and diode polarities are reversed.

Curve Tracer

Automatic transistor curve tracers do not require an external test circuit. The transistor is plugged into a receptacle on the curve tracer. The manual controls of the curve tracer are set to the required operating voltages and currents. The cathode-ray-tube indicator of the curve tracer then displays a graph (curve) of the transistor characteristics.

It is possible to display a whole family of characteristics, as in Fig. 13-3, by proper setting of the curve-tracer controls.

The automatic curve tracer whose operation was just briefly described is a completely self-contained unit (Fig. 13-6). Physically it resembles an oscilloscope, though the ordinary oscilloscope functions are not built into this instrument.

Separate curve-tracer units which can be used with oscilloscopes are available commercially. These also provide a display of transistor curves (Fig. 13-7a and b).

Oscilloscope cameras are available for photographing oscilloscope and curve-tracer displays. This makes it possible to obtain curves and families of characteristic curves for immediate use.

(a)

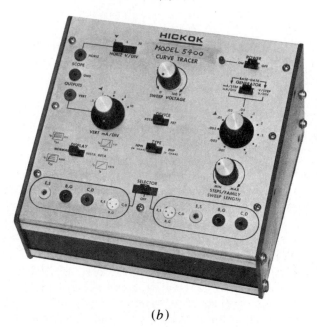

(b)

Fig. 13-7. (a) Curve-tracer unit used in conjunction with a general-purpose oscilloscope. (B & K) (b) Curve tracer used with general-purpose oscilloscope. (Hickok Teaching Systems)

Dual Power Supplies

In working with solid-state devices it is frequently necessary to use two independent power supplies. Dual power supplies, like the one in Fig. 13-8, offer a convenient source. Each of the supplies in Fig. 13-8 has its own controls. In addition there is a mode switch for selecting either one or both (A or B) of the independent supplies. In one mode it is possible to control both supplies from the A unit. In that mode the voltage output of each supply is exactly the same, but the polarity of each differs and A becomes a positive, B a negative supply. This feature is convenient when working with differential and operational amplifiers.

Fig. 13-6. Automatic transistor curve tracer. (Tektronix)

Fig. 13-8. Dual power supply. *(Hickok Teaching Systems)*

SUMMARY

1. Data on transistor characteristics are furnished by the manufacturer in tabular (chart) and graphic (curves) form.
2. Transistor data sheets, identified by number, include the following information: (*a*) mechanical, (*b*) descriptive, (*c*) maximum ratings, (*d*) electrical characteristics, (*e*) characteristic curves.
3. Transistors should not be operated beyond their rated wattage dissipation. For the CE configuration the product of V_{CE} and I_C is the collector dissipation of the transistor. Dissipation is specified for a specific operating temperature. If the operating temperature is higher than specified, dissipation must be derated by formula.
4. From the family of average collector characteristics (Fig. 13-3) it is possible to calculate the beta of a transistor.
5. Within the range where the relationship between I_C and I_B is linear, the beta of the transistor remains constant.
6. Average collector characteristics may be determined by any one of these methods:
   (*a*) *Point-by-point.* Figure 13-4 is the test circuit used to show how I_C varies with changes in V_{CE}. The data secured are plotted in the form of a graph.
   (*b*) *Oscilloscope display.* The test circuit of Fig. 13-5 is used to show on an oscilloscope a graph of I_C versus V_{CE} for some value of I_R for an NPN transistor.
   (*c*) *Curve tracers.* (1) Self-contained curve tracers come complete with an oscilloscope display unit. These generate a single curve or a whole family of curves. (2) Curve-tracer units which are used with general-purpose oscilloscopes are also available.

SELF-TEST

Check your understanding by answering these questions.

1. Maximum ratings are limiting values of _____ and _____ conditions which should not be exceeded.
2. The symbol V_{CEO} stands for _____ to _____ voltage, with _____ open.
3. The three different temperature characteristics which must be considered in operating a transistor are:
   (*a*) T_C, _____ temperature;
   (*b*) T_A, _____ temperature; and
   (*c*) T_J, _____ temperature.
4. Collector-cutoff current _____ is specified for a specific V_{CB} with the emitter open.
5. The static forward-current-transfer ratio of a transistor in the CE configuration is designated by the symbol _____ and represents the _____ _____ of a transistor.
6. The _____ of a transistor is the measure of the base's ability to control collector current.
7. Current gain of a transistor in the CE configuration may be determined from the family of _____ _____ characteristics.
8. In the graph of Fig. 13-2, for a V_{CE} greater than -1 V and $I_B = 0.1$ mA, _____ is practically independent of V_{CE}.
9. In Fig. 13-2, when $V_{CE} = -4$ V and $I_B = $ _____ mA, $I_C = 40$ mA.
10. In the test circuit of Fig. 13-3, the purpose of R_1 is to maintain a _____ _____ , as V_{CE} is varied by adjusting _____ .
11. In the test circuit of Fig. 13-4, the voltage applied to the vertical input of the oscilloscope is proportional to _____ _____ .
12. Collector current, in the circuit of Fig. 13-4, is limited to the _____ (*positive, negative*) alternation of collector-to-emitter voltage.

MATERIALS REQUIRED

- Power supply: Two variable low-voltage dc
- Equipment: Two multirange micromilliammeters (or 20,000-Ω/V VOMs); EVM; curve tracer
- Resistors: 100-, 470-Ω ½-W
- Semiconductors: 2N6004 with socket or equiv.
- Miscellaneous: 2500-Ω 2-W potentiometer; two SPST switches; other parts as required for extra-credit step 20

PROCEDURE

The procedure which follows requires some explanation. When changing ranges on current meters, readings may not seem to coincide from range to range. The meters, however, do read the actual current in the circuit. The discrepancy arises from changes in meter resistance when ranges are switched. The result is a change in total circuit resistance which affects the current in the circuit. Therefore, when changing ranges, it may be necessary to readjust the controls in the affected circuit to offset changes in meter resistance.

Another factor to consider is the interaction between the

collector and base circuit. It will be necessary to readjust the base-current control when collector voltage is varied, in order to hold I_B at a fixed value.

V_{CE} versus I_C

1. Connect the circuit of Fig. 13-4. Set V_{BB} at 1.5 V; V_{CC} at +3 V. S_1 and S_2 are open. Set R_1 to 0 V. M_1 and M_2 should be set on the highest milliampere range to protect the meters. The range is selected after power is applied. Check circuit connections before power is applied.
2. **Power on.** Adjust R_1 until low values of I_B (meter M_1) and I_C (meter M_2) are obtained to check circuit operation.
3. Adjust R_1 so that M_1 reads 10-μA current (I_B). Readjust R_1, when necessary during steps 4 and 5, to maintain $I_B = 10$ μA.
4. Adjust V_{CC} for 0 V. M_3 should read 0 ($V_{CE} = 0$). Read the value of I_C and record it in Table 13-1.
5. Adjust V_{CC} in turn to every value of V_{CE} shown in Table 13-1. Observe, and record in Table 13-1, the value of I_C for each value of V_{CE}.
6. Adjust V_{CC} for $V_{CE} = 0$ V. Set R_1 for $I_B = 20$ μA and maintain I_B at this value for steps 7 and 8.
7. Read value of I_C and record it in Table 13-1.
8. Adjust V_{CC} in turn to every value of V_{CE} shown in Table 13-1. Observe and record the value of I_C for each value of V_{CE}. Monitor I_B, and readjust R_1, if necessary, to maintain $I_B = 20$ μA.
9. Repeat steps 6 to 8 for all values of I_B shown in Table 13-1.
10. **Power off.** From the data in Table 13-1, plot the collector characteristic curves for the common-emitter connection of the 2N6004. Use graph paper. V_{CE} is the horizontal axis, I_C the vertical.

Curve Tracer—Tektronix Type (Self-Contained)

11. Set up the curve-tracer controls according to the manufacturer's instructions. Base-current steps should be set for 10 μA/step; collector-to-emitter volts at 5 V/horizontal div. The curve-tracer vertical axis should be set for 2 mA/div. The vertical axis will now read I_C, the horizontal axis V_{CE}, as in Fig. 13-9. Set the curve tracer for NPN-type transistor.
12. Insert a 2N6004 transistor into the test socket on the curve tracer. Readjust instrument controls for proper operation. A set of V_{CE} versus I_C curves should appear.
13. Photograph or draw the response curves, marking the current I_C calibration on the vertical axis, the voltage V_{CE} calibration on the horizontal axis, and the base-current I_B steps.
14. Now determine the ac beta ($\beta = \Delta I_C/\Delta I_B$) of the transistor, using the following procedure.
    (a) At some value of V_{CE}, say 15 V, erect a perpendicular to the horizontal axis, intersecting the I_{B2} and I_{B3} curves at A and B (Fig. 13-9).

TABLE 13-1. Test Data V_{CE} versus I_C (CE Connection): Low Base Current

| I_B, μA | I_C, mA | | | | | | | | |
| | V_{CE}, V | | | | | | | | |
	0	2.5	5	7.5	10	15	20	25	30
10									
20									
30									
40									
50									
60									

(b) From points A and B draw perpendiculars to the vertical axis, intersecting the axis at points C and D. Measure the number of graticule divisions between C and D and multiply this number by the current calibration factor (2 mA). The result is the value of ΔI_C. (EXAMPLE: If $CD = 2\frac{1}{2}$ div., then $\Delta I_C = 2\frac{1}{2} \times 2$ mA = 5 mA.)

(c) By our initial setting of the curve tracer, the difference between two successive base-current steps is 10 μA. Therefore, $\Delta I_B = 10$ μA.

(d) Substitute the values obtained in the formula for beta.

$$\beta = \frac{\Delta I_C}{\Delta I_B} = \frac{5 \times 10^{-3}}{10 \times 10^{-6}} = 500$$

Curve Tracer Requiring an Oscilloscope as Indicator

15. Following the manufacturer's instructions, connect the curve tracer to the oscilloscope. Set scope on Ext. sweep.
16. Calibrate the vertical axis of the oscilloscope for 2 mA/div and the horizontal at 5 V/div.

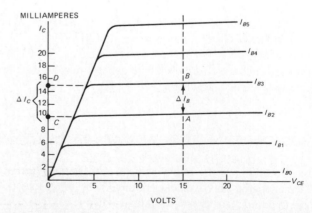

Fig. 13-9. Calculating ac β from family of collector characteristic curves.

17. Set the step selector at 10 μA/step, and sweep voltage at approximately 30 V.

18. Insert a 2N6004 into N-channel socket, or set the switch for NPN. Readjust curve-tracer controls, if necessary, to display the family of V_{CE} versus I_C characteristics of the transistor.

19. Now follow steps 13 and 14 and determine the ac beta of the transistor.

Extra Credit

20. Detail the procedure you would follow for displaying automatically on the oscilloscope without a curve tracer the V_{CE} versus I_C characteristic curve of a 2N6005 (PNP) transistor at $I_B = 40$ μA. Draw the circuit diagram. Connect the circuit and draw or photograph the characteristic curve. (HINT: See Fig. 13-5. How must this test circuit be modified for a PNP transistor?)

QUESTIONS

1. From the family of average collector characteristic curves, based on the data in Table 13-1, compute the value of beta in the vicinity $I_B = 40$ μA, $V_{CE} = 20$ V. Show all computations.

2. How do the experimental V_{CE} versus I_C characteristic curves compare with the published curves for the 2N6004? Explain any discrepancies.

3. Calculate the collector dissipation for the 2N6004 using Table 13-1 for the following values of I_B and V_{CE}:

I_B		V_{CE}	
30 μA	60 μA	10 V	10 V
30 μA	60 μA	30 V	25 V

Show a sample calculation.

4. Refer to the transistor data for the 2N6004.
(a) What is the typical h_{fe} when $V_{CE} = 1$ V and $I_C = 10$ mA?

(b) What is the maximum transistor dissipation for a case temperature up to 25°C?
(c) What is the maximum collector-to-emitter voltage which may be applied with the base open?

5. In the circuit of Fig. 13-4, what would be the effect of reversing the polarity of V_{BB}?

6. Draw the circuit diagram of a test circuit which could be used to determine the average collector characteristics of a PNP transistor in the CE configuration.

Answers to Self-Test

1. operating; environmental
2. collector; emitter; base
3. (a) case; (b) ambient; (c) junction
4. I_{CBO}
5. h_{fe}; current gain
6. beta
7. average collector
8. I_C
9. -0.5
10. constant I_B; V_{CC}
11. collector current
12. positive

SOLID-STATE DIODE AND TRANSISTOR TESTING

OBJECTIVES

1. To test solid-state diodes with an ohmmeter
2. To test solid-state diodes with go–no go circuits
3. To test transistors with an ohmmeter
4. To check transistors and solid-state diodes with a transistor tester

INTRODUCTORY INFORMATION

Diode Testing with an Ohmmeter

An electronics technician must be able to check the operating condition of all solid-state and other components found in electronic equipment, whether servicing that equipment, building new equipment, or working with laboratory models. Similarly a student, working in the school laboratory, must be able to test circuit components. If a circuit fails in the lab, a student should determine why, and if the reason is a component failure, find the defective part.

Out-of-circuit solid-state signal diodes and rectifiers can be checked, in most cases, with a conventional ohmmeter. First the polarity and voltage at the ohmmeter terminals are determined by measurement with a voltmeter across the ohmmeter leads. Then resistance measurements are made of the diode in both the forward and the reverse direction (see Experiment 1). The ratio of the *reverse*-resistance (R_R) to the *forward*-resistance (R_F) reading of most good diodes will be greater than 10:1. But in ohmmeter testing *caution* must be observed *to prevent damaging the diode and misleading the technician. The voltage at the terminals of the ohmmeter must not be higher than the rated voltage of the diode,* and meter lead polarity should be known.

Some points to remember:

1. If the reverse resistance of the diode is 0, or very low, the diode is shorted.
2. If the forward resistance of the diode is infinite or *very* high, the diode is open. However, when a diode shows low forward and high reverse resistance, it is seemingly good.
3. Ohmmeter checks of a diode *cannot* be made on the *low-power-ohms* function of an ohmmeter, because the voltage at the terminals of the meter is not high enough to *forward*-bias the diode.

Diode Testing with Go–No Go Circuits

High-Voltage Rectifiers

A dynamic go–no go test for silicon rectifiers operating at 120 V or higher is possible with the circuit in Fig. 14-1. The rectifier to be tested, D, is connected between terminals A and B of the test jig. A *good* rectifier gives these test indications: With switch S_2 *open* and S_1 *closed,* the 25-W bulb glows *dimly,* because it is operating on only the positive alternations of the 120-V input sine wave. (D acts as a *closed* switch on the *positive,* an *open* switch on the *negative* alternation.) When S_2 is closed, the bulb brightens.

An *open* rectifier does not permit the lamp to glow when S_2 is open. A *shorted* rectifier is just as bright whether S_2 is open or closed.

Low-Voltage Rectifiers

The test circuit of Fig. 14-2 checks low-current rectifiers. A 6-V battery or power supply is the power source. A double-pole double-throw (DPDT) switch acts as a polarity-reversing switch for the circuit. In position 1 of the switch shown in the figure, the diode is forward-biased. A good diode D permits current flow in the circuit, and the #49 pilot lamp glows. If the lamp does not light in position 1 of the switch, the diode is *open.*

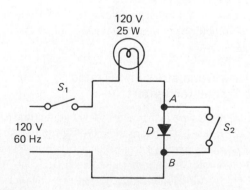

Fig. 14-1. Go–no go test for 120-V silicon rectifiers.

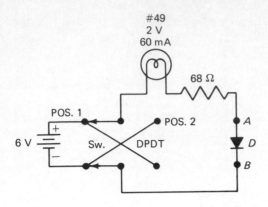

Fig. 14-2. Go–no go test for low-current rectifiers.

In position 2 of the switch, a *good* diode does not conduct and the lamp does *not* glow. If the lamp *does* glow in position 2 as well as in position 1, the diode is *shorted* and defective.

NOTE: In the absence of a reversing switch the test can still be made by manually reversing the polarity of the 6-V source in the test circuit.

Testing Transistors with an Ohmmeter

A conventional ohmmeter frequently helps a technician spot a defective transistor in an out-of-circuit test. Resistance tests are made between (1) emitter and base, (2) collector and base, and (3) collector and emitter. In testing resistance between any two terminals, the ohmmeter leads are first connected in *one* direction. Then they are reversed. In one of these lead positions the resistance between any two transistor terminals should be very high, 10,000 Ω or more. In the other lead direction, the resistance between the two terminals should be relatively low. Thus between emitter and base, and collector and base, the forward resistance should be 100 Ω or less. Between emitter and collector the low-resistance reading may be about 1000 Ω but will vary greatly.

If a transistor *fails* this test, it is defective. If it passes, it may still be defective. A transistor tester, a curve tracer, or operation of a transistor in a live circuit is a better test of a transistor than an ohmmeter.

CAUTION: As with diodes, the ohmmeter terminal voltage must not exceed the maximum voltage rating between any two terminals of the transistor. Nor must the current delivered by the ohmmeter be higher than the maximum current rating of any junction.

Checking Transistors and Diodes with a Transistor Tester

Transistor testers are instruments for checking transistors and solid-state diodes. There are three types of transistor testers: (1) quick-check in-circuit checker, (2) service-type tester, and (3) laboratory-standard tester. Each performs a unique function. In addition, signature-pattern checkers (curve tracers) are reliable indicators of transistor performance.

The in-circuit tester is used to check whether a transistor which has previously been performing properly in a circuit is still operational. The transistor's ability to "amplify" is taken as a rough index of its performance. Servicing with this type of tester indicates to a technician whether the transistor is dead or still operative. The advantage, of course, is that the transistor does not have to be removed from the circuit.

Service-type transistor testers usually perform three types of checks. Thus they test (1) forward-current gain, or beta, of a transistor, (2) base-to-collector leakage current with emitter open (I_{CO}), and (3) short circuits from collector to emitter and base. Some service testers include a go–no go feature. Some also provide a means of identifying transistor elements, if these are unknown. The tester in Fig. 14-3 has all these features and can check solid-state devices in and out of circuit.

Measurements with the service-type tester are relative rather than absolute. Even so, they are valuable to a technician in maintaining transistorized equipment.

Laboratory-standard transistor testers, or analyzers, are used for measuring transistor parameters dynamically under operating conditions. The readings they give are absolute. Among the important characteristics measured are (1) I_{CBO}, collector current with emitter open (common base), (2) ac beta (common emitter), and (3) R_{in} (input resistance).

Transistor testers have the necessary controls and switches for making the proper voltage, current, and signal settings. A meter with a calibrated "good" and "bad" scale is on the front panel.

Transistor testers are designed to check solid-state diodes as well as transistors. There are also testers for checking high-power transistors and rectifiers.

Fig. 14-3. In-circuit semiconductor analyzer. *(Hickok Teaching Systems)*

SUMMARY

1. Solid-state diodes may be tested with an ohmmeter by placing the meter leads across the diode to check its resistance, and then making a second resistance check of the diode with the leads reversed. A good diode will measure high resistance when it is reverse-biased, low resistance when it is forward-biased.
2. The ratio of the reverse-resistance reading to the forward-resistance reading of most good diodes is 10:1 or higher.
3. Rectifier diodes may be checked in go–no go test jigs (Figs. 14-1 and 14-2) which indicate whether the diode appears to be good, open, or shorted.
4. Transistor junctions may also be tested with an ohmmeter. When the junction is reverse-biased, the resistance between the two terminals should be high, 10,000 Ω or more. In the other position the resistance between the leads should be low.
5. In resistance-testing a diode or transistor be careful to avoid exceeding the maximum junction voltage or current by selecting the proper ranges of the ohmmeter.
6. Transistor testers check both diodes and transistors. There are in-circuit testers, service-type testers, and laboratory-standard testers.
7. With in-circuit testers the transistor need not be removed from the circuit. The tester gives a good/bad indication.
8. Curve tracers are also excellent devices for measuring out-of-circuit transistor performance.

SELF-TEST

Check your understanding by answering these questions.

1. A voltmeter, connected across the terminals of an ohmmeter, measures the internal _____ _____ of the meter.

2. The forward resistance of a nondefective diode measures _____ (high, low), the back resistance _____ (high, low).
3. The ratio of the back to forward resistance of a nondefective diode rectifier should be at least _____ : _____ .
4. In the circuit of Fig. 14-1 the brightness of the 25-W bulb does not change if S_2 is open or closed. D is a defective rectifier. _____ (true, false)
5. In Fig. 14-2 if the #49 bulb remains lit for either position 1 or 2 of the switch, then diode D is _____ (good, bad).
6. The forward resistance of the emitter-base junction of a transistor measures 10,000 Ω. The transistor is _____ (defective, nondefective).
7. Service-type transistor testers give relative rather than _____ values of transistor parameters.
8. If a transistor tester shows that a transistor is good, you can be certain that it is good. _____ (true, false).

MATERIALS REQUIRED

- Power supply: Low-voltage dc source
- Equipment: Transistor tester; VOM; EVM
- Transistors and solid-state diodes: A low-current signal diode and a silicon 120-V rectifier of which one is defective; two transistors, one defective
- Resistors: 68-Ω ½-W
- Miscellaneous: Two SPST switches; 1 DPDT switch; 25-W light bulb and socket; #49 pilot light and socket; fused line cord

PROCEDURE

1. Determine the polarity and voltage at the ohmmeter terminals by measuring with an EVM across the ohmmeter leads. Make a resistance test of a 120-V silicon rectifier, diode 1, by connecting the ohmmeter leads across the diode terminals. Now reverse the ohmmeter leads and again test the diode resistance. Enter the low-resistance measurement F_R in Table 14-1, column labeled "Forward-biased," the high-resistance reading R_R in column headed "Reverse-biased."
2. Calculate and record in Table 14-1 the reverse-to-forward-resistance ratio of the diode.
3. Indicate whether the diode is good or bad in the column labeled condition.
4. Repeat steps 1 to 3 for low-current signal diode 2.
5. Measure and record in Table 14-1, for transistor 1, the R_R and F_R resistance of the (a) E-to-B (emitter-to-base) junction, (b) C-to-B (collector-to-base) junction, and (c) E to C (emitter-to-collector).

6. Indicate in Table 14-1 whether the transistor checks good or bad.
7. Repeat steps 5 and 6 for transistor 2.

Diode Go–No Go Tests

8. Connect the test circuit of Fig. 14-1 and test silicon rectifier diode 1. (This is the same diode as 1 in Table 14-1.) Indicate in Table 14-2 whether this diode is good or bad.
9. Connect the test circuit in Fig. 14-2 and test signal diode 2. Indicate in Table 14-2 whether the diode is good or bad.

Transistor-Tester Checks

10. Read the instructions for using the transistor tester assigned to you. Record the transistor make and model

TABLE 14-1. Diode and Transistor Ohmmeter Tests

Component	Resistance, Ω		Ratio, R_R/F_R	Condition, Good or Bad
	Reverse-biased	Forward-biased		
120-V rectifier diode 1				
Signal (low-current) diode 2				
Transistor 1 E to B				
C to B			X	
E to C				
Transistor 2 E to B				
C to B			X	
E to C				

TABLE 14-2. Go–No Go Tests for Diodes

Component	Condition, Good or Bad
120-V Silicon rectifier diode 1	
Low-current signal diode 2	

TABLE 14-3. Transistor and Solid-state Diode Tests

	Manufacturer		Model Number	
Transistor Tester				
Tests which can be performed				
Silicon rectifier Diode 1				
Diode 2				
Transistor 1				
Transistor 2				

number in Table 14-3. List in the appropriate row of Table 14-3 the test functions which can be performed.

11. Test the two solid-state diodes you used previously and record the results in Table 14-3.
12. Test the two transistors you used previously and record the results in Table 14-3.

QUESTIONS

1. Do the ohmmeter tests (Table 14-1) confirm the go–no go tests (Table 14-2) for the diodes? If there are any discrepancies, explain.
2. Compare the results of the diode tests (Table 14-3) with the (a) diode ohmmeter tests (Table 14-1) and (b) diode go–no go tests (Table 14-2). Comment on any unexpected results.
3. Compare the results of the transistor ohmmeter tests (Table 14-1) with those using a transistor tester (Table 14-3). Comment on any unexpected results.
4. What precautions must be observed in resistance-checking a diode?
5. How does an in-circuit tester differ from an out-of-circuit tester?
6. What characteristics does a transistor checker test?

Answers to Self-Test

1. battery voltage
2. low; high
3. 10:1
4. true
5. bad
6. defective
7. absolute
8. false

(handwritten notes at top: "4-1-88 95 Act Questions 4-4-88")

COMMON-EMITTER AMPLIFIER BIASING AND GAIN

OBJECTIVES

1. To connect a transistor as a CE ac amplifier using voltage-divider bias
2. To measure the voltage gain of a CE amplifier
3. To observe the effect of an emitter bypass capacitor on amplifier gain.

INTRODUCTORY INFORMATION

The Transistor as an AC Amplifier

In Experiment 12 it was observed that in a transistor connected in a CE circuit, base current controlled collector current. Also, the increase in collector current was much greater than the increase in base current. This *current gain* of the transistor in the CE connection was designated beta (β). β was defined as follows:

$$\beta = \frac{\Delta I_C}{\Delta I_B} \qquad (V_{CE} \text{ const}) \qquad (15\text{-}1)$$

It should be noted that the bias conditions for which β was obtained were:

1. *The emitter-base junction was forward-biased* (and for this reason β was also defined as the *forward* current gain of the transistor in the CE connection).
2. *The collector-to-base junction was reverse-biased*.

These are the bias conditions used in amplifiers in audio, radio, and TV circuits.

Transistor amplifiers may be used to amplify dc or ac currents and/or voltages. In this experiment we shall be concerned with ac current *and* voltage amplifiers, for the CE amplifier accomplishes both.

Figure 15-1 is the circuit diagram of an NPN transistor connected as a grounded-emitter ac amplifier. Two batteries are used, V_{BB} for forward-biasing the base-emitter circuit, and V_{CC} for reverse-biasing the collector-emitter circuit. The input signal v_{in} is coupled by C_1 to the base; the output signal v_{out} is taken from the collector. R_1 limits the current in the base-emitter circuit. In conjunction with V_{BB}, R_1 determines the base-bias current or operating point. For large values of R_1 the base-bias current I_B may be approximately determined from the equation

$$I_B = \frac{V_{BB} - .6}{R_1} \qquad (15\text{-}2)$$

The relatively small value of input resistance of the base has been ignored in this approximation.

From our study of transistor characteristics, we know that a small increase in base current results in a relatively large increase in collector current. A small decrease in base current causes a relatively large decrease in collector current. Current in the base and collector circuits is therefore in phase. The extent of control of collector current depends on the β of the transistor in the common-emitter circuit.

The input-signal current coupled into the base circuit by C_1 is imposed on the reference-base (bias) current and causes the base current to increase or decrease. The current waveforms in Fig. 15-2 illustrate this condition.

Assume the bias current I_B is $+100$ μA and i_b is a sinusoidal, alternating-signal current. In one complete cycle the signal current varies from 0 to $+50$ μA, back to 0, then to -50 μA, and back to 0. Coupling the signal into the base causes a sinusoidal base current ($I_B + i_b$) to flow. For the cycle of input signal described above, the base current will vary, respectively, from $+100$ to $+150$ μA, back to $+100$ μA, then to $+50$ μA, and back to $+100$ μA.

The resulting sinusoidal variations of current in the base circuit cause in-phase sinusoidal variations of collector current if the amplifier is operated over the linear portion of the transistor characteristic. Assume that the collector current I_C is 2 mA when $I_B = 100$ μA. For the cycle of signal current i_b in Fig. 15-2, collector current $[I_C + \beta(i_b)]$ will vary, respectively, say from 2 to 3 mA, back to 2 mA, then to 1 mA, and

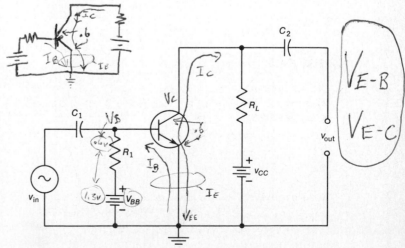

Fig. 15-1. NPN transistor connected as a grounded-emitter amplifier.

(handwritten notes bottom left:)
V_{EE}, V_{BB}, V_{CC} } SUPPLY VOLTAGES

V_E, V_B, V_C } VOLTAGES READ AT THE TRANSISTORS TERMINALS.

V_E } VOLTAGES DEAD

(handwritten box:) V_{BB} IS 1.3 BUT V_B IS .6

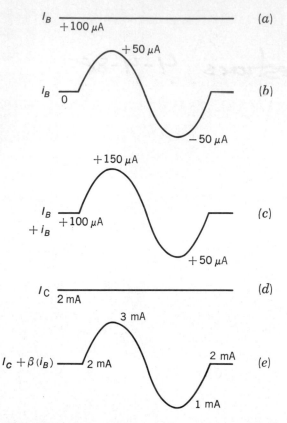

Fig. 15-2. Effect of signal current on base current and collector current in grounded-emitter amplifier.

a signal swing of 0.1 V p-p in the base will be amplified without distortion. If a larger signal is coupled into the base (that is, if the amplifier is overdriven), nonlinear operation will occur, resulting in a flattening of the signal on the peaks, as in Fig. 15-4. In this case, the peak of the positive alternation drives the collector into the nonlinear portion of its characteristic. The peak of the negative alternation drives the collector to cutoff.

If a PNP transistor is used in this grounded-emitter configuration instead of an NPN transistor, the effect of signal polarities is just the opposite. Thus, for the PNP transistor, i_c would decrease on the positive alternations of the input signal and increase on the negative alternations. As a result of the input-signal current i_b, an amplified-signal current i_c flows in the collector circuit. The output-signal voltage in the collector is given by the equation

$$v_{\text{out}} = i_c \times R_L \qquad (15\text{-}3)$$

Here R_L is the collector load resistor.

If the base-signal current i_b is sinusoidal and if the amplifier has been so designed that it will not distort i_c, the signal current in the output, the output voltage v_{out} is sinusoidal.

It should be noted that the transistor in Fig. 15-1 is basically a current amplifier. The signal-voltage gain which results is dependent on the current gain of the transistor, on the circuit arrangement, and on the circuit parameters.

Biasing Methods and Stabilization

It is evident from the preceding discussion that for an amplifier to provide *distortionless amplification,* the base must be biased properly so that the input signal operates over the linear portion of the transistor's characteristic. The manner in which a transistor is biased therefore determines the output signal it will produce for a given level input signal.

The CE amplifier lends itself readily to base biasing from a *single* power source, rather than from the two sources shown in Fig. 15-1. This, plus the fact that compared with a com-

back to 2 mA. In this circuit then, a 100-μA change of signal current in the base has caused a 2-mA change of collector current, for a current gain β of 20.

It is instructive to observe the effect of operating the amplifier over the nonlinear portion of its characteristic. Assume that in the NPN germanium transistor collector current I_C varies with base-to-emitter voltage V_{BE} in the manner shown by the characteristic curve (Fig. 15-3). This graph shows that linear variations in collector current occur only in the range of 0.1 to 0.2 base-to-emitter volt. If 0.15 V is maintained as the base-to-emitter-voltage operating point,

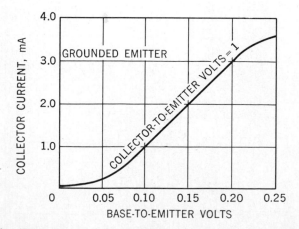

Fig. 15-3. Characteristic showing I_C as a function of V_{BE} for an NPN transistor.

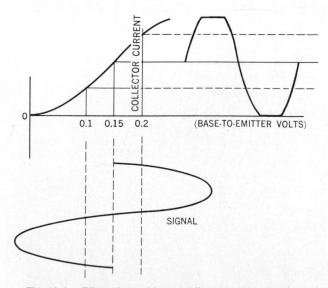

Fig. 15-4. Effect of overdrive on collector current waveform.

mon-base (CB) amplifier, the CE amplifier has a higher input impedance, a lower output impedance, and provides current, voltage, *and* power gain, makes it the most widely used *junction* transistor amplifier.

The circuit of Fig. 15-5 illustrates one means of biasing a CE amplifier by using a common power supply V_{CC}. This arrangement is possible because both the base and collector of the PNP transistor must receive a dc voltage which is negative with respect to the emitter. The dc voltage divider consisting of R_1 and R_{BE}, drawing current from the battery V_{CC}, determines the base bias (current). A common battery may also be used with an NPN transistor, but here the polarity of the battery must be reversed, since the collector and base must receive a positive voltage with respect to the emitter.

The circuit of Fig. 15-5 is not a practical means of biasing the CE amplifier, and it is not used. The reason is that the *stability* of the grounded-emitter amplifier is greatly affected by temperature changes. Moreover, variations in transistor characteristics affect amplifier performance. In the circuit shown, variations in transistor temperature and in transistor characteristics would change the operating point (the level of base-bias current), affecting in turn the gain and stability of the amplifier. Transistor runaway may occur from changes in operating temperature, resulting in the destruction of the transistor.

To protect the transistor against runaway and to stabilize the amplifier, a form of compensation is used which is known as *bias stabilization*. The circuit of Fig. 15-6 illustrates one type of stabilization. Resistor R_3 is connected in the emitter leg. The effect of this resistor is to compensate for any slight increase in collector current due to an increase in operating temperature or to variations in transistor characteristics. Thus an increase in collector current causes a larger voltage drop across R_3, with the polarity shown. The increased voltage developed across R_3 is in opposition to the forward bias in the base-emitter circuit and reduces the base current. This in turn reduces collector current, balancing the tendency of the collector current to increase in value. R_3, therefore, provides negative dc feedback, or degeneration to counteract the instability of the transistor.

Emitter Bypass Capacitor

Of interest in Fig. 15-6 is capacitor C_3, which is connected in parallel with emitter resistor R_3. If an ac signal were injected into the base of the transistor and C_3 were *not* present, the effect of R_3 would be to provide both dc *and* ac degeneration. For as collector current increased and decreased in step with the ac signal current in the base, an ac signal voltage would be developed across R_3, in phase with the signal voltage in the base. The effective base-to-emitter signal, which is the *difference* between the voltage on the base and the voltage on the emitter, would therefore become lower than the signal voltage on the base. Accordingly the transistor would "see" a *lower* input ac signal voltage, and the output signal in the collector would be lower, thus effectively *reducing the ac gain* of the amplifier.

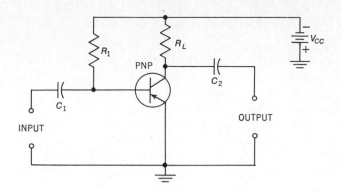

Fig. 15-5. Grounded-emitter amplifier using common power supply.

C_3 connected across R_3 provides another path for ac signal current. If the capacitive reactance of C_3 is very much lower than the resistance of R_3 for the ac signal frequency which the amplifier must process, then C_3 acts as a low-impedance path for the ac signal current. R_3 is then effectively bypassed, and the amount of ac degeneration is negligible. Signal gain is restored. R_3, however, still acts to provide dc bias stabilization.

A numerical example will illustrate this effect. Suppose at a certain frequency F, $X_{C3} = 100$ Ω and $R_3 = 1000$ Ω. Then $^{10}/_{11}$ of the emitter ac signal current will flow through C_3 and only $^1/_{11}$ through R_3. The signal voltage developed across R_3 is therefore only $^1/_{11}$ of the voltage which would have been developed if C_3 were not present. Bypassing the emitter resistor with C_3, in this instance, appreciably reduces the effects of signal degeneration. However, the direct current through R_3 is not affected.

If an amplifier is required to handle more than one frequency, then a value of the emitter bypass capacitor must be chosen that provides adequate bypassing for the *lowest* of all the frequencies. Then it will also be a good bypass for all the higher frequencies.

To determine the size of C_3, it is therefore necessary to know the lowest frequency $F_{\min}$ it is required to bypass and the value of the emitter resistor R_3 in parallel with it. C_3 is considered a good bypass if at $F_{\min}$

$$X_{C3} = \frac{R_3}{10} \qquad (15\text{-}4)$$

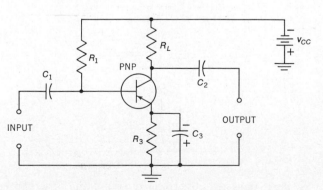

Fig. 15-6. Grounded-emitter amplifier with dc stabilization.

Voltage-Divider Bias

The most frequently used of all base-bias circuits is voltage-divider bias, illustrated in Fig. 15-7. Because an NPN transistor is connected here, a positive common supply V_{CC} is used. R_1 and R_2 constitute a voltage divider on the base. Because the base is made *positive* by this voltage divider, there is emitter-base and emitter-collector current flow. The polarity of voltage developed across emitter resistor R_3 is shown in Fig. 15-7. The bias voltage, which determines the base-bias current, is the potential difference between the base and emitter.

The other components in Fig. 15-7 perform the same function as in Fig. 15-6. C_1 couples the input signal to the base. C_2 connects the output signal to the stage which follows (not shown here). R_3 provides dc degeneration and compensates for current changes due to temperature and other variations. In a properly designed amplifier, the emitter resistor R_3 will compensate both for temperature changes and for variations in transistor β, since transistors with the same 2N number do have a variation in β. C_3 bypasses R_3 for the ac frequencies which the circuit must handle. R_L is the collector load resistor.

Figure 15-8 is a practical application of a grounded-emitter transistor audio amplifier. Of interest to us are the component values, the type of stabilization provided, and the manner in which power is applied. C_1 and C_2 are each 5-μF electrolytic coupling capacitors. The low input impedance of this amplifier requires a large coupling capacitor C_1. If a smaller capacitor were used, very little signal would be developed in the input circuit. Similarly, the value of C_2 must be large because of the low input impedance of the following stage. The values of R_1 (3300 Ω) and R_2 (22,000 Ω) were chosen to set proper operating bias. As a result of the divider action of R_1 and R_2 and of base-to-emitter current in the circuit including R_2 and R_3, 6.68 V is established at the base and 6.9 V at the emitter. This makes the emitter positive with respect to the base—the proper condition for forward emitter-base bias in a PNP transistor. A 2200-Ω resistor R_L is used as the collector load. The signal developed across this load is coupled by C_2 to the following stage (not shown).

R_3, a 1000-Ω resistor, provides stabilization; C_3, a 40-μF electrolytic capacitor, acts as an AF bypass. C_3 is large to

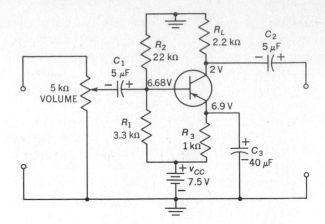

Fig. 15-8. PNP transistor used as an audio amplifier.

provide proper bypassing of R_3 at the lowest audio frequency which the amplifier must pass. V_{CC} is a 7.5-V battery whose negative terminal is grounded. This arrangement makes it possible to return the collector load conveniently to ground and permits the use of a PNP transistor from a single, positive power supply. The electrolytic capacitors have a low voltage rating because they operate in a low-voltage circuit. Because they are low-voltage capacitors, it is possible to keep their physical size small. Since the capacitors are electrolytic, they must be connected with the proper polarities in the circuit.

The amplifier of Fig. 15-8 is a small-signal amplifier and can handle a signal input somewhat greater than 0.1 V p-p without distortion. The polarities of C_1 and C_2 are chosen so that the dc potentials existing across their terminals will conform with the capacitor polarity markings.

Voltage Gain

The grounded-emitter amplifier is a current, voltage, and power amplifier. It is possible to determine experimentally the voltage gain of the amplifier by injecting a measured signal voltage into the input. The output-signal voltage is then measured (an oscilloscope or ac voltmeter may be used), and the ratio of output signal to input signal is the required voltage-gain figure. That is,

$$\text{Voltage gain} = \frac{v_{\text{out}}}{v_{\text{in}}} \qquad (15\text{-}5)$$

The amplifier must be operated over its linear region during this process.

It is possible to determine experimentally the range of linear operation of this amplifier. An audio signal, say, 1000-Hz sine wave, is injected into the input, and an oscilloscope is used to monitor the output at the collector. The attenuator on the signal generator is set at minimum at the start. The output of the generator is then increased, and the waveform is observed on the oscilloscope. The input signal is measured over the range where no output-signal distortion is evident on the oscilloscope. This measurement gives the range of signal input over which no distortion occurs.

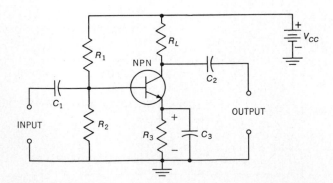

Fig. 15-7. Grounded-emitter amplifier using voltage-divider bias with dc stabilization

SUMMARY

1. Transistors connected as CE amplifiers are used in audio, radio, TV, and other applications.
2. The bias conditions for a CE distortionless amplifier are: (*a*) emitter-base junction forward-biased, and (*b*) collector-base junction reverse-biased.
3. The base-bias current I_B in a CE amplifier is the dc current in the base which establishes the operating point for the circuit and determines the steady-state collector current I_c.
4. An ac signal i_b injected into the base alternately increases and decreases the base current, as in Fig. 15-2. That is, the base current is the algebraic sum of I_B and i_b.
5. The amplifier collector current corresponding to the base-signal current is βi_b, and the total collector current is $I_C + \beta i_b$ (Fig. 15-2).
6. As long as the amplifier is operated over the linear portion of its characteristic, 0.1 to 0.2 V V_{BE} (Fig. 15-3), the output signal is a faithful reproduction of the input. If the amplifier is overdriven by too high an input signal, as in Fig. 15-4, or operated over the nonlinear portion of its characteristic (below 0.1 or above 0.2 V in Fig. 15-4), the output signal is distorted; that is, its wave shape is different from the input.
7. A practical base-bias circuit is shown in Fig. 15-6, where emitter resistor R_3 provides negative feedback or degeneration to overcome the tendency of the transistor to increase its current as its heat increases.
8. R_3 in Fig. 15-6, also stabilizes the circuit for different values of transistor β. It should be understood that the β of transistors of the same type (that is, the same JEDEC number) varies fairly widely. R_3 helps make the gain of the circuit fairly constant over this range of variation of β.
9. An even more practical and more widely used base-biasing circuit is that shown in Fig. 15-7. R_1 and R_2 provide voltage-divider bias for the transistor.

10. The voltage gain of the stage is defined as the ratio of the output-signal voltage to the input-signal voltage; that is,

$$\text{Gain} = \frac{v_{\text{out}}}{v_{in}}$$

11. A CE transistor amplifier has voltage gain, current gain, and power gain.

SELF-TEST

Check your understanding by answering these questions.

1. The voltage gain of the amplifier in Fig. 15-7 is 50. When C_3 is opened the gain of the amplifier should _____ (*increase, decrease, remain the same*).
2. If R_1 and R_2 in Fig. 15-7 are respectively 22 kΩ and 4.7 kΩ and V_{CC} is + 10 V, the dc voltage on the base will be _____ V (approx).
3. The circuit of Fig. 15-7 must amplify sine-wave signals in the frequency range 20 to 20,000 Hz. The design value of R_3 is 1200 Ω. The highest value X_C of C_3 which will act as a good bypass for this amplifier is _____ Ω at _____ Hz.
4. In an audio amplifier the collector to base must be _____ (*forward, reverse*)-biased.
5. The ac signal voltage measured at the base of a CE amplifier is 50 mV. The output-signal voltage measured at the collector is 2.5 V. The voltage gain of the amplifier is _____ .

MATERIALS REQUIRED

- Power supply: Variable low-voltage dc source
- Equipment: Oscilloscope; EVM; AF sine-wave generator
- Resistors: 560-, 1000-Ω, 8.2-, 18-, 220-kΩ ½-W
- Capacitors: Two 25-μF 50-V; 100-μF 50-V
- Semiconductors: 2N6004 (or equivalent) with appropriate socket
- Miscellaneous: SPST switch

PROCEDURE

Bias Stabilization

1. Connect the circuit of Fig. 15-9.
2. **Power on.** Measure current I_C in the collector circuit. Record it in Table 15-1. With an EVM measure base-to-emitter voltage V_{BE} and the collector-to-emitter voltage V_{CE}. Record these voltages in Table 15-1.
3. Connect an audio signal generator set at 1000-Hz minimum output to the input terminals of the amplifier. Connect the vertical input cable of an oscilloscope to the output terminals of the amplifier. Adjust the oscilloscope for proper viewing.
4. Set the output of the audio generator just *below* the point of distortion, so that the maximum undistorted sine wave appears. Measure the peak-to-peak amplitude of this output waveform and of the input waveform. Record the

results in Table 15-1. Measure also and record I_C, V_{BE}, and V_{CE}. Draw the input and output waveforms in Table 15-1.

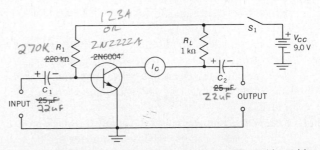

Fig. 15-9. Experimental grounded-emitter amplifier without bias stabilization.

TABLE 15-1. Bias Stabilization

Step	Current, mA I_C	Voltage, V V_{BE}	V_{CE}	Waveform Input	V p-p	Output	V p-p
2	5.6m	.663m	3.36	X	X	X	X
4	4.9m	.662m	3.45	∿	35m	∿	6
5	5.9m	.645m	3.22	∿	35m	∿	5.8
7	3.5m	.652m	3.45	X	X	—	0
9	3.6m	.65m	3.4	∿	54m	∿	5.6
10	3.7m	.632m	3.27	∿	54m	∿	5.4
11	3.4m	.65m	3.45	∿	13m	∿	1.44

5. Heat the transistor by holding a 60-W light close to it for about 2 min (until output waveform changes). Observe the waveform and its peak-to-peak amplitude. Draw and record them in Table 15-1. Measure I_C, V_{BE}, and V_{CE}. Record your measurements in Table 15-1.
6. **Power off.** Change the circuit to that shown in Fig. 15-10. Check all circuit connections.
7. **Power on.** Measure and record I_C, V_{BE}, and V_{CE}.
8. Connect an audio signal generator set at 1000-Hz minimum output to the input terminals of the amplifier. Connect the vertical input cable of an oscilloscope to the output terminals of the amplifier. Adjust the oscilloscope for proper viewing.
9. With the oscilloscope connected at the output of the amplifier, adjust the attenuator on the audio generator for the amplifier's maximum undistorted output. Draw the input and output waveforms in Table 15-1. Measure the peak-to-peak amplitude of the output and input waveforms. Record this data in Table 15-1. Measure also and record I_C, V_{BE}, and V_{CE}.
10. Heat the transistor by holding a 60-W light close to it for about 2 min. Observe both the waveform and its peak-to-peak amplitude. Draw the waveform and record its amplitude in Table 15-1. Measure and record I_C, V_{BE}, and V_{CE}.
11. Reduce the input signal to the smallest value consistent with undistorted output as observed on the oscilloscope. Measure I_C, V_{BE}, V_{CE}, and the peak-to-peak amplitude of the input- and output-signal voltages. Record them

and draw their input and output waveforms in Table 15-1.

Effect of Emitter Bypass Capacitor on Gain

12. With the oscilloscope connected across the output of the amplifier (Fig. 15-10), adjust the attenuator on the audio generator for 50 percent of the amplifier's maximum *undistorted output*.
13. Observe, measure, and record in Table 15-2 the peak-to-peak value of the input signal (base to ground) and output signal (collector to ground). Measure also and record the ac waveform from emitter to ground.
14. Calculate the ac gain of the amplifier and record in Table 15-2.
15. Do not change the attenuator setting of the audio generator. Remove C_3 from the circuit.
16. Repeat steps 13 and 14.

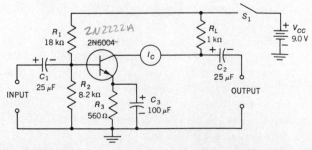

Fig. 15-10. Experimental grounded-emitter amplifier with bias stabilization.

TABLE 15-2. Emitter Bypass Capacitor and Gain

Step	Base	V p-p	Collector	V p-p	Emitter	V p-p	Gain
			Waveform (to Ground)				
13, 14	∿	25m	∪	2.75	∿	5m	110
15, 16	∿	35m	∪	60m	∿	35m	1.7

QUESTIONS

1. What is meant by bias stabilization? Why is it used?
2. How does heat affect transistor operation? Refer to your data in Table 15-1 to support your answer.
3. How can transistor amplifier operation be made relatively independent of small changes in operating temperature?
4. Is your answer to question 3 supported by the data in this experiment? Refer specifically to the data to support your answer.

5. What is the purpose of C_3 in Fig. 15-10?
6. What is the effect on transistor gain of removing C_3? Support your answer by referring specifically to the data in this experiment.

Answers to Self-Test

1. decrease   4. reverse
2. 1.76   5. 50
3. 120; 20

NO I_C
$V_B \downarrow$
$V_C \uparrow$

WHEN HEATED

$I \uparrow V \downarrow$

$(V_{BE} V_{CE})$

$I_B \uparrow$
$V_B \downarrow$
$I_C \uparrow$

11.6 V
NOT
12 V

SHORT   OPEN
R_{LB} R_{HB}

$V_B = \dfrac{R_{LB}}{R_{HB} + R_{LB}} \times V_{BB}$

$\dfrac{8.2}{18 + 8.2} \times 12 = 3.75v$

$V_E = V_B - .6 = 3.15v$

$I_E = \dfrac{V_E}{R_E} = \dfrac{V_E}{560\Omega} = 5.6mA$

$V_C = 6.5v$

	WORKING	SHORT E-B	OPEN E-B	OPEN C	SHORT BASE	SHORT ER	
	3.5	.33	3.6	.91	0	.6	0
	2.9	.33	0	.27	0	0	0
				.27			
	6.5v	(12) 11.6	11.6	11.6	11.6	45mV	11.6

E_R CONTROLS I_C

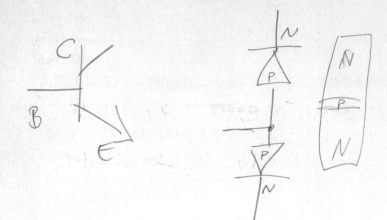

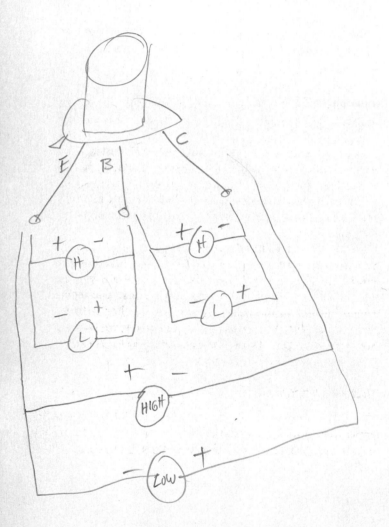

COMMON-EMITTER AMPLIFIER IMPEDANCE, POWER, AND PHASE RELATIONSHIPS

OBJECTIVES

1. To measure the input and output impedance of a CE amplifier
2. To determine the decibel power gain of a CE amplifier
3. To observe, with an oscilloscope, the phase of the input- and output-signal voltages in a CE amplifier

INTRODUCTORY INFORMATION

Input Impedance

The input impedance Z_{in} or R_{in} of an amplifier is defined as the ratio of input-signal voltage v_{in} and input-signal current i_{in}.

$$R_{in} = \frac{v_{in}}{i_{in}} \qquad (16\text{-}1)$$

R_{in} can therefore be determined experimentally by measuring v_{in} and i_{in} and substituting the measured values in Eq. (16-1).

An ac microammeter or milliammeter can be used to measure the input (base) current in an amplifier, or the circuit may be modified as in Fig. 16-1a to determine i_{in} by the voltmeter-ohmmeter method. A resistor R_x whose value is known or measured is connected in series with the output of the generator used to supply the signal to the amplifier. An ac

voltmeter measures the voltage v_x across R_x. The current i_x in R_x, which is the same as i_{in}, is then determined by Ohm's law.

$$i_{in} = i_x = \frac{v_x}{R_x} \qquad (16\text{-}2)$$

The signal voltage v_{in} measured between points B and C is the input to the amplifier. Z_{in} is then computed from the two measured values.

Another method is to connect a rheostat R_x instead of a fixed R_x between points A and B as in Fig. 16-1a. R_x is then varied until the voltage v_x across R_x equals v_{in}, the signal voltage measured from points B to C. R_x is next removed from the circuit and measured. The ohmic resistance of R_x is then equal to R_{in}.

The input impedance of a common-emitter amplifier may be increased by providing degenerative feedback to the input circuit. In Experiment 15 we observed that if the emitter resistor R_3 is unbypassed, degeneration occurs. The input-signal current in the base-to-emitter is smaller with R_3 unbypassed, for the same value of v_{in}, than it would be if C_3 were in the circuit. Hence, the value of R_{in} is higher when there is degeneration in the emitter circuit.

Output Impedance

The output impedance Z_{out} of the amplifier can be determined experimentally by the addition of rheostat R_{out} to the output circuit (Fig. 16-1b). The procedure is as follows. The

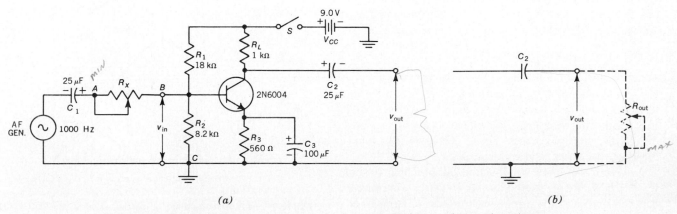

(a)   *(b)*

Fig. 16-1. Experimental common-emitter circuit to measure input and output impedance.

output signal v_{out} is first measured with no load. The rheostat load is then connected as shown and adjusted until a new output signal v_{out} is equal to one-half the original measured value of v_{out}. R_{out} is then removed from the circuit and its resistance measured. The measured value in ohms equals the output impedance Z_{out} of the amplifier.

In measuring the input and output impedance care must be taken to maintain an undistorted input and output signal.

Power Gain

The power gain of an amplifier is the ratio of output- to input-signal power.

$$\text{Power gain} = \frac{P_{out}}{P_{in}} \qquad (16\text{-}3)$$

Output and input power may be calculated when the input- and output-signal voltages and impedances are known. Thus

$$P_{out} = \frac{v_{out}^2}{R_{out}} \qquad (16\text{-}4)$$

$$P_{in} = \frac{V_{in}^2}{R_{in}} \qquad (16\text{-}5)$$

Substituting the values of P_{out} and P_{in} in Eq. (16-3), we have

$$\text{Power gain} = \left(\frac{v_{out}}{v_{in}}\right)^2 \times \frac{R_{in}}{R_{out}} \qquad (16\text{-}6)$$

The power gain of an audio amplifier is usually given in decibels (dB).

$$\text{Power gain (dB)} = 10 \log \frac{P_{out}}{P_{in}} \qquad (16\text{-}7)$$

Phase Relations

It can be demonstrated that in the common-emitter amplifier the output-signal voltage at the collector v_c is 180° out of phase with the input-signal voltage at the base v_{in}. Refer to the circuit in Fig. 16-2. The output voltage v_{ce} from the collector to the emitter (which is at ground potential) will vary inversely with the collector current i_c. This relationship can be demonstrated by means of Eq. (16-8), which is an application of Kirchhoff's voltage law to the output circuit.

$$V_{CC} = i_c \times R_L + v_{ce} \qquad (16\text{-}8)$$

Equation (16-8) states that the collector source voltage V_{CC} is equal to the sum of the voltage drop $i_c \times R_L$ across the load resistor R_L and the voltage v_{ce} from collector to ground. With V_{CC} fixed, as i_c increases, the voltage across R_L increases, the v_{ce} therefore decreases. As i_c decreases, the voltage across R_L decreases, and v_{ce} increases.

A numerical example will illustrate this principle. Assume $V_{CC} = 6$ V, $R_L = 1000$ Ω, and i_c varies sinusoidally between 3 mA maximum and 1 mA minimum. Assume 2 mA is the steady-state collector current with zero base signal.

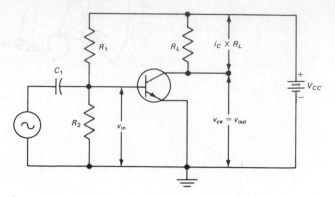

Fig. 16-2. Kirchhoff's voltage law applied to the output of the CE amplifier states that $V_{CC} = i_c R_L + v_{ce}$.

The output-signal voltage at the collector will vary from 3 to 5 V as in Fig. 16-3. It is evident that for an NPN transistor i_c and v_c are 180° out of phase.

For an NPN transistor as the input-signal voltage on the base v_{in} goes more positive, collector current increases; as v_{in} goes more negative, collector current decreases. Hence, in the circuit of Fig. 16-2, i_c is *in phase* with v_{in}. But we just established that v_c (the output signal) is 180° out of phase with i_c. Hence v_c is 180° out of phase with v_{in}.

The conclusion that v_c is 180° out of phase with v_{in} is equally true for a PNP CE amplifier.

It is possible to demonstrate this phase reversal experimentally using an oscilloscope and a sine-wave input with external sync/trigger as in preceding experiments. It is also possible to determine experimentally the phase relationship between the input and output signals in a CE amplifier using the test circuit shown in Fig. 16-4a. The diode rectifies the audio signal injected by the AF generator. The rectified waveform appearing across the 5000-Ω load is shown in Fig. 16-4b. This negative-going waveform is now injected into the

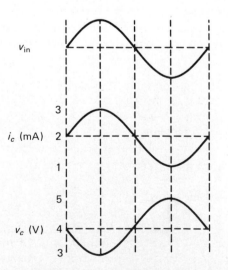

Fig. 16-3. In a common-emitter amplifier the collector voltage is 180° out of phase with the collector current and the input voltage.

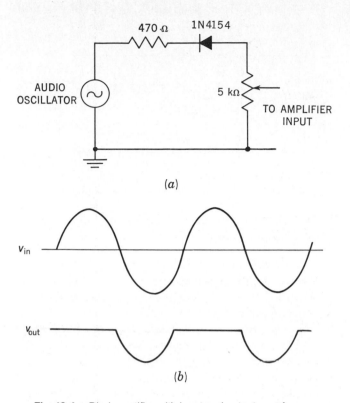

(a)

(b)

Fig. 16-4. Diode rectifier with input and output waveforms.

base of the grounded-emitter amplifier. The output waveform is then observed on an oscilloscope at the collector. A 180° phase reversal will appear on the oscilloscope as a signal reversal, that is, as a positive-going waveform. The amplifier must not be overdriven during this process.

SUMMARY

1. The input impedance Z_{in} or R_{in} of an amplifier is defined as

$$R_{in} = \frac{v_{in}}{i_{in}}$$

where v_{in} is the signal voltage on the base and i_{in} is the signal current in the base.

2. R_{in} may be determined experimentally by measuring v_{in} and i_{in} and computing their ratio.

3. R_{in} may also be computed by inserting a rheostat R_x in the base circuit as in Fig. 16-1, injecting an audio sine wave into the base, and adjusting R_x until the voltage v_x across R_x is equal to v_{in}.

4. The output impedance R_{out} of a CE amplifier may be measured experimentally by connecting a rheostat R_{out} (Fig. 16-1b) as a variable load in the collector circuit. R_{out} is adjusted until the load voltage v_{out} equals one-half the no-load voltage. The resistance of the rheostat then equals the output impedance of the amplifier.

5. The power gain of an amplifier is defined as

$$\text{Power gain} = \frac{P_{out}}{P_{in}} = \frac{(v_{out})^2}{(v_{in})^2} \times \frac{R_{in}}{R_{out}}$$

where R_{in} and R_{out} are the known or measured values of input and output impedance, respectively, of the circuit.

6. The power gain of an amplifier is usually stated in decibels (dB) as

$$\text{Power gain (dB)} = 10 \log \frac{P_{out}}{P_{in}}$$

7. In a CE amplifier the output-signal voltage at the collector is 180° out of phase with the input-signal voltage at the base.

SELF-TEST

Check your understanding by answering these questions.

1. In Fig. 16-1 R_x is adjusted until the voltage $V_{AB} = V_{BC}$. To find the input impedance of the amplifier it is necessary to measure the _____ of _____ .

2. In Fig. 16-1, $V_{in} = 0.5$ V$_{rms}$, $V_{AB} = 1$ V$_{rms}$, and $R_x = 1000$ Ω. The input impedance of the amplifier is _____ Ω.

3. The rms voltage measured at the collector of the amplifier in Fig. 16-1a is 4.6 V without load. When a 250-Ω load is connected across the output, the rms voltage measured at the collector load is 2.3 V. The output impedance of the circuit is _____ Ω.

4. An oscilloscope, externally triggered by the output signal at the collector of Fig. 16-1 and connected to observe the input signal v_{in}, is adjusted to show a single sine wave, whose positive alternation leads the negative alternation. When the oscilloscope is connected from collector to ground, the _____ alternation of the sine wave observed on the scope leads the _____ alternation by _____ degrees.

5. If the rectified output v_{out} of Fig. 16-4b is injected into the input of Fig. 16-1a, the waveform observed with an oscilloscope at the collector will be a _____ -going rectified signal voltage.

6. A CE amplifier has a voltage gain of 50, an input impedance of 1000 Ω, and an output impedance of 200 Ω. The power gain of this amplifier is _____ .

7. The decibel power gain of the amplifier in question 6 is _____ dB.

MATERIALS REQUIRED

- Power supply: Variable regulated low-voltage dc source
- Equipment: Oscilloscope; EVM; AF sine-wave generator
- Resistors: 470-, 560-Ω, two 1-, 4.7-, 8.2-, 18-kΩ ½-W
- Capacitors: Two 25-μF 50-V; 100-μF 50-V
- Semiconductors: 2N6004; 1N4154, or equivalent
- Miscellaneous: SPST switch; 5000-Ω 2-W potentiometer

Input Impedance

1. Connect the circuit of Fig. 16-1*a*. Note that R_x is a 1000-Ω resistor, not a potentiometer.
2. **Power on.** Adjust the AF sine wave generator for 1000 Hz and set the generator level (output) control for 70 percent of *maximum undistorted output*, v_{out}, as observed with an oscilloscope connected across the output.
3. With an oscilloscope, measure and record in Table 16-1 the peak-to-peak voltage of **(a)** v_{AC} across *AC*, **(b)** v_{BC} or v_{in} across *BC*, and **(c)** v_{out} in the output.
4. Compute v_x across R_x by subtracting v_{BC} from v_{AC}. Record in Table 16-1. Compute and record i_{in} and R_{in}. Show your computations.

Output Impedance

5. *Do not vary the input-signal level.* Connect a 5000-Ω rheostat R_{out}, as in Fig. 16-1*b*, across the output. Adjust R_{out} until the measured output signal v_{out} equals one-half the output measured in step 3**c**.
6. Remove R_{out} from the circuit. Measure and record its resistance. This is the value of the output impedance R_{out} of the amplifier.

NOTE: The output impedance of the amplifier is not a fixed quantity; it depends on the load resistance and transistor voltages.

Power Gain

7. Compute and record in Table 16-1 the voltage gain and power gain (in decibels) of the circuit under load. Show your computations.

Effect of Unbypassed Emitter Resistor

8. **(a)** *Do not change the level of the input signal.* Remove bypass capacitor C_3 from the circuit. With an oscilloscope observe the output signal v_{out}. Why has the output of the amplifier dropped so dramatically?
   (b) With C_3 still out of the circuit, increase the generator output until v_{out} equals 1 V p-p. Repeat steps 3 through 7.

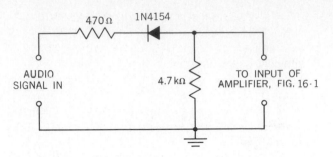

Fig. 16-5. Adding signal rectifier to experimental grounded-emitter amplifier.

Phase Relationship

9. **Power off.** Remove R_x from the circuit and connect points *A* to *B*. Connect the half-wave rectifier circuit shown in Fig. 16-5. The 1000-Hz signal from the generator is coupled to the input of the half-wave rectifier. The output of the half-wave rectifier is connected as the AF signal source for the CE amplifier in Fig. 16-1.
10. **Power on.** Reset the signal generator to that v_{in} is at the same *peak-voltage* level as in step 3**b**.
11. With an oscilloscope observe two cycles of the input signal waveform. Draw this waveform in Table 16-2.
12. Observe two cycles of the output waveform. Draw them in proper time phase with the input in Table 16-2.

Extra Credit

13. Explain in detail a method, other than the one used in this procedure, for determining the input impedance of a CE amplifier. Follow this procedure and record your results.
14. Explain in detail a method, other than the one used in this procedure, for verifying the phase relationships in the circuit of Fig. 16-1. Follow this procedure and record your results.

TABLE 16-2. Phase Relations in CE Amplifier

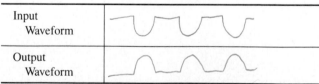

Input Waveform	
Output Waveform	

TABLE 16-1. CE Amplifier Impedance and Power Measurements

| | V p-p | | | | i_{in}, mA | R_{in}, Ω | R_{out}, Ω | Gain | |
Steps	v_{AC}	$\dfrac{v_{BC}}{v_{in}}$	v_{out}	v_x $\overline{v_{AC} - v_{BC}}$				Voltage	Power, dB
2–7	48m	24m	3	24m	.012m	2K	1K	125	44.9
8	.59v	.29v	1.53	.29v	.052m	5.6K	1.2K	1.8	11.9

QUESTIONS

1. (a) If you wished to measure v_x (voltage across R_x) directly, why would it be necessary to use a "floating" oscilloscope, that is, an oscilloscope whose case is not grounded to the electrical system?
   (b) Why is the use of a "floating" instrument generally not recommended?

2. What is the effect on input impedance of removing bypass capacitor C_3 in Fig. 16-1? Refer to your data to substantiate your answer.

3. How is the output of the half-wave rectifier (Fig. 16-5) helpful in determining the phase relationship between the input and output signals of a CE amplifier?

4. (a) What is the phase relationship between the input and output signals of a CE amplifier?
   (b) Was this relationship confirmed by the results of your experiment? Explain how.

5. Show the formula you use for computing the decibel gain of the amplifier in Fig. 16-1.

6. What is the phase relationship between the input-signal voltage and collector current in (a) NPN common-emitter amplifier? (b) PNP common-emitter amplifier?

7. Is the output impedance of a CE amplifier a fixed quantity? Confirm your answer by referring specifically to any substantiating data in this experiment.

Answers to Self-Test

1. resistance; R_x
2. 500
3. 250
4. negative; positive; 180
5. positive
6. 12,500
7. 41

TROUBLESHOOTING A CE AMPLIFIER

OBJECTIVES

1. To make a dynamic test which will determine if an ac amplifier is operating properly
2. To consider dc voltage and resistance norms at test points in an amplifier which is operating properly, and to draw inferences as to the nature of the trouble from voltage and resistance measurements in a defective amplifier
3. To troubleshoot a defective amplifier

INTRODUCTORY INFORMATION

Dynamic Test of a CE Amplifier

As used in this experiment an ac *amplifier* will be considered a device which develops, *without distortion,* a larger signal voltage in the output than it receives in the input. This basic definition of an amplifier suggests a means of testing its performance dynamically. Consider the CE amplifier shown in Fig. 17-1. Assume that the voltage gain of this amplifier is 50 and that it operates as a linear amplifier (that is, it introduces no distortion) for a signal input in the range 10 to 100 mV. Then, if a 50-mV sine wave is injected into the input terminals, a 2.5-V undistorted sine wave may be expected in the output. This constitutes a dynamic test for the amplifier. If the amplifier passes this signal-injection and signal-tracing test; that is, if the undistorted output signal is 2.5 V for a 50-

mV input, it may be assumed that every component in Fig. 17-1, including the power source, is operating satisfactorily.

If the exact voltage gain of the amplifier is not known, and if stage gain is not a concern in troubleshooting a specific defect, the signal-injection and signal-tracing technique can still be satisfactorily used. However, all that the technician would be concerned with is *distortionless amplification* of the stage, without specifying the stage gain.

The instruments which are used in this test are a sine-wave signal generator operating in the frequency range of the amplifier and an oscilloscope to observe the signal waveform and to measure the input- and output-signal voltages.

In the example above, it was assumed that the stage checked out perfectly. Suppose, however, that there is no signal at the output terminals of the amplifier for a specified signal input. It is still possible, using the dynamic signal-tracing method, to determine in the case of Fig. 17-1 whether capacitors C_1 and C_2 are *open*. The procedure is as follows. A sine-wave signal no larger than the amplifier can handle is injected into the input terminals of the amplifier and observed at these terminals with an oscilloscope. If the observed signal is normal, the oscilloscope probe is moved to point B (base) of the amplifier. The sine-wave signal at this point should be approximately the same as at the input terminals if the amplifier input is normal. If there is no signal at the base, two possible reasons exist. The first is that capacitor C_1 is *open*. The second is that the base terminal is shorted to ground

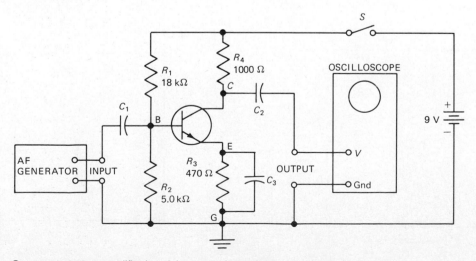

Fig. 17-1. One way to test an amplifier is to inject an ac input signal and observe the output signal with an oscilloscope.

(because of a defective transistor or other component in the base circuit). An open capacitor may readily be found by connecting the "hot" generator lead through a $0.1\text{-}\mu\text{F}$ capacitor to the base of the transistor and observing with an oscilloscope the output signal. If an output signal appears, this indicates that capacitor C_1 is open. If no signal appears at the output terminals of the amplifier, the oscilloscope probe is connected to the base. No signal indicates a short in the base circuit.

We may also determine if capacitor C_2 is open by signal tracing. Assume that the input circuit, including C_1 is found to be operating normally, but that there is no signal at the output terminals of the amplifier. The oscilloscope probe is then connected directly to the collector of Q. If a normal signal appears at the collector, but none exists at the output terminals, we know that C_2 is open.

DC Voltage Norms

The dynamic tests of an ac amplifier check ac operation and determine if the amplifier is operating properly. If no signal appears in the output, and capacitors C_1 and C_2 are found to be functioning properly, then the dc operation of the amplifier must be analyzed.

Figure 17-2 is the dc equivalent of the amplifier in Fig. 17-1. What steps should the technician take to determine which of the components or connections is defective? The first test is to determine if the transistor is good. A transistor tester may be used to check the transistor. *Power is turned off* and the suspected transistor is tested in the circuit or by removal from the circuit. The transistor is tested, and if found defective must be replaced.

If a transistor tester is not available, a substitution test is used. A known good transistor, of the same manufacturer's replacement number or JEDEC number if no special number is specified, is substituted for the suspected one. A dynamic signal-injection and oscilloscope test with power ON will determine if the transistor amplifier is operating normally. If it is, then the old transistor is defective and must be replaced. If the circuit is still not operating, the trouble is in a resistor, connections between resistors and transistor elements, or power supply.

DC Voltage Measurements

The next series of checks involves dc measurements of the power supply, and at the transistor elements. For these tests to be meaningful, the technician must know voltages in an identical circuit whose operation is normal. The voltages shown in Fig. 17-2 are for normal operation of this amplifier. Thus the power supply is 9 V. There is 1.2 V at the emitter, 1.9 V at the base, and 6.6 V at the collector. These voltages are all with respect to common (G). What is their significance?

The fact that 1.2 V is developed across the emitter resistor shows that *there is current in the emitter*. The 1.9 V at the base is close to the voltage which might be expected from the voltage-divider action of R_1 and R_2. By disconnecting the base of the transistor and measuring V_{BG} we get 1.96 V. This is slightly higher than the voltage V_{BG} measured at the base with the transistor in the circuit. The reduced voltage V_{BG} (1.90 V) is due to the loading action of the transistor, specifically the base circuit. The fact that the measured voltage V_{BG} = 1.90 V suggests that the effective resistance from point B to G must be less than R_2. Therefore there must be a current path from base to ground other than R_2. This other path is the emitter-base resistance (since there *is* base current) R_{EB}, in series with R_3, as in Fig. 17-3. The branch $R_{EB} + R_3$ is in parallel with R_2, and that is why the measured voltage V_{BG} is lower than the 1.96 V because of divider action alone, with the base open.

The 6.6 V measured at the collector in Fig. 17-2 shows that there is enough collector current to cause a 2.4-V drop across R_4, the collector resistor.

The measured voltages in Fig. 17-2 therefore show that the supply voltage is normal and that there is emitter current, collector current, and base current. It is possible from the measured values and the known values of emitter and collec-

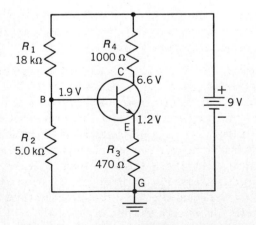

Fig. 17-2. Equivalent dc voltages for Fig. 17-1.

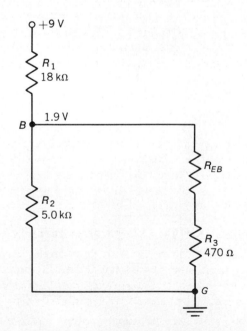

Fig. 17-3. Equivalent voltage divider that determines base voltage.

tor resistors to compute the dc currents in the emitter (I_E), collector (I_C), and base (I_B). For

$$I_E = \frac{V_E}{R_E} = \frac{1.2}{470} = 2.55 \text{ mA} \qquad (17\text{-}1)$$

$$I_C = \frac{V_C}{R_C} = \frac{2.4}{1000} = 2.40 \text{ mA} \qquad (17\text{-}2)$$

and

$$I_B = I_E - I_C = 2.55 - 2.40 = 0.15 \text{ mA} \qquad (17\text{-}3)$$

These results are consistent with our experience with transistor currents, for we demonstrated in a previous experiment that I_E is greater than I_C and that $I_B = I_E - I_C$ in a CE amplifier.

Let us analyze the voltages measured at the transistor elements in Fig. 17-2 another way. We know that the emitter is 0.7 V (1.9 − 1.2) more negative than the base, that is, the *forward bias* on the emitter to base is 0.7 V. Moreover, the collector is 4.7 V (6.6 − 1.9) more positive than the base, that is, the reverse bias on the collector to base is 4.7 V. These facts are consistent with the biasing requirements of a CE amplifier.

The voltages shown in Fig. 17-2 are normal for that amplifier using a silicon transistor for the circuit parameters shown. But amplifier parameters and transistor types differ, and the manufacturers of transistor equipment usually provide a chart of voltages for the circuit measured under normal operating conditions, or the voltage data are given on circuit diagrams. In the absence of such information, these norms are suggested as a guide for a *small-signal linear amplifier*.

Voltages are usually measured with respect to the common return (ground). The emitter-base of a transistor amplifier should be forward-biased. The range of bias which may be expected depends on the signal level which an amplifier is designed to handle and on the transistor used. In a small-signal linear amplifier using silicon transistors, the bias voltage may vary from 0.65 to 0.75 V (approx). In an NPN transistor the emitter should be negative relative to the base; in a PNP the emitter should be positive with respect to the base.

The collector-base junction of a transistor should be reversed-biased. Thus in an NPN amplifier the collector is positive; in a PNP, it is negative relative to the base. The collector-base voltages depend on circuit design and on the supply voltage. They range from approximately half the supply voltage to almost the full battery voltage. For example, if a 9-V battery is used, the collector-base voltage range may be from 5 to 8.9 V.

Inferences from DC Voltage Measurements

DC voltage readings are used to draw inferences of proper or improper functioning in transistor circuits. To show this, we will assume certain abnormal voltages in the circuit of Fig. 17-2 and analyze the possible causes of these voltages.

1. $V_C = 9$ V. The full supply voltage measured at the collector indicates that the collector circuit is not drawing any current. If the transistor is assumed to be good, the trouble could be (*a*) an open in the emitter circuit (R_3 or any of the connecting wires); (*b*) an open in the base circuit (open connection between the junction of R_1 and R_2 and the base); (*c*) base-emitter short circuit so that there is no forward bias on the emitter-base junction; (*d*) base short-circuited to ground so that there is no forward bias on the emitter-base junction.

2. $V_C = 0$ V. Possible troubles include (*a*) open collector circuit (including R_4 or the connective wiring), and (*b*) collector shorted to ground.

3. $V_E = V_C = 2.9$ V. Collector to emitter shorted.

4. $V_E = 0$ V. This condition would indicate that (*a*) there is no current flowing in the emitter, or (*b*) the emitter is short-circuited to ground. It would *not* indicate an open emitter circuit, for if there were an open in the emitter circuit, connecting the voltmeter from emitter to ground would complete the circuit and the meter would indicate a voltage.

Caution on Voltage Measurements

Associated components or the transistor itself can be damaged during voltage measurements. Care must therefore be exercised to prevent the voltmeter probes from causing short circuits between transistor terminals, or between high-voltage sources and components carrying a low-voltage rating. Moreover, voltmeters with a high input impedance should be used to prevent circuit loading. For transistor-voltage measurements, EVMs or 20,000-Ω/V VOMs are preferable to low-input-impedance meters.

Resistance Measurements

Resistance measurements in transistor circuits, *always made with power turned off*, are helpful in determining defective components. However, if a conventional ohmmeter is used, the readings obtained may be highly misleading because the energizing battery in the ohmmeter applies a voltage to the circuit under test which may cause current flow in the transistor. This parallel path through the transistor will affect readings in the external circuit.

CAUTION: Be sure you know the voltage and polarity of the leads and meter resistance before circuit testing, or you may damage components in the circuit.

Suppose, for example, it is desired to measure the resistance of R_2, from base to ground, in Fig. 17-4. Assume that the ohmmeter M is connected in the circuit so that the positive terminal of the battery is applied to the base, and the negative terminal to ground. Since this is the condition for forward bias in the emitter-base junction of this NPN transistor, current flows in this section. Therefore the resistance from emitter to base R_{EB} in series with R_3 is effectively in parallel with R_2 and affects the resistance measured by the ohmmeter. The reading obtained does not give an accurate measure of R_2 in the external circuit.

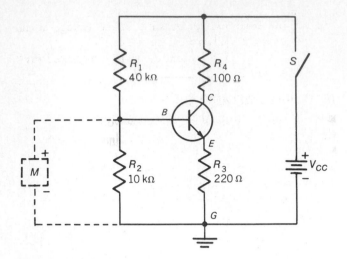

Fig. 17-4. Effect of ohmmeter and transistor on resistance measurements in external circuit.

To avoid such ambiguous measurements, the *low-power-ohms* function of an EVM should be used. This function is especially designed to produce such low voltages at the ohmmeter terminals that transistor or diode junctions cannot be forward-biased. Therefore a low-power-ohms meter measures the *actual* external resistances at selected test points in a transistor circuit and not parallel transistor or diode junction resistances.

Refer to the circuit of Fig. 17-4. What resistances may be expected if measurements are made at the transistor terminals with respect to common G, assuming that the low-power-ohms function of an EVM is used and that the circuit is operating properly? The resistance measured at the emitter E to G should be 220 Ω and at the base, point B to G, 10,000 Ω. The resistance from collector C to G would be the sum of R_4, R_1, and R_2; in this case 50,100 Ω. These values, then, are the standard or norm for the circuit of Fig. 17-4 (with switch S open). Measured values would be compared with these norms.

The most obvious resistor defects are opens, which can be spotted very easily. For example, if the resistance measured from E to G is infinite (∞), then either R_3 is open, or the connective wiring between the upper terminals of R_3 and the emitter is open, or the wiring between the lower terminal of R_3 and G is open. A resistance check across the terminals of R_3 will reveal if the resistor is good. If it is, continuity checks of the wiring to E and to G will reveal which is open.

Similarly, a resistance check from B to G indicates if the resistance in the base circuit is normal (10,000 Ω). If there is an open, further measurements will reveal if R_2 or the connective wiring from the terminals of R_2 to the base B or to common G is open.

If the resistance from point C to G is infinite and the resistance in the base circuit measures normal, then either R_4, R_1, or the connective wiring is open. Resistance checks of R_4 and R_1 and continuity checks will reveal where the defect is.

SUMMARY

1. The order in which troubleshooting checks are made in a CE amplifier is:
   (*a*) A *dynamic* signal-injection, signal-tracing check to determine if the amplifier is normal.
   (*b*) If it is not, the transistor is tested.
   (*c*) If it is normal, dc voltage measurements are made at the transistor elements with respect to common and compared with amplifier norms.
   (*d*) Other suspected components in the external circuit may be further isolated by resistance measurements at the transistor elements in the external circuit.

2. A dynamic test of the ac amplifier (Fig. 17-1) involves injecting a sine-wave signal at the amplifier input and observing, with an oscilloscope, the waveform at the output. If this test shows that the output waveform is undistorted and that the amplifier has the required gain, then the amplifier and all its components may be assumed to be operating normally.

3. The transistor may be checked with a tester or by substituting for it a transistor (of the same JEDEC number) known to be good.

4. Voltage and resistance checks are made, in troubleshooting a circuit, in an attempt to isolate a defective component.

5. In a normally operating small-signal amplifier the range of voltages at the collector will vary from one-half of V_{CC} to a voltage only slightly lower than V_{CC}. The measured voltages should be compared with the manufacturer's listing of normal voltages.

6. Voltage checks should be made with a high-ohms-per-volt meter; that is, the meter should have a high input impedance. In measuring voltages care should be exercised not to short between elements of the transistor.

7. In making resistance checks, the low-power-ohms function of an EVM should be used.

SELF-TEST

Check your understanding by answering these questions.

1. The gain of the CE amplifier in Fig. 17-1, when operated as a linear amplifier, is 100. A 10-mV signal at the input will appear as a _____ -V signal at the output, under normal operation.

2. A sine wave injected into the base of the transistor (Fig. 17-1) results in a normal output. When the generator leads are moved to the input, no signal appears in the output. The most probable cause of trouble is an

   _____ .

3. The voltage measured at the emitter (with respect to common) of the transistor in Fig. 17-2 is 0 V; the voltage at the collector is 9 V. The voltage measured at the base is 1.96 V. The most probable cause of trouble is a

   _____ _____ .

4. In a normally operating circuit, like that in Fig. 17-4, if $V_{CC} = 10$ V, the dc voltage measured from B to G will be less than _____ V.

5. In the circuit shown in Fig. 17-4, a technician measures 10.0 V at the collector. $V_{CC} = 12$ V. The collector current is _____ mA.

6. For the amplifier in question 5, the voltage at the emitter will be greater than _____ V.

7. If R_4 in Fig. 17-4 is open, a resistance check from points C to G will measure _____ Ω.

8. If the emitter is shorted to the base in Fig. 17-2, a resistance check from the base to ground will show approximately _____ Ω.

9. For the conditions in question 8, the voltage at the collector will measure _____ V.

10. If R_1 in Fig. 17-2 is open, the voltage measured from E to G will be _____ V.

MATERIALS REQUIRED

- Power supply: Variable regulated low-voltage dc
- Equipment: Oscilloscope; EVM; AF sine-wave generator
- Resistors: 560-, 1000-Ω; 8.2-, 18-kΩ $1/_2$-W
- Capacitors: Two 25-μF 50-V; 100-μF 50-V
- Semiconductors: 2N6004 or equivalent
- Miscellaneous: SPST switch; defective components for the circuit of Fig. 17-5

PROCEDURE

Dynamic Check of Amplifier Operation

1. Connect the circuit of Fig. 17-5. Inject a 30-mV peak-to-peak sine wave into the input. Observe with an oscilloscope, measure, and record in Table 17-1 the peak-to-peak output of the amplifier. Indicate also if the amplifier operation is or is not normal.

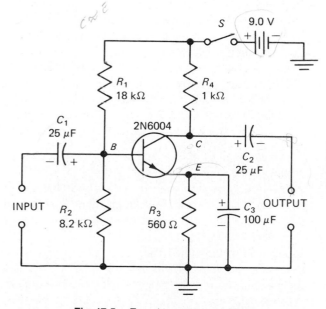

Fig. 17-5. Experimental ac amplifier.

2. Observe, measure, and record in Table 17-1 the peak-to-peak voltage of the signal at the base and collector of the amplifier.

3. With the generator still connected to the input of the amplifier, remove C_1. Observe, measure, and record in Table 17-1 the voltage of the waveform at the input, base, collector, and output of the amplifier. Is amplifier operation normal?

4. Replace C_1 and remove C_2 from the circuit. Observe, measure, and record the voltage of the waveform at the input, base, collector, and output. Is amplifier operation normal?

DC Voltage Measurements

5. Remove the signal generator from the input to the amplifier. Estimate and record in Table 17-2 the normal dc voltages at the emitter, base, and collector of the transistor. Assume 4.0 mA of emitter current.

6. Measure with respect to common (ground) the voltages at the emitter, base, and collector. Record your results in Table 17-2.

7. **Power off.** Remove R_3, the emitter resistor. Estimate the voltages at the transistor elements for this condition and record these values in Table 17-2.

8. **Power on.** Measure and record in Table 17-2 the voltages at the transistor elements.

TABLE 17-1. Dynamic Checks

Step	Condition	Input	Base	Collector	Output	Amplifier Operation
1, 2	Normal	30m	30m	2.6	4.6	NORMAL
3	C_1 open	44m	0	0	0	ABNORMAL
4	C_2 open	34m	30m	4.5	0	NORMAL

(handwritten note above table: ? Should be close to the same)

TABLE 17-2. DC Voltages

Expect voltage to go down

Element	Voltage (Normal)		Voltage (Emitter Open)		Voltage (Base Open)		Voltage (Collector Open)	
	Estimated	Measured	Estimated	Measured	Estimated	Measured	Estimated	Measured
Emitter	3.15	2.9	0	0	0	0	0	.27
Base	3.75	3.5	3.75	3.75	3.75	3.6	3.75	.91
Collector	6.375	6.5	12	12	12	12	12	Vcc

9. **Power off.** Replace the emitter resistor. Disconnect the lead from the base to the voltage divider, leaving the base open. Estimate and record in Table 17-2 the voltages at the elements of the transistor for this condition.

10. **Power on.** Measure and record the voltages at the elements of the transistor.

11. **Power off.** Reconnect the base to the voltage divider. Remove R_4, the collector resistor, from the circuit. Estimate the voltages at the transistor elements for this condition and record in Table 17-2.

12. **Power on.** Measure and record the voltages at the transistor elements.

13. **Power off.** Replace the collector resistor.

Resistance Measurements

14. Estimate the resistance values at the elements of the experimental amplifier and record your values in Table 17-3.

15. **Power off.** Set your EVM to the low-power-ohms function. Measure and record in Table 17-3 the resistances from the base, emitter, and collector to common (*G*).

TABLE 17-3 Resistance Measurements

Element	Resistance to Ground (G)	
	Estimated	Measured
Emitter	580	.56K
Base	8.2K	8.22K
Collector	27K	27K

Troubleshooting the CE Amplifier

16. Your instructor will insert trouble into the amplifier (Fig. 17-5). Troubleshoot the circuit, keeping a step-by-step record of each check that you make in the order performed. Record these checks in the standard troubleshooting report which will be used for all servicing procedures. When you have found the trouble, correct it and notify your instructor, who will put another trouble into the circuit, which you will again service, using a separate troubleshooting report to record your results. Service as many troubles as time permits.

QUESTIONS

1. Why is a *dynamic* test of an amplifier's operation more meaningful than static dc voltage and resistance checks?

2. The technique of signal tracing can be used to isolate trouble in a two- (or more) transistor (stage) amplifier to a specific transistor stage. Explain how this technique may be used to determine which transistor stage is defective in a three-stage amplifier. The transistors are connected so that the output of the first is capacitively coupled to the input of the second, and the output of the second is capacitively coupled to the input of the third.

3. Compare the estimated and measured voltages in Table 17-2 for an amplifier operating normally. Explain any discrepancies.

4. Interpret and explain the voltages at the elements of the transistor with the emitter open (Table 17-2).

5. Interpret and explain the voltages at the elements of the transistor with the collector resistor open (Table 17-2).

6. Is an understanding of the operation of a transistor amplifier helpful in troubleshooting it? Why?

7. Why is it necessary to use the low-power-ohms function of an EVM, rather than the conventional ohms function, in measuring resistance in the external circuit of a transistor amplifier?

8. Assume a short-circuited base to emitter in Fig. 17-5. Determine the voltages with respect to common which you would measure at the elements of the transistor. Show your computations.

Answers to Self-Test

1. 1
2. open C_1
3. defective transistor
4. 2
5. 20
6. 4.4
7. infinite
8. 470
9. 9
10. 0

95 AC
Quest. Due
4-19-88

18

EXPERIMENT

THE EMITTER FOLLOWER
(COMMON-COLLECTOR AMPLIFIER)

OBJECTIVES

1. To measure the input and output impedances of an emitter follower
2. To measure the power gain of this amplifier
3. To observe the phase relationship between the input and output signal voltage

INTRODUCTORY INFORMATION

Emitter Follower

The grounded or common-collector amplifier, more commonly known as the *emitter follower*, is a transistor circuit whose voltage gain is approximately unity, exhibits current and power gain, and has high input impedance and low output impedance. The impedance characteristic of this amplifier makes it useful for impedance-matching applications.

Figure 18-1a is the circuit diagram of an emitter follower using two power sources, V_{BB} and V_{CC}. A variation of this circuit is shown in Fig. 18-1b, which shows an emitter follower using a single power source. The input ac signal v_{in} is coupled by capacitor C_1 to the base. The load resistor R_2 is connected in the emitter, and the output-signal voltage v_{out} is

developed across this *unbypassed* emitter resistor. The collector is connected directly to V_{CC}, and it is at ac common or ground, because capacitor C_2 acts as a low-impedance bypass for the collector (hence the name *grounded* collector). C_2 can be an actual capacitor connected from collector to ground or the output filter capacitor in the V_{CC} supply. Since the input signal appears between base and ground (collector), and the output signal appears between emitter and ground (collector), it is evident that the collector (ground) is common both to the input and output circuits. The name "common collector," by which this amplifier is also known, is derived from this characteristic. Why this circuit is also called an "emitter follower" will become evident soon.

In the circuit of Fig. 18-1a V_{CC} serves as the collector power source. Its positive terminal is connected to the collector because an NPN transistor is used. If a PNP type were in the circuit, the battery polarity would be reversed. Also, in Fig. 18-1a V_{BB} acts as the bias source. In Fig. 18-1b V_{CC} is the source both for the collector and for base bias.

A distinguishing characteristic of the common- or grounded-collector circuit is that the emitter resistor R_2 is not bypassed for alternating current by a capacitor. Hence an output ac signal voltage appears across R_2 when an ac signal voltage is applied to the base.

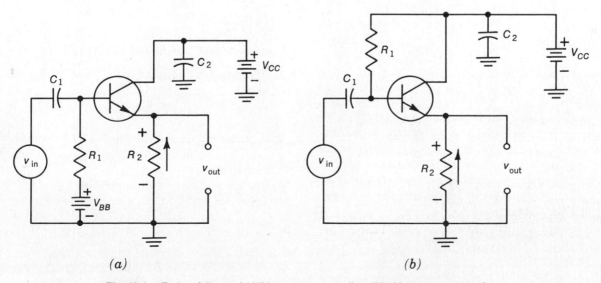

Fig. 18-1. Emitter follower. *(a)* With two power supplies; *(b)* with one power supply.

109

Phase Relations

To establish the phase of the input and output signals we observe that on the positive alternation of the input signal v_{in}, base current increases for this NPN transistor. As a result *both* collector and emitter currents increase. In the *external* emitter circuit current flows through R_2 in the direction shown by the arrow. The instantaneous emitter voltage v_{out}, which is equal to $i_{out}R_2$, therefore becomes more positive relative to ground than it was prior to the ac signal input. Hence as the base voltage goes positive, the emitter voltage follows. Similarly, as the base voltage goes negative on the negative alternation of v_{in}, the emitter voltage v_{out} becomes less positive and hence more negative. Therefore the input and output signals are in phase in an emitter-follower circuit. The fact that the phase of the output signal in the emitter *follows* the phase of the input signal gives rise to the name "emitter follower." Since the phase of the signal voltage in the emitter is the same as that in the base, and since the input-signal voltage to the circuit is the *difference* between the signal voltage on the base and that on the emitter, the effect of the unbypassed emitter resistor is to provide degenerative or negative feedback to the circuit. The amplifier therefore "sees" a lower effective input signal (between base and emitter) than v_{in}.

Impedance and Gain

A mathematical analysis of a common-collector amplifier, operating in the linear mode, leads to rather complex equations from which voltage gain, current gain, power gain, and input and output impedance may be calculated. However, the following approximate equations can be given as a basis for design:

$$A_v = \text{voltage gain} = 1 \tag{18-1}$$

$$A_i = \text{current gain} = \frac{1}{1 - \alpha} \tag{18-2}$$

$$A_p = \text{power gain} = \frac{1}{1 - \alpha} \tag{18-3}$$

$$\text{Input resistance } R_{in} = \frac{R_L}{1 - \alpha} \tag{18-4}$$

$$\text{Output resistance } R_{out} = r_e + (1 - \alpha)(R_G + r_b) \tag{18-5}$$

In Eq. (18-5), r_e and r_b are the emitter and base resistances, respectively, in the T equivalent model of a transistor; that is, they are two of the T parameters. R_G is the internal resistance of the signal source. The approximate formulas are valid only for the circuits from which the formulas were derived and are based on certain assumptions which relate to such typical values of T parameters as

$$r_b = 500 \ \Omega$$
$$r_e = 30 \ \Omega$$
$$r_c = 1.5 \text{ megohms (M}\Omega\text{)}$$
$$\alpha = 0.975$$

It should be noted further that the input resistance of the common-collector stage will be affected not only by α and R_L but also by any resistance which connects to the base. Thus, if R_1 is a low-valued resistance, its shunting effect on the circuit will affect R_{in}.

The voltage gain and input and output impedances of the emitter-follower amplifier may be determined by the methods previously employed in the common-emitter and common-base amplifiers. The power gain may be calculated from the experimentally determined values of R_{in}, R_{out}, v_{in}, and v_{out}. Thus

$$P_{in} \text{ (power in)} = \frac{v_{in}^2}{R_{in}} = i_{in}^2 R_{in}$$

$$P_{out} \text{ (power out)} = \frac{v_{out}^2}{R_{out}} = i_{out}^2 R_{out}$$

$$\text{Power gain} = \frac{P_{out}}{P_{in}}$$

SUMMARY

1. The common-collector (CC) amplifier is also known as the grounded collector and the emitter follower.
2. In the CC amplifier the collector is returned directly to V_{CC} and is at ac ground.
3. The input impedance of a CC amplifier is high; the output impedance is low.
4. The CC amplifier exhibits no voltage gain. Its voltage gain is approximately 1, or *unity*.
5. The CC amplifier exhibits current and power gain.
6. The input signal to a CC amplifier is injected into the base; the output signal is developed in the emitter.
7. The emitter of a CC amplifier is *unbypassed* for ac.
8. The phase of the output signal in the emitter is the same as that in the base.

SELF-TEST

Check your understanding by answering these questions.

1. The emitter-follower amplifier in Fig. 18-1 is also known as the _____ or _____ .
2. In Fig. 18-1a $v_{in} = 4$ V. v_{out} is approximately _____ V.
3. A negative-going signal appears on the emitter when a _____ (*negative/positive*)-going signal is applied to the base.
4. The input voltage to an emitter follower is 6 V. Its input impedance is 50,000 Ω. The input power is _____ mW.
5. In the amplifier in question 5, the output power is 0.1 W. The power gain of the circuit is _____ .
6. In the amplifier in questions 4 and 5 the output impedance is _____ Ω (approx).

MATERIALS NEEDED

- Power supply: Variable regulated low-voltage dc source
- Equipment: Oscilloscope; EVM; sine-wave generator
- Resistors: 3300-, 12,000-Ω; 470-kΩ ½-W

- Capacitors: 25-μF 50-V, 100-μF 50-V
- Semiconductors: 2N2102 or equivalent
- Miscellaneous: Three SPST switches; 500-Ω potentiometer

PROCEDURE

Voltage Gain

1. Connect the circuit of Fig. 18-2. **Power off.** Switch S_1 is *closed*, S_2 *open*. Set AF generator at *zero* output. Connect an oscilloscope across points DF and adjust it for proper viewing. The oscilloscope will be used to measure the peak-to-peak value of the signal at specified points in the circuit.
2. **Power on.** Slowly bring up the gain of the AF generator until a 150-mV undistorted output signal v_{out} appears across points DF in the emitter circuit. With the scope measure v_{out}. Record it in Table 18-1.
3. Measure and record the input signal voltage v_{in} (points AC). Compute and record the voltage gain v_{out}/v_{in}.

Input Impedance

4. *Open switch* S_1. Increase AF generator output until v_{in} is at same level as in step 3. With the oscilloscope "floating," measure, and record in Table 18-1, the signal voltage v_{AB} across points AB. Compute the input-base signal current i_{in} by substituting v_{AB} in the formula

$$i_{in} = \frac{v_{AB}}{R_{AB}} = \frac{v_{AB}}{12,000} \quad \text{13mV}$$

Show your computations. Record i_{in} in Table 18-1.

5. Compute the input resistance R_{in} substituting for i_{in} and v_{in} in the formula

$$R_{in} = \frac{v_{in}}{i_{in}} \quad \frac{150m}{1.08u} = 138.4k$$

Show your computations. Record R_{in} in Table 18-1. Compute also the input power $i_{in}{}^2 R_{in}$. Record it in Table 18-1.

Output Impedance

6. *Close switch* S_1. Reduce generator output until v_{out} measures 100 mV. S_2 is still open. Measure, and record in Table 18-2, the output voltage v_{out} (across DF).
7. *Close* S_2. The load resistor R_L is now in the circuit. Adjust R_L until the output voltage with load is one-half the value of v_{out} measured in step 6. Record $v_{out}/2$. *Open* S_2. Measure and record the resistance of R_L. This is the value of output impedance R_{out} of the circuit. Compute also the output power $v_{out}{}^2/R_{out}$, W. Record this wattage in Table 18-2.

Power Gain

8. Compute and record power gain using the formula

$$A_p = \frac{P_{out}}{P_{in}}$$

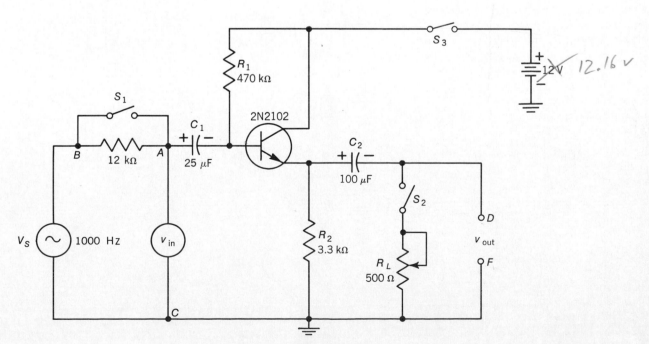

Fig. 18-2. Experimental emitter follower.

TABLE 18-1. Voltage Gain, Input Impedance, and Input Power

v_{out}, V	v_{in}, V	$Gain = \dfrac{v_{out}}{v_{in}}$	v_{AB}, V	i_{in}, A	R_{in}, Ω	$i_{in}{}^2 R_{in}$, W
150mV	150mV	1	16mV	1.33uA	112.78KΩ	199.5 mW

TABLE 18-2. Output Impedance, Output Power, and Power Gain

v_{out}, V	$\dfrac{v_{out}}{2}$, V	$R_L = R_{out}$, Ω	$\dfrac{v_{out}{}^2}{R_{out}}$, W	Power Gain
100m	50m	52.1Ω	191.9uW	29.8dB

TABLE 18-3. Phase Relations

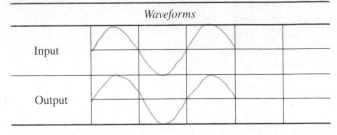

Waveforms	
Input	
Output	

Phase Relations

9. Following the procedure outlined in previous experiments, determine the phase relationship between the input and output waveforms. Draw the waveforms in proper time phase in Table 18-3.

Extra Credit

10. Explain how you would determine an approximate value for the α of the transistor used in this experiment.
11. Find α following your procedure.

QUESTIONS

1. Comment on your experimental data concerning the voltage gain of the emitter follower.
2. How does the input impedance of the common-collector amplifier compare with that of the common emitter? Explain why they differ.
3. How will the size of the bias resistor R_1 in Fig. 18-1 affect the input impedance (resistance) of the circuit?
4. How will the size of the emitter resistor R_2 in Fig. 18-1 affect the input impedance of the circuit?
5. What is the relationship between the phase of the input and output waveforms in the emitter follower?

6. Explain in detail the method you used to determine phase relations in the experimental circuit. How was your oscilloscope triggered/synchronized for this check?
7. Assume that $v_{out}/v_{in} \cong 1$. Write an approximate formula for the power gain of an emitter follower.
8. Explain in detail how you determined the input impedance of the experimental amplifier.

Answers to Self-Test

1. grounded; common collector
2. 4
3. negative
4. 0.72
5. 139
6. 360

LOAD-LINE ANALYSIS OF A TRANSISTOR AMPLIFIER

OBJECTIVES

1. To construct a maximum dissipation curve
2. To construct a dc load line in a CE amplifier and to verify the predicted operating conditions of the amplifier

INTRODUCTORY INFORMATION

Transistor amplifiers must be operated over the linear portion of their characteristic when it is necessary to reproduce the input signal without distortion. It is therefore necessary to choose carefully the operating point, the characteristics of the transistor, and the associated circuit components in considering the design of a linear amplifier.

In transistors the operating point is determined by the dc voltages chosen for collector, emitter, and base. Hence, in designing an amplifier, the engineer or technician should consult the transistor characteristic curves and other listed data.

Collector-Dissipation Curves

Figure 19-1 is a typical family of collector characteristic curves for a transistor connected in the common-emitter configuration. These curves show the behavior of the transistor at 25°C ambient temperature, over a range of collector currents and voltages, for specified values of base or bias currents. There is a certain restriction, however, on the por-

tions of the curve over which the transistor may be operated. This restriction relates to the maximum collector dissipation permissible at a specified temperature level. For example, the 2N649 NPN alloy junction transistor is limited to a collector dissipation of 100 mW when operating at an ambient temperature of 25°C. The permissible operating region may be shown in the manufacturer's data sheets by a graph superimposed on the family of collector characteristics. If a maximum dissipation curve is not included in the manufacturer's data, the technician can easily plot it on the family of characteristics, as in Fig. 19-2. The curve represents 100 mW at every point; that is, the coordinates of every point on this curve are selected so that the product of

$$V_{CE} \times I_C = 100 \text{ mW} \qquad (19\text{-}1)$$

For example, the 2-V 50-mA point satisfies this condition, as does the 10-V 10-mA point, etc. The curve is, of course, one branch of the hyperbola defined by Eq. (19-1). The transistor must be operated to the left of this curve.

DC Load Line

The performance of a transistor amplifier can be predicted graphically by means of a load line. The effect of an input on the output signal for specified operating conditions, assuming no ac load, can then be determined from the load line.

Refer to Fig. 19-3, which shows the output circuit of a grounded-emitter amplifier. By Kirchhoff's voltage law the

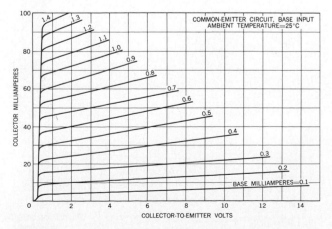

Fig. 19-1. Average collector characteristics of a 2N649. *(RCA)*

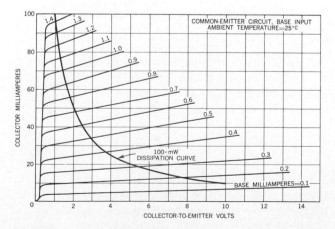

Fig. 19-2. A 100-mW dissipation curve superimposed on collector characteristics.

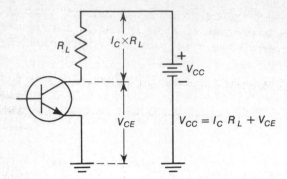

Fig. 19-3. DC voltages in the output of a grounded-emitter amplifier.

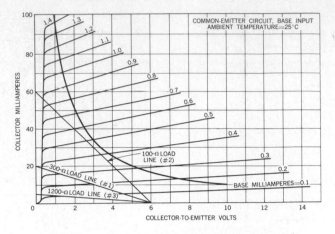

Fig. 19-4. Load line for specified values of R_L and V_{CC}.

relationship between the dc supply voltage V_{CC}, the voltage drop $I_C \times R_L$ across R_L, and the voltage V_{CE} from collector to emitter is

$$V_{CC} = I_C \times R_L + V_{CE} \qquad (19\text{-}2)$$

If V_{CC} and R_L are known, the load line may be drawn through the two limiting points on the line:

$$\left(V_{CE} = 0, I_C = \frac{V_{CC}}{R_L} \right) \qquad \text{on the collector current axis}$$

$$(I_C = 0, V_{CE} = V_{CC}) \qquad \text{on the collector voltage axis}$$

A numerical illustration will show how the coordinates of the limiting points are found. Assume in Fig. 19-3 that V_{CC} = 6 V and R_L = 300 Ω. Equation (19-2) may therefore be written as

$$6 = I_C(300) + V_{CE} \qquad (19\text{-}3)$$

Now when $V_{CE} = 0$, $I_C = 6/300 = 20$ mA. When $I_C = 0$, $V_{CE} = 6$V. The line drawn through these two points (1) is shown in Fig. 19-4.

What would be the effect on the line of increasing or decreasing the value of R_L, all other conditions remaining equal? Again a numerical example will indicate the answer. First, if R_L is decreased in value to 100 Ω, the load line (2) coordinates are

$$V_{CE} = 0, \quad I_C = \frac{6}{100} = 60 \text{ mA}$$

and

$$I_C = 0, \quad V_{CE} = 6 \text{ V}$$

This line (2) has a common point with load line 1, namely, $V_{CC} = 6$, $I_C = 0$, but lies above it. Again if the value of R_L is increased to say 1200 Ω, the coordinates of the load line (3) are

$$V_{CE} = 0, \quad I_C = \frac{6}{1200} = 5 \text{ mA}$$

and

$$I_C = 0, \quad V_{CE} = 6.0 \text{ V}$$

Load line 3 shares a common point with lines 1 and 2 but lies below line 1. Note that in all three cases the load line is to the left of the 100-mA maximum-dissipation curve.

The general method of drawing the load line is clear as

long as the values of R_L and V_{CC} are known. Suppose the value of R_L is not known, but the operating point of the transistor amplifier is known. (The operating point may be defined by the quiescent conditions of the amplifier, that is, by the values of V_{CE} and I_C with no signal present.)

If the operating point Q is $V_{CE} = 5.25$ V, $I_C = 19$ mA, and the supply voltage $V_{CC} = 10$ V, the load line is drawn through Q and through the point $I_C = 0$, $V_{CE} = 10$, as in Fig. 19-5. Note that the load line intersects the collector milliampere axis at the point $V_{CE} = 0$, $I_C = 40$ mA. The value of R_L may now be easily found, for $I_C = V_{CC}/R_L$ when $V_{CE} = 0$. Therefore

$$R_L = \frac{V_{CC}}{I_C} \qquad (19\text{-}4)$$

Substituting the values $V_{CC} = 10$ V, $I_C = 40$ mA in Eq. (19-4) gives

$$R_L = \frac{10}{40 \times 10^{-3}} = 250 \text{ Ω}$$

A 250-Ω collector load resistor would permit the amplifier to be operated at the required point Q.

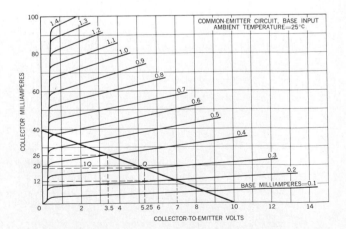

Fig. 19-5. Load line drawn if the operating point Q and the supply voltage are known.

Predicting Amplifier Operation from the Load Line

From the load line it is now possible to predict the manner in which collector current will vary with changes in base current. The intersection of the load line with any one of the family of collector curves may serve as the operating point. The best operating point is determined by circuit requirements. The operating point selected in the last example, V_{CE} = 5.25 V, I_C = 19 mA, lies on the intersection of the load line with the I_B = 0.3-mA characteristic curve. This means that a base current of 0.3 mA will cause a current of 19 mA to flow in the collector circuit of a 2N649 transistor. The transistor operates from a 10-V source with a load resistor of 250 Ω, connected in the grounded emitter configuration of Fig. 19-3. What would be the effect on collector current of increasing the base current in this circuit to, say, 0.4 mA? The answer is obtained by noting the point at which the load line intersects the 0.4-mA characteristic. In this case I_C = 26 mA, an increase of 7 mA, and V_{CE} = 3.5 V. If I_B is reduced to 0.2 mA, the collector current goes down to 12 mA, a decrease of 7 mA, and V_{CE} = 7.0 V. For the operating point and circuit chosen, we see that changes in collector current are linearly related to changes in base current between base-current limits of 0.1 to 0.3 mA.

From the load line also we can find the current gain for the amplifier. We previously defined current gain as the ratio of output-current change to input-current change. That is,

$$A_i - \frac{\Delta I_{\text{out}}}{\Delta I_{\text{in}}} \qquad (19\text{-}5)$$

In the example cited we see that

$$\Delta I_{\text{out}} = 26 - 12 = 14 \text{ mA}$$
$$\Delta I_{\text{in}} = 0.4 - 0.2 = 0.2 \text{ mA}$$

Substituting in Eq. (19-5), we find

$$A_I = \frac{14}{0.2} = 70$$

SUMMARY

1. The family of average collector characteristics (Fig. 19-1) shows how collector current I_C varies with base current I_B and collector voltage V_{CE}, in a specific transistor operating at a specified temperature.
2. This family of curves does not, however, indicate the maximum wattage W that the collector can dissipate. For the graphs to be useful, the maximum dissipation curve of the transistor must be drawn on the collector characteristics (Fig. 19-2).
3. A dc load line is a straight line drawn on the family of characteristic curves. Amplifier performance may be predicted from this line.
4. To draw the load line it is assumed that V_{CC} and R_L are known. The end points of the line may then be calculated as follows: (a) The coordinates of the point on

the collector-current axis are: V_{CE} = (0, I_C = V_{CC}/R_L; (b) the coordinates of the point on the collector-voltage axis are: V_{CE} = V_{CC}, I_C = 0.
5. If V_{CC} = 10 V and R_L = 250 Ω, the load line is shown in Fig. 19-5. Now, if the amplifier whose load line is shown in Fig. 19-5 is operating with 0.3-mA base (bias) current, the intersection of the load line with the 0.3-mA curve is called the *operating point Q* of the amplifier.
6. From the load line it is possible to predict the output-current and output-voltage variations for a specified base signal input.
7. From the load line (see statement 5) we can also find the current gain of the amplifier.

SELF-TEST

Check your understanding by answering these questions.

1. In Fig. 19-1, the characteristic curve which is identified by 0.4 is the graph showing the relationship between _____ and _____ of the 2N649 transistor, where the base current is maintained at _____ mA.
2. In Fig. 19-2, the intersection of the 100-mW dissipation curve with the 0.5-mA (base-current) curve occurs at the point where V_{CE} = _____ V and I_C = _____ mA.
3. In question 2, the product of V_{CE} and I_C is _____ mW.
4. The product of the coordinates of every point on the collector-dissipation curve in Fig. 19-2 is _____ mW.
5. The load line is the line defined by the equation V_{CC} = I_C × R_L + V_{CE}, where V_{CC} and R_L are fixed quantities, and I_C and V_{CE} are variable. If V_{CC} = 20 V and R_L = 500 Ω, the end points of the load line are _____ V, 0 mA on the V_{CE} axis, and 0 V, _____ mA on the I_C axis.
6. In Fig. 19-5, assume the operating point of the amplifier is at 0.2-mA base current. At this point, the collector voltage is _____ V; collector current is _____ mA.

MATERIALS REQUIRED

- Power supply: Variable regulated low-voltage dc
- Equipment: EVM; 0–50-µA and 0–10-mA dc ammeters (or equivalent ranges on a multimeter)
- Resistors: 1- and 68-kΩ ½-W
- Semiconductors: 2N6004 or equivalent
- Miscellaneous: One SPST switch; 10,000-Ω 2-W potentiometer; equipment and components as required for extra-credit question

PROCEDURE

1. On Fig. 19-6, the family of collector characteristics of the transistor type 2N6004, draw a 100-mW collector-dissipation curve.
2. On Fig. 19-6, draw the load line for the common-emitter amplifier of Fig. 19-7. In this circuit $R_L = 1$ kΩ and the collector supply voltage $V_{CC} = 10$ V. Show your computations for the load-line end points.
3. Find the intersection of the load line and the 15-μA base-current curve. Identify this point on the load line as Q_1, operating point 1, and record in Table 19-1 the I_C and V_{CE} coordinates of Q_1.
4. From the load line determine and record the values of I_C and V_{CE} when **(a)** $I_B = 10$ μA; **(b)** $I_B = 20$ μA; **(c)** $I_B = 25$ μA; **(d)** $I_B = 30$ μA. Identify as Q_2 the intersection of the load line and the 25-μA base-current curve.
5. Construct the circuit of Fig. 19-7. Switch S_1 is open, and R_1 is set for minimum resistance.
6. *Close S_1.* Set R_1 so that the base current measures 10 μA. Record this base-current value in column "I_B" under the heading "Load-line Value, Measured." Measure I_C, V_{CE}, and V_{EB}. Record the results under the heading "Load-line Value, Measured" in the respective columns. Repeat for each I_B in Table 19-1.
7. Using the measured values, compute the collector-current gain of the transistor operating at Q_1 for a total base-current change of 10 μA. Record your results in Table 19-2. Show your computations. Compute also the voltage gain at the collector for the same conditions. Record the data in Table 19-2.
8. Repeat step 7 for Q_2, operating point 2.

Extra Credit

9. Explain in detail the procedure you would use to determine experimentally the maximum ac input current and voltage the experimental amplifier in Fig. 19-7 can handle without distortion, using a 1000-Hz sine-wave input. Show the circuit diagram. Explain also how to measure ac current and voltage gain.
10. Connect the circuit, measure the gain at Q_1 and Q_2, and record the data in a table.

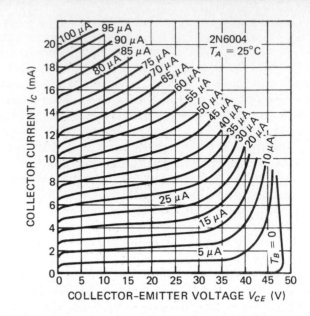

Fig. 19-6. Common-emitter collector characteristics for the 2N6004. *(GE)*

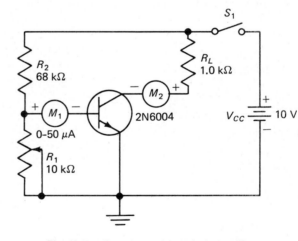

Fig. 19-7. Experimental transistor amplifier.

TABLE 19-1. Load-line Measurements

Load-line Value, Computed			Load-line Value, Measured			
I_B, μA	I_C, mA	V_{CE}, V	I_B, μA	I_C, mA	V_{CE}, V	V_{EB}, V
10						
15						
20						
25						
30						

TABLE 19-2. Load-line Gain Calculations

Operating Point	Collector-current Gain $= \dfrac{I_C}{I_B}$	Collector-voltage Gain $= \dfrac{V_{CE}}{V_{EB}}$
Q_1 (15 μA)		
Q_2 (25 μA)		

QUESTIONS

1. Explain the significance of the maximum dissipation curve.
2. What is the purpose of a dc load line?
3. What information is needed to draw a dc load line?
4. How do the load-line values in Table 19-1 compare with the measured values? Explain the differences.
5. Is the operation of the experimental circuit in Fig. 19-7 linear over the range of 10-μA input base current at Q_1? At Q_2? Explain.
6. Compare the current and voltage gains of the experimental circuit at (a) Q_1, Q_2.

7. If the current gains at Q_1 and Q_2 are not the same, explain why.

Extra Credit

8. Comment on the relationship between the alternating and direct current and voltage gains of the experimental circuit.

Answers to Self-Test

1. V_{CE}; I_C; 0.4
2. 3; 33⅓
3. 100
4. 100
5. 20; 40
6. 7; 12

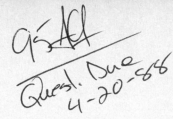

CASCADED TRANSISTOR AMPLIFIER

OBJECTIVES

1. To determine the range of linear operation of an *RC*-coupled two-stage amplifier
2. To observe the phase relationships at the input and output of each stage in the amplifier

INTRODUCTORY INFORMATION

When two amplifiers are connected in such a way that the output signal of the first serves as the input signal to the second, the amplifiers are said to be connected in *cascade*. Amplifiers, either transistor or vacuum-tube, may be operated in cascade to extend the gains possible with single-stage amplifiers.

Theoretically there are nine possible transistor cascade arrangements of which the most common is the grounded-emitter to grounded-emitter configuration. As you will recall, grounded-emitter amplifiers exhibit high voltage, high current, and high power gains. They are used in sound-reproducing systems as audio amplifiers, in TV receivers as video (picture) amplifiers, and in many other applications.

Coupling Methods

Transformer Coupling

Transformers are frequently used in coupling amplifier stages. Transformers make it possible to match the output impedance of the first stage to the input impedance of the next. Proper impedance matching ensures maximum transfer of power from one stage to the next.

Figure 20-1 is the circuit diagram of a two-stage transistor amplifier using transformer coupling. The input signal is

coupled by T_1 into the base of Q_1. C_1 is used to isolate the secondary of transformer T_1 from the dc base-bias circuit. Base bias and bias stabilization are provided by the voltage-divider arrangement of R_1 and R_2 across the battery V_{CC} and resistor R_3 in the emitter. C_2 bypasses R_3 to prevent degeneration. T_2 couples the output signal of transistor Q_1 to the base circuit of Q_2. In the circuit of Q_2, the components $C_3, C_4, R_4, R_5,$ and R_6 have the same functions, respectively, as $C_1, C_2, R_1, R_2,$ and R_3, their counterparts in Q_1. The amplified output of Q_1 is thus fed to Q_2, where additional amplification occurs.

RC Coupling

Other coupling methods are used. Thus there are direct-coupling, *RC*-coupling, and impedance-coupling arrangements. Figure 20-2 shows an *RC*-coupled cascaded amplifier. Capacitors C_1 and C_3 couple the signal into Q_1 and Q_2, respectively. C_5 is used for coupling the signal from Q_2 to its load. In other respects, component functions in stages Q_1 and Q_2 are the same as the equivalent components in the transformer-coupled amplifier in Fig. 20-1.

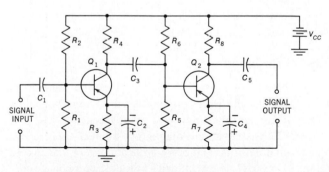

Fig. 20-2. *RC*-coupled transistor amplifier.

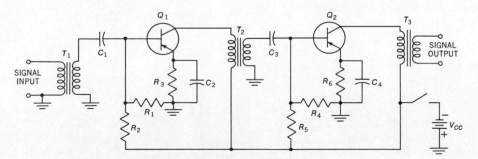

Fig. 20-1. Cascaded transistor amplifier with transformer coupling.

If the operation of coupled amplifiers is considered, a complicating factor appears. The addition of a second stage may alter the characteristics of the first stage and thus affect the level of signal fed to the second stage. For example, if C_3 in Fig. 20-2 were open, R_4 would act as both the ac (signal) load and the dc load for Q_1. When C_3 is connected as shown (see Fig. 20-2), R_4 is no longer the ac collector load for Q_1. Instead, the ac load now consists of R_4 in parallel with R_5, R_6, and the input resistance R_{in} of Q_2 (Fig. 20-3). The ac collector load resistance, R_L, is therefore smaller in value than R_4. In this analysis it is assumed that the reactance of C_3 is negligibly small at the frequency of the input signal and that the battery V_{CC} presents a very-low-impedance path for the ac signal; that is, V_{CC} acts like an ac short circuit for the signal.

The signal voltage developed at the collector of Q_1 equals $i_c \times R_L$. An effect of the reduced value of ac load resistance is the reduction in signal voltage v_{1out} at the collector of Q_1. It is the signal voltage v_{1out} which determines the value of signal current flow in the input base circuit of Q_2. It is evident then that the ac plate load resistance of Q_1, one of the factors which determines the signal current coupled to the base of Q_2, is changed by RC coupling of cascaded amplifiers.

The preceding analysis considered the signal voltage v_{1out} as the source of signal current i_{b2} for the base of Q_2. Another view is that the collector signal current i_{c1} supplies the base current i_{b2}. Refer again to Fig. 20-3. It is evident that not all the collector current i_{c1} is fed to the base. However, since the base input resistance R_{in} is very low, a large percentage of collector signal current will be coupled to the base.

A combination of RC and transformer coupling may be used effectively. Figure 20-4 is the circuit diagram of a transistor amplifier in a combination coupling arrangement. Stage Q_1 is RC-coupled to Q_2. Stage Q_2 is transformer-coupled to its load. The bias arrangements for stages Q_1 and Q_2 are conventional.

Direct Coupling

Direct coupling is also used in cascaded transistor amplifiers. An advantage of direct coupling is the savings possible in components and the improvement in frequency response. Figure 20-5 is the circuit diagram of a direct-coupled two-stage amplifier. Bias and bias stabilization of Q_1 are conventional. Bias of Q_2 is determined by the collector voltage of Q_1 and the emitter voltage of Q_2. The values of R_4 (in the collector of Q_1), of R_5, and of the operating point of Q_1 must

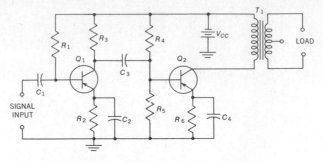

Fig. 20-4. *RC* and transformer coupling.

be so chosen that the collector of Q_1, and hence the base of Q_2, is negative with respect to the emitter of Q_2. This establishes forward bias for Q_2. Bias stabilization for Q_2 depends on the direct connection between Q_2 and Q_1. Thus, if an increase in operating temperature occurs, the collector currents of Q_1 and Q_2 will increase. An increase in collector current of Q_1, however, will cause the voltage at the collector of Q_1 to become less negative, and hence more positive. This will make the base of Q_2 more positive and will reduce the forward bias of Q_2, thus reducing collector current in Q_2. Effectively then, the direct-coupled circuit of Fig. 20-5 provides bias stabilization of Q_2.

Direct coupling is also possible using PNP and NPN transistors as in Fig. 20-6. PNP and NPN transistors exhibit a property known as complementary symmetry; that is, the polarity of a signal necessary to increase current in one type is the opposite of that necessary to increase current in the other. Thus in stage Q_1 a PNP transistor is used. Forward bias is provided conventionally, and the base is negative with respect to the emitter. The circuit parameters have been so chosen that the collector voltage of Q_1 is more positive than the emitter voltage of Q_2. Since the base of Q_2 is directly connected to the collector of Q_1, the condition for forward bias of Q_2 is met; that is, the emitter is negative with respect to its base. On the positive alternation of a signal in the input of Q_1, collector current in Q_1 decreases (because it is a PNP type), resulting in a negative-going signal at the collector of Q_1. This polarity of signal decreases the forward bias on Q_2 and causes a decrease in collector current in Q_2 and a positive-going signal at the collector of Q_2.

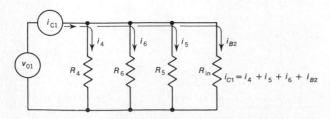

Fig. 20-3. AC collector load of Q_1.

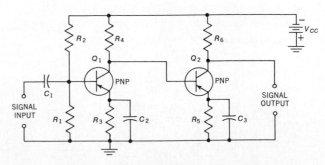

Fig. 20-5. Direct-coupled transistor amplifier.

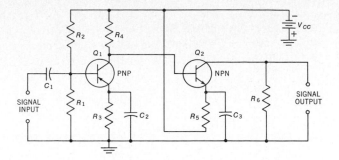

Fig. 20-6. Direct-coupled transistor amplifier using complementary symmetry of PNP and NPN transistors.

As for the phase of the signal, the input and output signals in the circuit of Fig. 20-6 are in phase. This is also true of the amplifier in Fig. 20-5.

Linear Operation

Two or any number of amplifiers operated in cascade may be considered as a single amplifier with a single input and a single output. Figure 20-6 shows such a two-stage amplifier. A cascaded amplifier of any number of stages is shown in block diagram form in Fig. 20-7.

When two or more amplifiers are operated in cascade, the characteristics of the total unit must conform to the requirements of the application. For example, if two or more transistor amplifiers in cascade constitute an audio amplifier, the amplifier must be operated over its linear characteristic for distortionless reproduction of sound.

An oscilloscope may be used to test linear operation. An audio sine-wave generator is used as the signal source. The output of the amplifier is monitored with an oscilloscope. To determine the range of linear operation, the input-signal level is increased from 0 to just below the point of distortion (clipping) in the output. The maximum generator signal which does not introduce distortion is thus determined and may be measured.

In this experiment you will study the operation of an *RC*-coupled, two-stage audio amplifier.

SUMMARY

1. Two amplifiers are said to operate in cascade when the output signal of the first serves as the input signal to the second.
2. Amplifiers may be coupled in a variety of ways. These include *RC* coupling, direct coupling, transformer coupling, and impedance coupling.

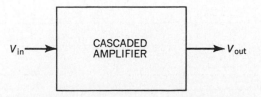

Fig. 20-7. Block diagram of a cascaded amplifier.

3. Transformers are used when it is necessary to match the output impedance of the first stage to the input impedance of the second, thus ensuring a maximum transfer of power from the first to the second stage. Figure 20-1 is an example of transformer coupling.
4. Figure 20-2 is an example of an *RC*-coupled amplifier.
5. In coupling two amplifiers the effect of the second-stage input on the ac load resistance of the first stage must be considered. Thus in Fig. 20-2, the *ac* load for Q_1 consists of the parallel combination of R_4, R_5, R_6, and R_{in} (for Q_2), as in Fig. 20-3. The effect of this "reduced" ac load in Q_1 is to reduce the level of ac output signal at the collector of Q_1.
6. Combination-coupling arrangements may be used between amplifiers. For example, in Fig. 20-4, Q_1 and Q_2 are *RC*-coupled, but the output of Q_2 is transformer-coupled to its load.
7. Direct-coupled stages (Fig. 20-5) using PNP to PNP or NPN to NPN transistors are frequently found in transistor circuitry. One advantage is the elimination of coupling components (e.g., transformers, capacitors). Another advantage is the improvement in frequency response.
8. Direct-coupled stages using PNP to NPN or NPN to PNP are also common (Fig. 20-6). This coupling arrangement makes use of the property called *complementary symmetry*. In complementary symmetry, the polarity of a signal necessary to increase current in one type (say PNP) is the opposite of that necessary to increase current in the other type (NPN).

SELF-TEST

Check your understanding by answering these questions.

1. Coupling between Q_1 and Q_2 (Fig. 20-1) is accomplished by means of _____ _____ .
2. In Fig. 20-2 _____ _____ couples the signal from the output of Q_1 to the input of Q_2.
3. In Fig. 20-2 the ac load in the output of Q_1 is the resistor R_4. _____ (*true*, *false*)
4. Direct-coupled transistor amplifiers may use only transistors. _____ (*true*, *false*)
5. In a two-stage complementary-symmetry amplifier, if the first stage uses a PNP transistor, the second stage must use a(n) _____ transistor.
6. A sine wave is injected into the input of Fig. 20-6. The signal in the output is a clipped wave. The cascade amplifier is said to be operating _____ (*linearly*, *nonlinearly*).
7. By _____ (*increasing*, *decreasing*) the drive signal at the input of the amplifier in question 6, the output waveform can be made sinusoidal, in a properly designed cascade audio amplifier.
8. An oscilloscope is used to check the waveforms at the _____ and _____ of an amplifier to determine if it is operating linearly.

MATERIALS REQUIRED

- Power supply: Variable regulated low-voltage dc source
- Equipment: Oscilloscope; EVM; 0–25-mA dc milliammeter; AF generator
- Resistors: 100-, 470-, 560-, 1000-Ω; 8.2-, 10-, 18-, 33-kΩ ½-W
- Capacitors: Two 25-μF 50-V; two 100-μF 50-V
- Semiconductors: 2N6004; 2N2102 (or equivalent)
- Miscellaneous: 500-Ω 2-W potentiometer; two SPST switches

PROCEDURE

1. Connect the circuit of Fig. 20-8. An AF signal generator set at 1000-Hz *minimum* output is used as the signal source. Power switch S_1 is *open*, S_2 is *closed*. The volume-control center arm is set at minimum, point B. M_1 is a milliammeter connected to measure total dc current in the circuit.

2. Set the output of the power supply at *9 V*, as measured with the EVM. Close S_1, applying power to the circuit. Monitor the dc supply and maintain its output at 9 V throughout the experiment.

3. Connect an oscilloscope calibrated for voltage measurement across the volume control. Set the output of the generator at 50 mV. Now connect the oscilloscope at the collector of Q_2, test point 5 (TP 5). Slowly turn up the volume control just below the point where the sine wave starts distorting. The volume control is now set for the maximum input signal which the circuit can handle without distortion. Leave the control set at this level.

NOTE: If the circuit is unstable (oscillates), bypass the collector of Q_1 with a 0.1-μF capacitor. This stops oscillation but also reduces high-frequency response.

4. Measure with the oscilloscope, and record in Table 20-1, the peak-to-peak signal voltage at every test point shown in Fig. 20-8.

NOTE: You probably will not be able to read, directly, the signal level at TP 1. You will determine the level at TP 1 indirectly in step 6.

Measure also, and record in Table 20-1, the dc voltage at every test point and the total current I_T as read on M_1.

5. *Open* S_2. Again measure and record the peak-to-peak signal level and dc voltage at every test point.

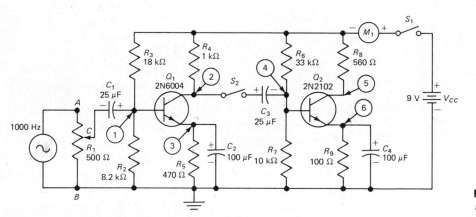

Fig. 20-8. Experimental *RC*-coupled audio amplifier.

TABLE 20-1. Cascade Amplifier Measurements

Test Point	Signal, V p-p		DC, V		I_T, mA	
	S_2 Closed	S_2 Open	S_2 Closed	S_2 Open	With Signal	Without Signal
2	48m	152m	4.8	4.8	12	12
3	1m	0 ^.	2.07	2.07		
4	48m	0	1.25	1.25	R_{CB}, Ω	R_{AB}, Ω
5	4.7	0	4.95	4.95		
6	17m	0	.71	.71	12	500
1	1.2m	2m	2.48	2.48		

Measuring Signal Voltage at TP 1

6. *Remove* the signal generator from the circuit. Do *not* vary the setting of the volume control. **Power off.** *Disconnect* the *center arm* of the volume control from the circuit. Measure the resistance from the center arm (point *C*) to ground (point *B*). Record it in Table 20-1. Measure also and record the total resistance of the control, points *A* to *B*. Compute and record the input signal in millivolts delivered at TP 1 by substituting the measured values R_{CB} and R_{AB} in the formula

$$v_{in} = \frac{R_{CB}}{R_{AB}} \times 50 \text{ mV} \quad \frac{12}{500} \times 50m = \boxed{1.2mV}$$

(You will recall that 50 mV is the signal applied across the volume control in steps 2 to 4.)

7. Reconnect the volume control in the circuit and set it for zero output. **Power on.** Measure the total current I_T, without signal. Record it in Table 20-1.

8. Remove the signal-generator leads from the volume control.

Extra Credit

9. Explain the procedure you would use to determine the range of linear operation at 1000 Hz of each amplifier stage in Fig. 20-8, when the amplifiers are not operated in cascade.

10. Following the detailed procedure in step 8, measure the linear range of operation of Q_1 and Q_2 and record these data.

QUESTIONS

1. What is the purpose of cascading amplifiers?
2. Is the single-stage amplifier as effective as the two-stage amplifier? Justify your answer by referring to the data.
3. Does the input circuit of Q_2 have any effect on the signal level at the collector of Q_1? Describe this effect, and explain it. Refer to your data.
4. Does the experimental procedure suggest a method for isolating the trouble to Q_1 or Q_2 in a dead amplifier such as that in Fig. 20-8? Explain the procedure.
5. Explain the level of signal at test points 3 and 6 in the experimental circuit.
6. What is the range of linear operation of the total amplifier? Refer to your data.

7. What is the voltage gain of the total amplifier? Show your computations. How is the total voltage gain related to the individual voltage gains of Q_1 and Q_2?
8. Comment on the total dc current in the circuit, with and without signal.
9. Is there any apparent change in dc voltage level at test points 4, 5, or 6 with S_2 open or closed? Why?
10. List three methods for coupling amplifiers in cascade.

Answers to Self-Test

1. transformer T_2
2. capacitor C_3
3. false
4. true
5. NPN
6. nonlinearly
7. decreasing
8. input; output

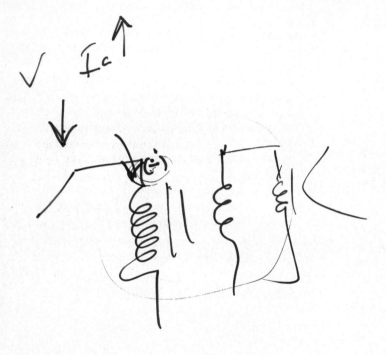

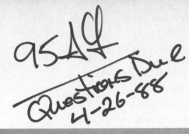

21
EXPERIMENT

THE LOUDSPEAKER AND THE CLASS A AUDIO POWER AMPLIFIER

OBJECTIVES

1. To become familiar with the loudspeaker and its operation
2. To become familiar with the output transformer and its operation
3. To connect and operate a speaker in the output stage of a class A audio power amplifier

INTRODUCTORY INFORMATION

The Dynamic Speaker

The loudspeaker, or simply speaker, changes electric audio energy into sound. The speaker is a *transducer*, one in a family of devices which converts energy from one form into another.

The speaker in greatest use today is the permanent-magnet (pm) dynamic speaker. The electrostatic speaker may be found as part of a speaker system in hi-fi installations. Electromagnetic dynamic speakers are no longer used, although they were very popular early in the development of audio systems. Dynamic speakers are similar in operation. They differ in the method used to obtain the stationary or static magnetic field.

Figure 21-1 shows the mechanical construction of a pm speaker. A permanent magnet concentrates a magnetic field at the pole pieces of a highly permeable housing. The pole pieces are very close together to obtain an intense magnetic field. A voice coil cemented to the speaker cone is freely suspended between the magnetic poles. A flexible membrane called the *spider* is attached to the voice-coil form and cemented to the speaker frame. The spider centers the voice-coil form between the speaker poles to keep it from rubbing against them. The flared end of the cone is flexibly attached to the speaker frame.

The permanent magnet of a pm speaker is made of a mixture of aluminum, nickel, and cobalt and is called an *alnico* magnet.

The audio-signal currents fed to the voice coil set up a moving magnetic field about the voice coil. This interacts with the fixed magnetic field and results in a vibratory motion of the voice coil and hence of the speaker cone. The rate of vibration of the speaker cone is determined by the frequency of the audio current. The amplitude of vibration, i.e., how far the cone moves, depends on the amplitude of audio current. The speaker cone moves the air mass surrounding it, producing sound.

Connections from the voice-coil ends are brought to insulated solder terminals on the speaker frame. When output transformers are used to match the impedance of the last audio amplifier to that of the speaker, the transformer may be mounted on the speaker frame, but more commonly it is mounted on the amplifier chassis. Leads from the secondary of the output transformer are then connected to the voice coil at the solder terminals on the frame.

A continuity check may be made on the voice coil of a dynamic speaker. The resistance is very low, usually 3 or 4 Ω, although in some solid-state amplifier-driven speakers it may be as high as 32 Ω. The continuous motion of the voice coil may cause the coil wire to break, resulting in an "open" voice coil. This would check as an infinite resistance.

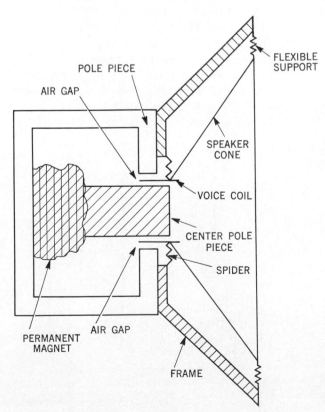

Fig. 21-1. Permanent-magnet loudspeaker construction.

125

In checking the resistance of a voice coil, in a *nondefective* speaker, an interesting effect may be noted. As the ohmmeter leads are placed across the coil (ohmmeter should be set on its lowest range), the cone moves, and a click is heard. When the leads are removed, the cone moves back to its rest position and again a click is heard. If the ohmmeter leads are reversed, the cone moves in the opposite direction.

A $1^1/_2$-V battery may be used instead of the ohmmeter for the click check, which is an indication that the speaker is operating. In some systems where two or more speakers are used, the battery test will also show if speakers are properly connected. The cone movement determines proper connection to the driving source. All cones should move in the same direction, using the same battery polarization.

Permanent-magnet speakers are rated for their size, power-handling ability, and voice-coil impedance. For hi-fi installations, the frequency response of the speaker is also specified. A 5-in speaker is one whose maximum cone diameter is 5 in. Large-speaker cones can move greater quantities of air, thus producing louder sound. More driving power is required, however. Hence large speakers have higher wattage ratings. Table-model receivers use small speakers. A 5-in speaker in a table-model radio can handle approximately 3 W of power.

Defective speakers should be replaced by identical speakers or by speakers with equivalent ratings.

Output Transformer

Dynamic speakers used in radio receivers are low-impedance devices. One way in which a low-impedance device is connected to the audio power amplifier from which it derives its driving power is through an output transformer. The high-impedance primary of the output transformer is connected in the high-impedance collector circuit of the transistor; the low-impedance ~~primary~~ SECONDARY is connected to the low-impedance voice coil of the speaker, as in Fig. 21-2. This ensures a maximum transfer of audio power from the output transistor to the speaker. The audio output transformer, wound on an iron core, is a voltage-stepdown, current-stepup device.

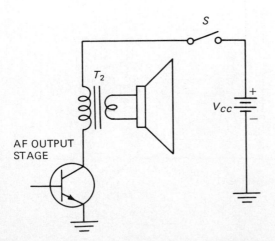

Fig. 21-2. Audio output transformer matches the high impedance of the collector circuit to the low impedance of the speaker.

TABLE 21-1. EIA Color Code for Audio Output Transformers

Transformer Winding Lead	Color
Primary:	
Collector	Blue
$V+$ or V_{CC}	Red
Collector (push-pull only)	Blue, brown, or white
Secondary:	
High side	Green or yellow
Low side	Black

The output impedance of a transistor varies with the transistor type and with the circuit arrangement. Hence, to match a particular output transistor stage to a given speaker requires an output transformer with the proper turns ratio. A variety of output transformers are manufactured to match the various combinations of output circuits and speakers. In replacing an output transformer in an audio system, an identical transformer or one with similar characteristics should be used.

The EIA color code for the leads of an audio output transformer is shown in Table 21-1. If there is any question about which winding is primary, which secondary, check the resistance of the windings. The resistance of the secondary winding is much lower than that of the primary. There is a wide range of resistances of the windings of output transformers. For example, one manufacturer lists the range of primary resistance for a particular series of output transformers from 600 to 2.3 Ω. The range of secondary resistance is listed as 25 to 0.25 Ω.

Note that in current audio-system design the output transformer is frequently eliminated in favor of direct speaker drive. In this type of system, the speaker acts as the load in the output of an emitter-follower power amplifier.

Class A Power Amplifier

Preceding experiments have dealt with small-signal high-gain voltage amplifiers. In an audio amplifier, the end product of the system is the sound produced by the speaker. As we have seen, sound is caused by the vibration of the cone of a speaker. To cause this vibration audio power is required. This power is the audio current and voltage delivered to the voice coil of a speaker by the final amplifier in the audio system. The last stage in an audio system is therefore called the *power stage* or *power amplifier*. It is usually a large-signal stage.

There are various classes of power amplifiers, including classes A, AB, B, and C. We shall be concerned here with the class A linear power amplifier. A class A amplifier is one whose emitter-to-base remains forward-biased during the entire input signal. If, in addition, the output signal is an exact reproduction of the input, the amplifier is said to be linear. A linear class A power amplifier therefore does not distort or change the shape of the signal waveform and delivers audio power to a load.

Figure 21-3 is the circuit diagram of a power amplifier Q, which receives its input signal from a preamplifier or driver. An output transformer T delivers audio power to the speaker.

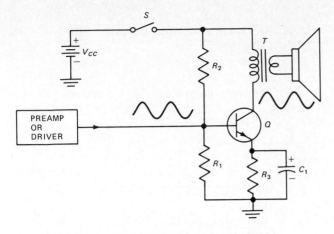

Fig. 21-3. Class A power output audio amplifier.

R_1 and R_2 provide voltage-divider forward bias to Q. R_3 is required for bias stabilization, and C_1 is a bypass capacitor for R_3, to prevent degeneration of the audio signal. Proper operation of this stage as a class A amplifier requires the bias to be properly set so that an undistorted input signal delivered by the driver will be amplified without distortion by Q. As previously noted, for maximum transfer of audio power to the speaker, transformer T must match the output impedance of Q to the input impedance of the speaker.

Figure 21-4 is the circuit which will be used in this experiment to demonstrate an RC-coupled two-stage audio amplifier, including a 2N2102 power output amplifier. The design of small-signal amplifier Q_1 is very similar to that of the first stage in Fig. 20-8 (Experiment 20). Q_2 in Fig. 21-4 is a class A small-signal power amplifier similar in circuit arrangement to that in Fig. 21-3.

Test points in Fig. 21-4 include the bases, emitters, and collectors of Q_1 and Q_2 and the secondary of output transformer T. Assume that the audio amplifier is operating normally and a sine-wave test signal is coupled to the input. Signal-tracing this circuit with an oscilloscope will reveal a sine wave at test points 1, 2, 4, 5, and across 7–8, the secondary of T_2. Points 1 to 6 are also dc test points. The dc

voltages measured at the bases (1, 4) and emitters (3, 6) should show that for these NPN transistors, each base is positive relative to its emitter (for forward bias of the base-to-emitter circuit). The dc voltage at the collector of Q_1 is approximately one-half of V_{CC}. The voltage at the collector of Q_2 is approximately equal to V_{CC}, because of the low primary resistance of T.

When a sine wave or other audio signal is applied to the volume control, and when the control is set for the proper input level, sound should be heard from the speaker.

SUMMARY

1. A speaker converts electric audio-frequency signals into sound.
2. The permanent-magnet (pm) speaker is used almost exclusively in sound-reproducing systems. It is called a pm dynamic speaker.
3. The parts of a pm speaker, shown in Fig. 21-1, include a frame with pole pieces at the end, a permanent alnico magnet, a center pole piece on which the voice coil rides, and a flexible cone to which the voice-coil assembly is physically attached.
4. Audio-frequency currents, flowing in the voice coil, set up changing magnetic fields which interact with the fixed magnetic field of the speaker magnet. This interaction causes the voice coil to move back and forth, driving the speaker cone back and forth. The rate at which the speaker cone moves is determined by the audio frequency of the signal. How far it moves is determined by the amplitude of the signal. The motion of the speaker cone creates sound waves.
5. The resistance of the voice coil of a speaker is usually very low, about 3 to 4 Ω.
6. PM speakers are rated for their physical size, power-handling capability, magnet size, voice-coil impedance, and frequency response.
7. An audio output transformer is a voltage-stepdown, current-stepup device. It is specifically designed to match a

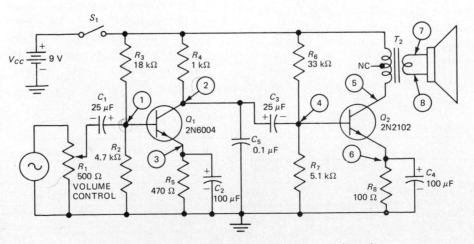

Fig. 21-4. Audio amplifier with output transformer and speaker.

certain output impedance to a specified speaker impedance. A defective output transformer in an audio amplifier must therefore be replaced by an identical unit or by one with identical characteristics.

8. The resistance in the primary winding of an output transformer is much greater than the resistance in the secondary winding.

9. A class A linear transistor amplifier is one whose emitter-to-base section remains forward-biased during the entire input signal and which amplifies the input signal without distortion.

10. A class A audio power amplifier delivers audio power to a load.

fer _____ _____ from the audio output stage to the _____ .

6. An audio output transformer is an impedance-_____ device.

7. The resistance of the voice coil of a speaker is normally about _____ to _____ Ω.

8. The resistance of the primary of an output transformer is _____ (*greater, lower*) than the resistance of the secondary.

9. A rough test to determine if a speaker is working is to place an ohmmeter across the _____ _____ .

10. In question 9, if the speaker is working, a _____ will be heard.

SELF-TEST

Check your understanding by answering these questions.

1. The purpose of a loudspeaker is to _____ electric energy into _____ .
2. Modern sound systems use _____ speakers.
3. The voice coil of a speaker carries the _____ frequency currents when the speaker is actuated.
4. The sound that issues from a loudspeaker is created by the movement of the _____ , which is driven by the motion of the _____ _____ .
5. The purpose of an audio output transformer is to trans-

MATERIALS REQUIRED

- Power supply: Variable regulated low-voltage dc source
- Equipment: Oscilloscope; EVM; AF generator
- Resistors: 100-, 470-, 1000-Ω; 4.7-, 5.1-, 18-, 33-kΩ, ½-W
- Capacitors: Two 25-μF 50-V; two 100-μF 50-V; 0.1-μF
- Semiconductors: 2N6004; 2N2102 (or equivalent)
- Miscellaneous: 500-Ω 2-W potentiometer; SPST switch; AF output transformer Kelvin type 175-45 or equivalent; pm loudspeaker (3.2-Ω); audio source (e.g., microphone or record player).

PROCEDURE

Speaker

CAUTION: *Carefully examine the pm speaker which you have received. The cone is fragile and can be easily damaged.* If possible, the speaker should be kept in a small enclosure to protect it.

1. Measure and record in Table 21-2 the resistance of the voice coil (listen for the click).
2. Observe the direction, forward or backward, in which the speaker cone moves when the ohmmeter (or $1\frac{1}{2}$-V battery) leads are placed across the voice coil. Repeat with ohmmeter leads reversed.

Output Transformer

3. Examine the output transformer you have received. Identify the primary leads. Observe that the primary is center-tapped. *The tap will not be connected in this experiment.* Measure and record in Table 21-2 the total resistance of the primary winding, the resistance from the center tap to each primary connection, and the resistance of the secondary.
4. Connect an AF sine-wave generator set at 1000 Hz across the primary. Connect a 3.2-Ω speaker across the secondary. With an oscilloscope measure the AF signal

TABLE 21-2. Speaker and Transformer Measurements

Speaker	Voice-coil Resistance, Ω 7.2 〜						
Transformer	Resistance, Ω				V p-p		Turns Ratio
	Primary Top to CT	Primary Bottom to CT	Full Primary	Secondary	v_p	v_s	a
	19.6	20.2	41	.3	4	330m	12.12

TABLE 21-3. Audio Amplifier Measurements

Test Point	1	2	3	4	5	6	7–8 (Voice Coil)
Signal volts (peak-to-peak)	2m	41m	2m	40m	3.3	8m	275m
DC volts	1.8	6.6	1.2	.98	9 v	.36 v	0 v

DC current in collector Q_2: (mA)	With signal	Without signal
	3.3	3.3

voltage across the primary. Adjust the output of the generator for 4 V p-p (v_p).

5. With the oscilloscope measure the peak-to-peak voltage across the secondary (v_s) and record it in Table 21-2.

6. Compute the turns ratio a of the transformer by substituting the measured values of v_p and v_s in the formula:

$$a = \frac{v_p}{v_s} = \frac{4}{330m} = 12.\overline{12}$$

Record in Table 21-2.

Audio Amplifier

7. Connect the circuit of Fig. 21-4. The AF generator, at 1000 Hz, is connected across the volume control. Set volume at 0. Switch S_1 is OFF.

8. Adjust the output of the power supply to 9 V. *Close S_1* (**power on**). Do you hear any sound?

9. With an oscilloscope observe the signal voltage at the collector of Q_2 to ground. Adjust the volume control for the *maximum undistorted* sine wave. What do you hear?

10. With the oscilloscope measure and record the peak-to-peak signal voltage at every test point listed in Table 21-3.

11. Measure also and record the dc voltage with respect to ground at the test points shown in Table 21-3.

12. Measure and record the dc current in the collector of Q_2, with and without signal.

13. Remove the AF signal generator and replace it with a microphone or record player. Adjust the volume control for comfortable sound level. With the oscilloscope signal-trace the amplifier and observe the waveforms at the base and collector of Q_1 and Q_2. Observe if the pattern is stable or if it varies with sound pitch and sound level.

QUESTIONS

1. From the test data in Table 21-3, calculate the turns ratio a of the output transformer.

2. How does the value a determined in question 1 compare with the value of a in Table 21-2? Explain any difference.

3. Using the measured voltage across the secondary (Table 21-3) and assuming the impedance of the speaker is 3.2 Ω, what is the audio current (i_s) in the voice coil? Show your calculations.

4. Using the value a determined in question 1 and the value i_s calculated in question 3, what is the audio current in the primary of the output transformer? Show your calculations.

5. From your measurements and calculations can you determine the impedance of the primary of the output transformer? Explain.

6. What is the voltage gain of Q_1 in the circuit? Show your calculations.

7. Which of your measurements indicates if the dc current rating of the primary of the transformer (rated at 4 mA) is exceeded?

8. How can you determine if there is an (a) open voice coil in a speaker, (b) open winding in a transformer?

9. Refer to the dc data in Table 21-3. Are the base-to-emitter sections of Q_1 and Q_2 forward-biased?

10. Refer to the signal data in Table 21-3. Are Q_1 and Q_2 operated as linear amplifiers in this experiment? If they were not, what would happen to the 1000-Hz test sine waveform in the circuit?

11. Is it possible to overdrive the amplifiers? How?

Answers to Self-Test

1. convert; sound
2. pm
3. audio
4. cone; voice coil
5. maximum power; speaker
6. matching
7. 3; 4
8. greater
9. voice coil
10. click

PUSH-PULL POWER AMPLIFIER

OBJECTIVES

1. To define class B operation
2. To connect and signal-trace a push-pull audio power amplifier

INTRODUCTORY INFORMATION

Class B Operation

High-power audio systems require more audio power than a single output stage can provide. One solution is to use two or more transistors connected in push-pull. Push-pull amplifier circuits are operated in either class B or class AB.

In the preceding experiment we learned that in class A operation the emitter-base section of the transistor is forward-biased for the entire period of the input signal. Current flows for 360° and the output is undistorted, as in Fig. 22-1a.

In class B operation, the emitter base of a transistor is forward-biased by the signal during one-half of the input signal and reverse-biased during the other half. The collector-current waveform of a class B circuit appears as in Fig. 22-1b. Note that current flows for approximately 180° and is cut off during the remainder of the cycle.

Class AB operation lies between class A and class B. Current flows for more than 180° but less than 360°, as in Fig. 22-1c. It is clear from the current waveforms that if a single transistor operating into a resistive load were biased class B

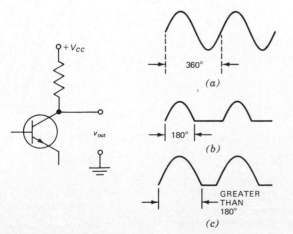

Fig. 22-1. Current waveforms. (a) Class A; (b) class B; (c) class AB.

or AB, signal distortion would occur. A push-pull circuit eliminates the distortion which would occur with a class B or AB single-transistor power amplifier.

Push-Pull Amplifier

When biased for class AB or class B operation, push-pull output amplifiers can handle a signal amplitude approximately twice as large as that of a conventional class A power amplifier. For this reason class B and class AB output stages can deliver more power than a single-ended class A stage.

A push-pull output stage is shown in Fig. 22-2. Q_2 and Q_3 are output transistors connected as common-emitter amplifiers in a balanced circuit. T_2, an input transformer, couples Q_1, called the *driver*, to stages Q_2 and Q_3.

The bases of Q_2 and Q_3 are connected to the ends of the CT secondary of T_2. Hence the bases receive two signals, equal in amplitude but 180° out of phase. For a sine-wave input, the base of Q_2 will be positive when the base of Q_3 is negative. When Q_2 swings negative, the base of Q_3 goes positive. As a result, when current flows in the collector of Q_2, no current flows in Q_3, and vice versa.

The collector currents of Q_2 and Q_3 flow in opposite directions through the primary of T_3, the output transformer. Suppose the magnetic field that the collector current of Q_2 sets up about the primary is expanding when collector current in Q_2 is increasing. At the same time collector current in Q_3 is decreasing, and the resulting magnetic field is moving in the same direction as the field arising from Q_2. Thus the fields aid and induce a larger emf in the secondary than either could alone.

Q_2 and Q_3 are two medium-power transistors, operated close to class B, with just enough forward bias (supplied by R_5 and the combination of R_6 and R_7) to cause a small collector idling current to flow in Q_2 and Q_3 which prevents crossover distortion. The waveform in Fig. 22-3 shows the distortion which occurs when transistors in a push-pull stage are dc-biased to cutoff. If there is cutoff (zero) bias on the base-emitter of a silcon transistor, no current flows in the transistor until the input-signal voltage rises to about 0.7 V. Therefore there is a period of time—when the signal is rising from 0 to 0.7 V—that the transistor will not conduct. If the transistors of a push-pull stage are biased at cutoff, the current waveform will appear as in Fig. 22-3. The gaps in current, periods t_1 and t_2, represent the time when the signal polarity is changing, and when the signal is crossing over to

131

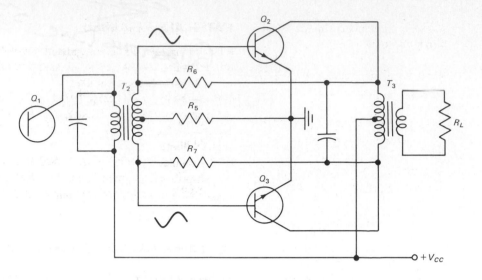

Fig. 22-2. Push-pull audio output amplifier.

activate one transistor while the other is turning off. No current flows during the times t_1 and t_2. The resulting current waveform is a distortion of the input waveform. Therefore, to eliminate crossover distortion, the transistors in a push-pull stage are not biased class B (at cutoff), but are forward-biased slightly so that a small collector current flows even in the absence of any signal. However, they are biased close enough to cutoff to designate their mode of operation as class B.

The emitters of Q_2 and Q_3 are returned directly to ground. The collectors are returned to $+V_{CC}$ through their load windings on the primary of T_3.

The low level of forward bias on Q_2 and Q_3 keeps the transistor *on* and its collector currents *very low* in the absence of signal. When a signal is applied, Q_2 and Q_3 conduct alternately on each half-cycle of the incoming signal. The average collector current is therefore low without signal and much higher in the presence of a signal.

The audio power delivered to a resistive load R may be computed, if the signal voltage across the load is known, by using the equation

$$P = \frac{V^2}{R} \qquad (22\text{-}1)$$

where V is the rms voltage, given in volts, and R is the load resistance, measured in ohms.

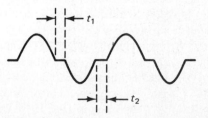

Fig. 22-3. Crossover distortion caused by biasing push-pull transistors at cutoff.

SUMMARY

1. High-power audio systems normally use push-pull outputs dc-biased either class B or AB.
2. In class B operation the emitter-base section of a transistor is dc-biased at cutoff. In class AB it is biased ON, but the bias is between class B and class A.
3. In class B operation the input signal causes collector current to flow for approximately 180°, during the interval when the *signal forward-biases* the base-to-emitter section.
4. In class AB operation collector current flows more than 180° but less than 360°.
5. Push-pull output amplifiers, rather than single-ended class A output amplifiers, are used in high-power audio systems because push-pull stages can deliver more power to a load than a single-ended stage, all other things being equal.
6. The power P delivered by the output stages to a resistive load R (Fig. 22-2) can be calculated using the equation

$$P = \frac{V^2}{R}$$

where V is the rms voltage measured across R, and R is the resistance in ohms of the load.

SELF-TEST

Check your understanding by answering these questions.

1. The output amplifiers in Fig. 22-2 are called _____-_____ amplifiers.
2. In a class A amplifier current flows during the _____ _____ of the input signal.
3. In a class B amplifier current flows for _____ ° of the input signal.

4. Push-pull amplifiers are not completely dc-biased at cutoff, but close to cutoff, in order to prevent _____ .

5. The average dc current in the collector circuits of a class B push-pull amplifier is _____ _____ with signal than without.

6. The voltage divider R_5–R_6 provides some _____ _____ for the base-emitter section of _____ .

7. The rms voltage across the secondary of T_3 (Fig. 22-2) is 1.5 V. $R_L = 3.2~\Omega$. The output power to the load is _____ W.

MATERIALS REQUIRED

- Power supply: Variable regulated low-voltage dc
- Equipment: Oscilloscope; AF sine-wave generator; EVM; 0–100-mA dc milliammeter
- Resistors: 5-, 680-, 820-Ω; 2.2-, 8.2-, two 18-kΩ, $^1/_2$-W
- Capacitors: 0.02-μF, 0.1-μF, 25-μF 50-V, 100-μF 50-V
- Semiconductors: 2N6004; two 2N2102 with heat sinks (or equivalent)
- Miscellaneous: SPST switch; 500-Ω 2-W potentiometer; push-pull transformers: input type Kelvin 155-08, output type Kelvin 175-45 or equivalents; 3.2-Ω speaker

PROCEDURE

1. Connect the circuit of Fig. 22-4. M_1 is a 0–100-mA meter connected to measure the combined collector currents of Q_2 and Q_3. Check all circuit connections. S is *open*. Output of power supply is set at 0.

2. *Close S*, applying power to the circuit. Gradually increase power-supply output (V_{CC}) until M_1 reads about 2 mA. V_{CC} will be about 9 V; Q_2 and Q_3 will be biased close to cutoff.

3. Connect an oscilloscope, set for proper viewing, across R_L, the output load resistor. Connect a sine-wave generator set at 1000 Hz across R_1, the volume control. Generator output is at minimum level; R_1 at maximum.

4. You will observe a sine wave on the oscilloscope as the signal-generator attenuator is advanced. If, however, the output of the audio generator at its minimum setting overloads the amplifier, the signal on the oscilloscope will be distorted. In that case reduce the input to the amplifier by readjusting R_1 until the distortion just disappears. Next, if there is crossover distortion, gradually increase V_{CC} until this distortion disappears (V_{CC} will be about 9 V).

5. If initially the sine wave on the oscilloscope was undistorted, increase the output of the generator for maximum undistorted output.

   The audio generator and R_1 are now set to give maximum undistorted output at 1000 Hz. *Do not change their settings.*

6. Measure, and record in Table 22-1, the dc voltage with respect to ground at each of the test points (1 to 8) in Fig. 22-4.

7. With an oscilloscope, observe and measure the peak-to-peak amplitude of audio signal at TP 1 to 9. Record these in Table 22-1.

8. Observe and record the collector current I_C in Q_2 and Q_3 with signal applied as in step 4. Measure and record in Table 22-1.

9. Reduce R_1 to minimum (zero signal). Observe and record I_C in stages Q_2 and Q_3 with no signal. Also measure and record their bias.

10. Replace R_L, the load resistor, with a 3.2-Ω speaker. Substitute an audio source, such as a record player or microphone, for the signal generator. Vary volume

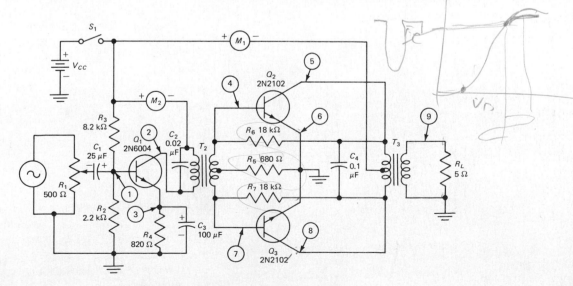

Fig. 22-4. Experimental push-pull amplifier.

TABLE 22-1. Transistor Push-pull Amplifier Checks

Test Point		DC (to Ground), V	Signal, V p-p
B	1	2.39	2m
C	2	10.5	7
E	3	1.76	2m
B	4	.39	.9
C	5	10.5	20
E	6	0	10m
B	7	.39	.9
C	8	10.5	20
E	9	0	3

	I_C, mA	Bias Q_2	Bias Q_3
With signal	44	22	23
Without signal	7.6m	3.1	1.9

control from minimum to maximum undistorted level. Observe if there is more audio power than with the single-ended class A amplifier in Experiment 21.

Extra Credit

11. Explain in detail how you would set oscilloscope triggering/sync to observe the phase of the signal waveforms at TP 2, 4, 7, 5, and 8 with relation to the signal at TP 1 Fig. 22-4).
12. Following the procedure of step 11, observe the waveforms at TP 1, 2, 4, 7, 5, and 8. Record these in a special table labeled "Phase Relations in a Driver and Push-pull Output Stage."

QUESTIONS

1. Explain the difference in operation between class A and class B push-pull amplifiers.
2. Refer to the data in Table 22-1. Are Q_2 and Q_3 in Fig. 22-4 operated class B? Explain why.
3. What are the advantages of operating an amplifier class B push-pull?
4. In Fig. 22-4 compute the power developed in R_L for maximum undistorted signal. Show computations.
5. Interpret the meaning of the signal-voltage readings (Table 22-1) at test points 3 and 6.
6. Explain the polarity of C_1 (Fig. 22-4).
7. Explain the difference, if any, in collector current in Fig. 22-4, with and without signal. Refer to your data in Table 22-1.

8. Do the results of your experiment indicate that the class B push-pull amplifier can deliver more power than the class A output amplifier? Explain why.
9. Explain the differences between overload distortion and crossover distortion. Illustrate with waveforms.
10. When does overload distortion occur? Crossover distortion?

Answers to Self-Test

1. push-pull
2. entire period
3. 180
4. crossover distortion
5. much greater
6. forward bias; Q_2
7. 0.703

COMPLEMENTARY-SYMMETRY PUSH-PULL AMPLIFIER

OBJECTIVES

1. To observe the operation of a class B push-pull complementary-symmetry audio amplifier
2. To measure the dc bias, current, and waveforms in this amplifier

INTRODUCTORY INFORMATION

Complementary Symmetry (Two Power Supplies)

The class B push-pull audio power amplifier in Experiment 22 required the use of an input and output transformer. These transformers are not needed in a class B push-pull audio power amplifier employing complementary symmetry.

The complementary-symmetry circuit uses two transistors with identical characteristics. However, one transistor is a PNP, the other an NPN. Figure 23-1a shows an idealized complementary-symmetry push-pull amplifier. Q_2 is an NPN, Q_3 a PNP transistor, each connected as an emitter follower, with the emitters connected together. The load R_L in the emitter circuit is common to Q_2 and Q_3. The collector of Q_2 goes to $+V_{CC}$, a positive supply. The collector of Q_3 receives its *dc* voltage from $-V_{CC}$, a negative supply. The bases of Q_2 and Q_3, connected together, receive the input signal from some external circuit.

Assume that the dc bias keeps Q_2 and Q_3 just cut off. Now observe what happens when a sine wave is applied to the input of this amplifier. During the positive alternation (1) the base of Q_2 is driven positive relative to its emitter, turning on Q_2, the NPN transistor. Q_2 remains on during this positive alternation (1), and the current waveform in Q_2 is shown in Fig. 23-1b. During the positive alternation, Q_3, a PNP transistor, remains reverse-biased and is cut off. During the negative alternation (2), Q_3 is forward-biased by the signal and turns on while Q_2 is cut off. Figure 23-1b shows that current in Q_3 is opposite in direction to current in Q_2. This is so because Q_2 is an NPN, Q_3 a PNP transistor. The arrows in Fig. 23-1a show the direction of electron current in the external circuit of Q_2, Q_3, and the load R_L. The voltage developed across R_L is a sine wave, like the input. Since the actions of Q_2 and Q_3 complement each other and the circuit is symmetrical, the arrangement in Fig. 23-1a is called *complementary symmetry*. Complementary-symmetry circuits can be common emitter as well as common collector, and they may be operated class A as well as class B.

Complementary-symmetry amplifiers require very careful design to prevent thermal runaway and destruction of the power transistors. Unbalance or leakage in power transistors can cause multiple failures. Diode stabilization is frequently employed as thermal compensation. These conditions are mainly found in direct coupling from driver to output stages.

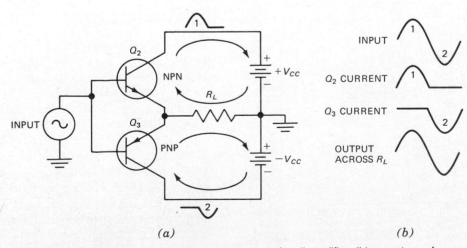

Fig. 23-1. *(a)* Idealized complementary-symmetry push-pull amplifier; *(b)* current waveforms.

Complementary Symmetry (One Power Supply)

The circuit in Fig. 23-1 utilized two power supplies, equal in voltage but opposite in polarity, with a common ground. The circuit of Fig. 23-2 uses a single supply for a push-pull complementary-symmetry amplifier. The symmetry of the circuit is maintained by two equal voltage dividers, R_1 and R_2. The top divider provides just enough forward bias for Q_2, and the bottom divider enough for Q_3, to give each transistor a low idling current to eliminate crossover distortion.

The arrows in Fig. 23-2 show the path for this idling current in the external circuit of Q_2 and Q_3. Notice that the path for idling current includes Q_3, Q_2, and the power supply. Since Q_2 and Q_3 are assumed to have identical characteristics, point D is the dc voltage midpoint of the circuit, that is, $V_{AD} = V_{DG} = V_{CC}/2$. Similarly C is the dc voltage midpoint in the voltage divider between points A and G, assuming that both R_1s are equal in resistance, both R_2s are equal in resistance, and Q_2 and Q_3 have identical characteristics. Therefore a dc voltmeter connected between points C and D would measure 0 V because C and D are at the same dc potential with respect to ground (G). By voltage-divider action of top resistors R_1 and R_2, point B_1 is more positive than point C (and D). Hence the base of Q_2 (B_1) is positive relative to its emitter (D), and this provides forward bias for Q_2 (NPN) to develop a low idling current. Similarly point B_2 is negative with respect to C (and D). Hence the base of Q_3 is negative relative to its emitter, providing forward bias for Q_3 (PNP).

The ratio of R_1 to R_2 is critical in setting the forward bias for a low idling current in Q_2 and Q_3.

The input signal is coupled by C_1 to the bases of Q_2 and Q_3 and is the same signal on each base. Q_2 is forward-biased by the positive alternation of the signal. The resulting current flow develops the positive alternation across R_L. Q_3 is forward-biased by the negative alternation of the signal and develops the negative alternation across R_L. The operation of Fig. 23-2 is, in all other respects, the same as that of Fig. 23-1 which uses two power supplies.

Complementary-symmetry push-pull class B amplifiers are frequently used in the output stage of high-power audio amplifiers. In such a system the speaker voice coil acts as the load and takes the place of R_L. The circuit may be designed to eliminate capacitor C_2; the load is then directly connected between emitter and ground. This is possible because of the low output impedance of the emitter-follower design.

SUMMARY

1. A complementary-symmetry amplifier uses two transistors, a PNP and an NPN, with identical electrical characteristics.
2. A completely symmetrical circuit arrangement biases both transistors equally, so that each permits the same current to flow, in the absence of an input signal.
3. A complementary-symmetry push-pull amplifier may use one or two separate power supplies. It may be biased class B, as in Figs. 23-1 and 23-2, or it may be operated class A.
4. The bias circuit of a class B push-pull complementary-symmetry circuit must provide a small idling current, in the absence of an input signal, to prevent crossover distortion.
5. The design of the complementary-symmetry audio output stage may permit direct connection of the speaker voice coil as the load in the output of the amplifier. This arrangement is possible where the amplifiers are designed as emitter-follower circuits.
6. Use of a complementary-symmetry class B audio output stage permits the advantages of class B operation while eliminating costly input and output transformers.

SELF-TEST

Check your understanding by answering these questions.

1. In the idealized class B circuit of Fig. 23-1, good design requires that Q_2 and Q_3 both be biased at cutoff in the absence of a signal. _____ (*true, false*)
2. A dc voltmeter connected across CD in Fig. 23-2 will measure _____ V, theoretically.
3. The dc voltage across AC in Fig. 23-2 equals _____ , if the circuit is completely balanced.
4. The _____ alternation of the input signal causes current to flow in Q_3.
5. Q_2 and Q_3 in Fig. 23-2 are connected as _____-_____ amplifiers.
6. In the absence of signal (Fig. 23-2) the base of Q_2 is _____ (*positive, negative*) with respect to the emitter.

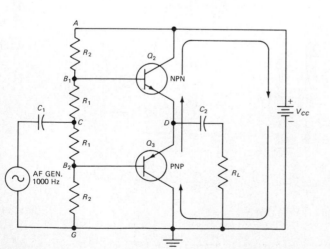

Fig. 23-2. Complementary-symmetry push-pull class B amplifier using one power supply.

7. In Fig. 23-2 the dc voltage across Q_2, V_{AD}, is _____ (*equal to, smaller than, greater than*) the voltage across Q_3, V_{DG}.

8. In the circuit of Fig. 23-2, transistors Q_2 and Q_3 may be interchanged without affecting circuit performance. _____ (*true, false*)

MATERIALS REQUIRED

- Power supply: Two variable regulated low-voltage dc sources

- Equipment: Oscilloscope; EVM; AF sine-wave generator; 0–100-mA milliammeter
- Resistors: 5-, two 100-, 470-, three 1000-, 1200-Ω; 4.7-, 10-kΩ ½-W
- Capacitors: 0.002-μF, 25-μF 50-V; four 100-μF 50-V
- Transistors: 2N2102 with heat sink; 2N4036 with heat sink; 2N6004 (or equivalents)
- Miscellaneous: Two SPST switches; 3.2-Ω speaker; 500-Ω 2-W potentiometer; audio source (record player or microphone)

PROCEDURE

Single Power Supply Circuit

1. Connect the circuit of Fig. 23-3. S_1 is *open*. Volume control is set for minimum output. Sine-wave signal generator, at 1000 Hz, is at minimum output. V_{CC} is at 0 V.

2. *Close S_1*. **Power on**. *Gradually* increase the output of V_{CC} until M_1 measures about 2 to 3 mA. (This is the idling current for Q_2 and Q_3.) Measure V_{CC}. It should be about 15 V.

3. Connect an oscilloscope across R_L, TP 8 to ground. Increase volume control output (and sine-wave generator level, if necessary) for the maximum undistorted signal across R_L.

NOTE: If the output signal shows evidence of crossover distortion, gradually increase V_{CC} until the distortion disappears. Recheck *idling current* in Q_2 and Q_3, *without signal*. It must not exceed 5 mA.

4. With maximum signal measure and record in Table 23-1 V_{CC}, and the dc voltages with respect to ground at TP 1 to 8. Measure also and record the current in Q_2 and Q_3.

5. With the oscilloscope, measure and record in Table 23-1 the peak-to-peak signal voltage at TP 1 to 8. Observe that Q_2 and Q_3 current is increasing. Why?

6. Remove the input signal. Again measure, without signal, and record in Table 23-1 the dc voltages at TP 1 to 8. Measure and record the idling current in Q_2 and Q_3. NOTE: If current is higher than original idling current, let current stabilize for about 3 to 5 minutes (min) without signal, or by shutting off power. Then measure idling current.

7. Replace R_L with a 3.2-Ω speaker. Replace the generator with an audio source (record player or microphone). Adjust volume control for comfortable level. Check the quality of sound.

8. **Power off.**

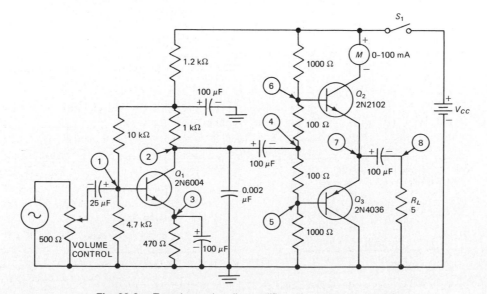

Fig. 23-3. Experimental audio amplifier using one power supply.

TABLE 23-1. Measurements in Audio Amplifier—One Power Supply

| Test Point | DC, V | | Signal, V p-p | Current, mA, in Q_2 and Q_3 |
	With Signal	Without Signal		
V_{CC}			X	
1				
2				With Signal
3				
4				
5				
6				Without Signal—Idling
7				
8				

Two Power Supply Circuit

9. Modify the circuit, as in Fig. 23-4. Two power supplies are used.

10. S_1 and S_2 are *open*. **Power off.** $+V_{CC}$ and $-V_{CC}$ at 0 V. Volume control at minimum output. Generator, at 1000 Hz, is at minimum output.

11. *Close* S_1 and S_2. **Power on.** NOTE: In the power-supply adjustments which follow, it will be necessary to maintain the two supplies at the same voltage level. If a dual power supply with a single adjustable control is available, $+V_{CC}$ and $-V_{CC}$ will be automatically equal. If separate supplies are used, the voltages must be checked with voltmeters to see that they are equal.

12. Gradually increase the output of the two supplies, maintaining $+V_{CC}$ and $-V_{CC}$ equal in output, until the idling current as measured by M_1 is about 2 to 3 mA. Measure $+V_{CC}$ and $-V_{CC}$. They should be about 7.5 V each.

13. Repeat step 3.

14. Repeat step 4. Measure and record in Table 23-2 both $+V_{CC}$ and $-V_{CC}$ with respect to ground.

15. Repeat step 5.

16. Repeat step 6.

17. Repeat step 7.

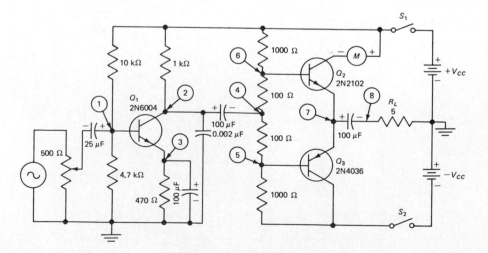

Fig. 23-4. Experimental audio amplifier using two power supplies.

TABLE 23-2. Measurements in Audio Amplifier—Two Power Supplies

| Test Point | DC, V | | Signal, V p-p | Current, mA, in Q_2 and Q_3 |
	With Signal	Without Signal		
$+V_{CC}$			X	
$-V_{CC}$			X	
1				With Signal
2				
3				
4				
5				
6				Without Signal
7				
8				

QUESTIONS

Refer specifically to the data in Tables 23-1 and 23-2, whenever possible, in answering these questions.

1. In the experimental amplifier of Fig. 23-3, without signal, what was the base-emitter bias on Q_2? Was this forward or reverse bias? Explain.
2. In Fig. 23-3, without signal, what was the base-emitter bias on Q_3? Was this forward or reverse bias? Explain.
3. In Fig. 23-3, do your measurements confirm that the dc voltage, without signal at TP 4 and 7 with respect to ground, equals $V_{CC}/2$? If not, why not? (HINT: Was the circuit absolutely balanced?)
4. What was the voltage gain of Q_1 in Fig. 23-3? Show your computation.
5. In Fig. 23-3, compare the dc collector current in Q_2 and Q_3 with and without signal and explain the difference, if any.

6. In Fig. 23-2, why is the adjustment of V_{CC} so critical in determining the idling current?
7. How does the experimental amplifier in Fig. 23-4 differ from that in Fig. 23-3?
8. Compare the effectiveness of the amplifiers in Figs. 23-3 and 23-4 in delivering a signal to the load.
9. Would driver Q_1 in Fig. 23-4 operate with S_1 ON and S_2 OFF? Explain.
10. Why do transistors Q_2 and Q_3 require heat sinks, when operated with signal, under full load?

Answers to Self-Test

1. false
2. 0
3. $V_{CC}/2$
4. negative

5. emitter-follower (or common-collector)
6. positive
7. equal to
8. false

FREQUENCY RESPONSE OF
AN AUDIO AMPLIFIER

OBJECTIVES

1. To measure the frequency response of an audio amplifier
2. To measure the effect of negative feedback on the frequency response of an audio amplifier

INTRODUCTORY INFORMATION

Frequency Response

The range of sound frequencies that the average person can hear is 30 to 15,000 Hz (approx). An audio amplifier should be able to amplify *equally* the signal voltages in this range. Because of price and other practical considerations, many audio amplifiers do not permit the entire range of audio frequencies to pass equally. Low frequencies (up to 500 Hz) and high frequencies (beyond 5000 Hz) may be attenuated.

An important characteristic of an audio amplifier, then, is the output voltage that it provides at each of the frequencies within the audio range for a given level of input signal. This characteristic can be determined by using an audio sine-wave generator as the signal source and an oscilloscope for observing and measuring the input and output signal. The gain is then computed at each frequency.

One method for checking the gain of an amplifier at a specific frequency is to introduce a measured signal voltage at that frequency into the amplifier and measure the output signal voltage.

$$\frac{v_{out}}{v_{in}} = \text{gain} \qquad (24\text{-}1)$$

The frequency response of an amplifier is a plot of gain or output versus frequency at many frequency points. When these points are connected smoothly, we have a frequency response curve.

Figure 24-1 represents an amplifier (which may consist of one or more stages), the signal source v_{in}, and the output signal voltage v_{out} (which is coupled to the vertical input terminals of an oscilloscope). If the output of the audio generator were constant over its entire frequency range, it would be possible to set v_{in} at some value, say 1 V p-p, and simply monitor the output v_{out} with an oscilloscope. However, the output of the audio oscillator may vary under load with frequency. Hence it is necessary to measure the genera-

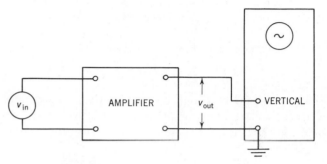

Fig. 24-1. Measuring output of amplifier.

tor output under load for every test frequency. This is the value of v_{in} used in the gain formula.

Another method that can be used is to maintain the output of the audio generator at a fixed value of v_{in}, say 1 V p-p. This again requires the use of the oscilloscope for monitoring v_{in} for every frequency setting of the generator and adjusting the output of the generator to keep v_{in} at the desired level.

If a 1-V p-p signal is used for v_{in}, the value of v_{out} is also the gain of the amplifier because

$$\text{Gain} = \frac{v_{out}}{v_{in}} = \frac{v_{out}}{1} = v_{out} \qquad (24\text{-}2)$$

In plotting the response curve of an audio amplifier on graph paper, values of v_{out} are plotted vertically as the ordinates of the graph. The corresponding frequency settings are the abscissas.

Extending the Frequency Response of a Transistor Amplifier by Negative Feedback

Degenerative, or negative, feedback is employed in transistor amplifiers to stabilize the amplifier and extend its frequency response. *These desirable characteristics, however, are accomplished at the cost of a reduction in amplifier gain.*

One method of introducing degenerative feedback is to leave unbypassed the emitter resistor in a grounded-emitter amplifier (Fig. 24-2). The signal voltage developed across the emitter resistor is in phase with the applied signal voltage in the base. The signal which the amplifier "sees" is the difference between the input signal and the signal developed across the emitter resistor. The unbypassed emitter results in increased input and output impedance and in lower stage gain.

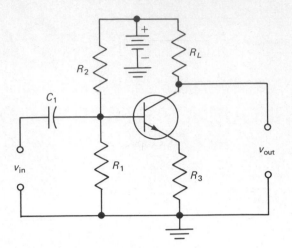

Fig. 24-2. Unbypassed emitter resistor provides negative feedback.

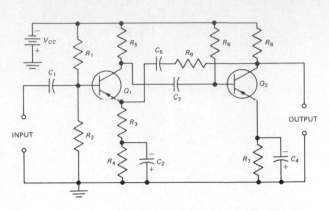

Fig. 24-4. Amplifier using two-stage degenerative feedback.

Figure 24-3 is a variation of Fig. 24-2. Here R_3, part of the emitter circuit resistance, is unbypassed to provide degenerative feedback. DC bias stabilization is provided by R_4 in parallel with C_4.

Many feedback combinations provide degenerative feedback. Another method is shown in Fig. 24-4. This is the circuit of a two-stage amplifier in which negative feedback is accomplished by coupling a portion of the output signal from the collector of stage Q_2 to the emitter of Q_1. The signal fed back is in phase with the degenerative signal developed across the unbypassed emitter resistor R_3. The amplitude of signal fed back to the emitter is determined by the ac divider consisting of R_9, C_5, and R_3. Hence the total input signal that the base-to-emitter circuit "sees" is the difference between the input signal v_{in} and the sum of the in-phase feedback voltages across R_3.

Figure 24-5 is another two-stage transistor amplifier employing degenerative feedback. In this push-pull amplifier, feedback is taken from the secondary of T_2, the output transformer, and coupled by R_8 and C_5 to the base of Q_4. The amplitude of signal feedback is determined by the ac divider

consisting of R_8, C_5, and the input impedance of Q_4. The proper terminal of the secondary of T_2 must be selected for feeding back a signal 180° out of phase with the input signal which appears at the base of Q_4. This type of feedback is frequency-selective and may be used to accentuate some desired frequencies in the audio spectrum.

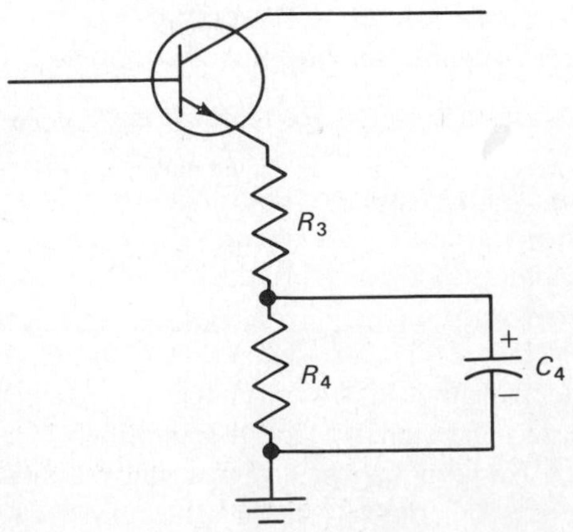

SUMMARY

1. The frequency response of an audio amplifier describes how faithfully the amplifier reproduces the range of audio frequencies.
2. Frequency response is the characteristic which indicates what output voltage the amplifier develops for the frequencies within the audio range, for a fixed level of input signal.
3. An audio sine-wave generator is used as a signal source, and an oscilloscope acts to measure the input and output voltages.
4. The frequency response is a graph of amplifier gain versus frequency, where gain is v_{out}/v_{in}.
5. *If the input signal to the amplifier is kept constant*, then the frequency response can also be a graph of v_{out} versus frequency.
6. In plotting the response curve, values of v_{out} or gain are the ordinates of the graph and values of frequency are the abscissas of the graph.
7. The frequency response may be improved or extended by introducing negative feedback.
8. Negative feedback extends the frequency response but reduces the gain of an amplifier.

SELF-TEST

Check your understanding by answering these questions.

1. The block diagram in Fig. 24-1 illustrates one method of connecting test equipment to check the ___FREQENCY___ ___RESPONSE___ of an amplifier.
2. If a positive-going signal is applied to the base of the amplifier in Fig. 24-2, the signal developed across R_3 will be ___POSITIVE___.

Fig. 24-3. Combination of dc stabilization circuit and degenerative feedback.

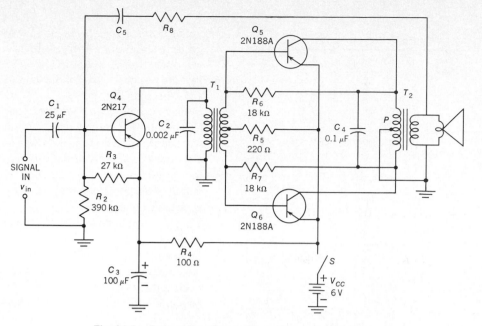

Fig. 24-5. Push-pull audio amplifier with negative feedback.

3. In Figs. 24-2 and 24-3, the signal voltage which the transistors amplify is the difference in amplitude between the signal on the _____ BASE _____ and the signal developed across _____ R₃ _____ .

4. In Fig. 24-4, to provide degenerative (negative) feedback, R_9 and C_5 must couple a signal to R_7 which has the *same* phase as the signal developed across R_3. _____ (*true*, false)

5. The test instruments used in checking the frequency response of an amplifier are an _____ OSCIL, _____ and _____ SINE - WAVE GEN. _____ .

6. In a frequency response curve, the horizontal axis is _____ FREQ. _____ , the vertical axis is _____ SIG OUT _____ or _____ GAIN _____ .

7. The range of frequencies most people can hear is about _____ 20 30 _____ to _____ 20K 15Ω _____ Hz.

8. It is possible to use a sine-wave generator as the signal source in making a frequency response check of an amplifier, even though the generator output is *not* constant over the range of test frequencies. _____ (*true*, false)

MATERIALS REQUIRED

■ Power supply: Variable regulated low-voltage dc source
■ Equipment: Oscilloscope; AF sine-wave generator; EVM
■ Resistors: 5-, 100-, 150-, 330-, 560-Ω; 1-, 4.7-, 5.1-, 8.2-, 10-, 18-, 33-kΩ ½-W
■ Capacitors: Two 25-μF 50-V; two 100-μF 50-V; 0.0047-μF, 0.047-μF
■ Transistors: 2N2102; 2N6004 (or equivalent)
■ Miscellaneous: 500-Ω 2-W potentiometer; SPST switch; output transformer—Kelvin 175-45 (or equivalent)

PROCEDURE

1. Connect the circuit of Fig. 24-6. S_1 is *open*. Set the output of an AF sine-wave generator at 0 and the volume control for minimum; that is, arm C is set at point B. Connect an oscilloscope from the collector of Q_2 to ground.

2. *Close* S_1. **Power on.** *Close* S_2. Tune the frequency of the generator to 1000 Hz, and increase the output until a 60-mV sine wave v_{AB1} is applied across AB. Now gradually increase the volume control setting until a 4-V p-p output sine wave $v_{1\text{out}}$ is measured at the collector of Q_2. Do not change the setting during the remainder of this experiment.

NOTE: If a 60-mV signal is maintained across AB and if the setting of the volume control is not changed, the signal measured between point C and ground will remain constant within the range of audio frequencies delivered by the generator.

3. Set the frequency of the generator at 30 Hz. Measure the signal across AB, and readjust generator output, if necessary, until the signal level reaches 60 mV as in step 2.

4. Measure and record the output signal voltage $v_{1\text{out}}$ from the collector of Q_2 to ground.

5. Repeat steps 3 and 4 for every frequency setting of the generator as shown in Table 24-1.

TABLE 24-1. Frequency-response Data

Frequency, Hz	v_{AB1}: 60 mV v_{1out}, p-p No Degeneration	v_{AB2} .8v v_{2out}, p-p With Degeneration
30	.14	.47
40	.21	.64
60	.4	.85
100	.42	1.3
200	2.2	1.8
400	3.4	1.9
600	3.7	2
1000	4 V	2 V
2000	4	2.1
3000	4	2.1
4000	4	2.1
5000	3.9	2.1
6000	3.9	2
8000	3.9	2
10,000	3.8	2
12,000	3.8	2
15,000	3.8	2
Step 5	R_{CB}, Ω 8.1	R_{AB}, Ω
20K	3.7	2
50K	3.1	1.75
100K	2.5	1.35
500K	.8	
1M	.27	
10M	.02	

Power off. (*Open S_1.*) *Open S_2.* Measure and record the resistance from C to B. Measure and record the resistance from A to B.

6. Plot the response curve of the amplifier on graph paper, and label it 1.

7. Connect C_2, the 25-μF emitter bypass capacitor, across R_6, leaving R_5 unbypassed, introducing degenerative feedback into the first stage. **Power on.** *Close S_1 and S_2. Do not vary volume control from its original setting.* Set the generator frequency at 1000 Hz, and adjust the output of the generator until a 2-V p-p output sine wave is observed at collector of Q_2. Measure the signal voltage V_{AB2} as it appears across AB and record it in Table 24-1.

8. Maintaining the output of the generator at V_{AB2}, measure the output v_{2out} at each frequency shown in Table 24-1. Record the data in the appropriate column. Graph the response curve 2 on the same axes as graph 1.

Extra Credit

9. Explain *in detail* how you would measure the frequency response of the amplifier in Fig. 24-7 (**a**) without degeneration, (**b**) with degeneration. *Where* would you measure the output?

10. Following your detailed procedure, measure the frequency response of the amplifier under conditions **a** and **b** specified in step 9. Record the data in a special table. Graph the data on the same axes used for graphs 1 and 2. Label the graphs 3 and 4, respectively.

11. Measure the effects on frequency response of removing C_5 from the circuit.

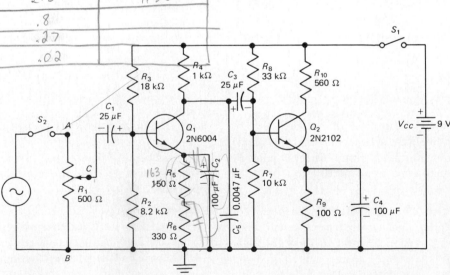

Fig. 24-6. Circuit used in determining frequency response of two-stage audio amplifier.

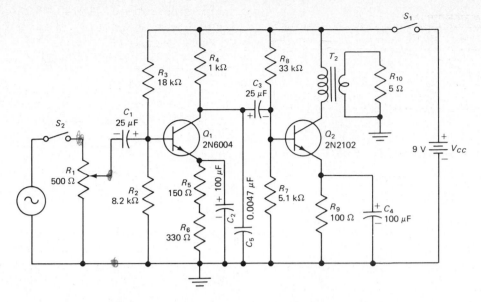

Fig. 24-7. Circuit used in extra-credit experimental procedure.

QUESTIONS

1. What is meant by the "frequency response" of an amplifier?
2. How do coupling capacitors C_1 and C_3 in Fig. 24-6 affect the frequency response of the amplifier? Why?
3. Why is it necessary to maintain a constant-level input signal across the volume control?
4. Compute the total gain of the amplifier at 1000 Hz (*a*) without degeneration, (*b*) with degeneration. Show computations. Comment on any difference in gain.
5. Compare the frequency response of the amplifier with and without degeneration. Refer to your graphs.

Extra Credit

6. In Fig. 24-7 list the *linear* components, that is, those parts which respond equally to the whole range of audio frequencies.

7. In Fig. 24-7 list the *nonlinear* components, that is, those parts which *do not* respond equally to the whole range of audio frequencies.
8. What is the effect, if any, on the frequency response of removing C_5 from the circuit? Why?

Answers to Self-Test

1. frequency response
2. positive
3. base; R_3
4. true
5. oscilloscope (or EVM); sine-wave generator
6. frequency; signal output; gain
7. 30; 15,000
8. true

TROUBLESHOOTING AN AUDIO AMPLIFIER

OBJECTIVES

1. To isolate a defective stage in an audio amplifier by signal-tracing
2. To isolate a defective component by voltage and resistance checks

INTRODUCTORY INFORMATION

Logical Troubleshooting Procedure

Locating a trouble in a defective electronic device such as a multistage audio amplifier requires a logical, systematic procedure. An effective first step in such a procedure is to assume that there is just *one* defect causing the trouble. The technician tries to locate that defect in the simplest and fastest way. When the trouble has been found and corrected, amplifier operation should be restored to normal. If it is not, then the technician must seek a second possible source of trouble; then if necessary, a third, etc. The steps in this logical procedure are:

1. *Listen to the amplifier's performance.* For this, an audio source such as an AF sine-wave generator or a record player is needed. The symptoms, coupled with past knowledge and experience, may lead to a rapid diagnosis.
2. *Inspect the audio amplifier for obvious defects*, such as open connections, improper connections, burned resistors, broken parts, or overheated parts. Burned-out resistors and transformers have a characteristic odor. *Determine the reason for the failure of a defective part before replacing it.*
3. *Measure the supply voltage* to determine if that is the source of trouble.
4. *Isolate the defective stage* by signal-tracing the amplifier. When the defective stage is found, the trouble must be in one or more of the components in that stage.
5. *Isolate the defective part.* DC voltage checks are made in the defective stage and compared with the rated values. If no ratings are available, the technician should be able to determine approximate values from knowledge of the circuit. Refer to Experiment 17 for a discussion of voltages which may be expected in an amplifier. Differences between rated and measured values must be analyzed for inferences as to the defective part. The suspected part can

then be tested or a known good part substituted to verify if the trouble has been located. DC voltage checks are supplemented by resistance checks in the suspected circuit. See Experiment 17 for a discussion of resistance norms.
6. *Correct the defect.* The replacement part must meet the electrical and physical specifications of the original. In amplifier servicing, use the manufacturer's recommended replacement parts.

Isolating a Defective Stage by Signal-Tracing

Figure 25-1 is a block diagram of a three-stage audio amplifier which includes a predriver, driver, and complementary symmetry output amplifier. The dynamic test points for signal-tracing this amplifier are numbered 1 to 7. The detailed block diagram does not show the circuitry, but it does identify the transistor elements where the test points are. Assume that the amplifier is dead (no sound is heard in the speaker) and that we wish to isolate the defective stage by signal-tracing. The procedure which follows describes the process.

1. Connect an AF generator to the input of the amplifier. A 1000-Hz signal set at a moderate level acts as the signal source.
2. With an oscilloscope, trace the progress of the signal from the input (base of Q_1) to the output (voice coil of the speaker).
3. If the amplifier were operating normally, we would expect to find the 1000-Hz sine wave at each test point 1 to 7. The approximate levels (amplitudes) of the sine wave are:

TP 1   Low level, determined by generator output.
TP 2   Higher peak-to-peak signal than at TP 1 (because of gain of Q_1).
TP 3   The signals at TP 2 and TP 3 should be approximately the same level.
TP 4   Higher peak-to-peak signal than at TP 3.
TP 5   Same level signal as at TP 4.
TP 6   Same level signal as at TP 4.
TP 7   We have assumed a complementary-symmetry, emitter-follower output stage. In this type of circuit the output sine wave at the emitter will be somewhat lower in level than the signal at TP 5 and 6.

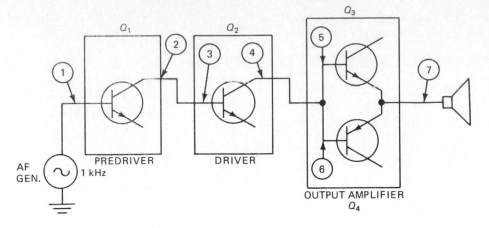

Fig. 25-1. Detailed block diagram of three-stage audio amplifier showing dynamic signal-tracing test points.

Circuit variations from the amplifier in the block diagram must be treated individually, depending on the actual circuit used. For this reason, the technician should examine the diagram of the circuit being signal-traced to identify the dynamic test points and know what to expect at each.

4. If the amplifier were defective (completely dead), it is clear that the signal would disappear at some point in the circuit. Suppose, for example, that the signal is normal up to TP 3 but disappears at TP 4. A logical inference is that either the driver stage (Q_2) is defective or a defective output stage (Q_3–Q_4) is loading Q_2 and preventing it from functioning properly. If possible, unload Q_3–Q_4 and check the signal again at TP 4. If it is *now* normal, the trouble is in Q_3–Q_4. But if there is still no signal at TP 4, the trouble is in Q_2. In this *signal-tracing* process, the trouble is isolated to the defective stage. Then it is possible to examine the suspected stage to determine the defect.

Isolating the Defective Part by Measurement

Assume that trouble has been isolated to the driver Q_1 in Fig. 25-2. A 60-mV sine wave is measured at the base TP 1, but there is no signal at the collector, TP 2. A dc voltage check at the base, emitter, and collector of Q_1 may suggest the trouble.

Normal dc voltages at the test points are: collector $+6.6$ V, base $+1.8$ V, emitter $+1.2$ V. Assume the measured voltages in the defective amplifier are: collector $+9$ V, base $+1.9$ V, emitter 0 V. The transistor is obviously not conducting, since there is no voltage drop across the emitter or collector resistors. The measurements suggest a defective transistor. Replacing Q_1 with a good 2N6004 transistor should clear up the trouble.

A further check of the transistor is possible, before replacement, with an in-circuit transistor checker or with an in-circuit curve tracer.

In the trouble just discussed it was very evident that the

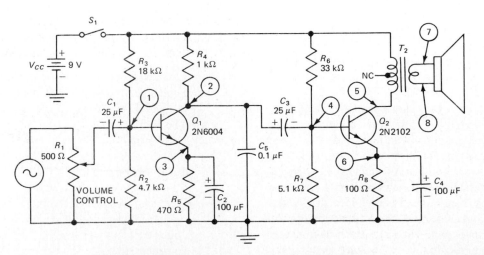

Fig. 25-2. Isolating the defective stage by signal-tracing.

transistor was defective. More subtle troubles frequently require both voltage and resistance checks to find the bad part.

NOTE: Solid-state audio amplifiers are generally more complex than the circuit in Fig. 25-2, using some form of dc bias and ac stabilization. These feedback circuits add a new dimension in troubleshooting. Chain-reaction failures may occur, and feedback paths must be opened to isolate trouble.

SUMMARY

1. Troubleshooting an electronic device, such as an audio amplifier, requires a logical, systematic procedure.
2. A troubleshooting assumption which helps simplify the process is that there is just *one* defect. When that defect is found, if the circuit is still not operating properly, the technician looks for another defect, etc., until all the troubles have been found and corrected.
3. The steps in a logical troubleshooting procedure for an audio amplifier are:
   (a) Listen to the amplifier for a clue as to the trouble.
   (b) Inspect the amplifier for obvious defects such as broken, burned, or open parts.
   (c) Measure the dc supply voltage to be sure that it is normal.
   (d) Isolate the defective stage in a multistage amplifier by signal-tracing, that is, by following with an oscilloscope the progress of the signal from the input to the output.
   (e) When a discontinuity in the signal occurs, we assume that everything is working properly through the last test point where a normal signal was found. The trouble must be in the stage where the signal disappeared, or in the loading of that circuit by the next stage.
   (f) Voltage and resistance checks of the suspected stage should lead to the defective part.
4. An in-circuit transistor tester or an in-circuit curve tracer is helpful in locating defective transistors.

SELF-TEST

Check your understanding by answering these questions.

1. The amplifier in Fig. 25-2 is dead. An inspection of the circuit does not reveal any obvious defects. The first measurement to make is the _____ _____ .
2. The amplifier in Fig. 25-2 is dead. In signal-tracing the circuit a normal waveform appears at TP 2, but there is no waveform at TP 4. A possible cause of trouble is a(n) _____ _____ .
3. Trouble has been isolated to Q_2 (Fig. 25-2). A dc voltage check of Q_2 shows the following: collector 0 V, base 1.2 V, emitter 0 V. The most likely cause of trouble is: (a) defective 2N2102, (b) open R_8, (c) open primary in T_2, (d) primary to secondary short in T_2?
4. Volume in the amplifier in Fig. 25-2 is very low. Signal-tracing reveals that a normal signal exists up to TP 4. There is a sine wave at TP 5 but its amplitude is very low. The sine wave at TP 6 is almost equal in amplitude to the signal at TP 4. DC voltage measurements are all normal. The most likely cause of trouble is: (a) defective 2N2102, (b) open secondary in T_2, (c) open R_8, (d) open C_4?
5. The amplifier in Fig. 25-2 is dead. A check shows that there is $+9$ V at the output of the power supply but 0 V at the junction of R_3, R_4, R_6, and the primary of T_2. A likely cause of trouble is: (a) primary of T_2 is shorted to the frame, (b) defective S_1, (c) defective Q_2, (d) shorted C_5?
6. Trouble in Fig. 25-2 has been isolated to Q_1. A resistance check to ground at the elements of Q_1 shows the following: collector 0 Ω, base 4700 Ω, emitter 470 Ω. The most likely cause of trouble is: (a) shorted C_5, (b) open C_5, (c) open R_4, (d) shorted R_4?

MATERIALS REQUIRED

- Power supply: Variable regulated low-voltage dc
- Equipment: Oscilloscope; EVM; AF sine-wave generator; transistor tester
- Miscellaneous: Commercial low-power audio amplifier and speaker with defect; circuit diagram for above

NOTE: If a commercial audio amplifier is not available, the student may connect any of the audio amplifiers (including Fig. 25-2) used in previous experiments. The instructor will insert defects in the amplifier for troubleshooting purposes.

PROCEDURE

Your instructor will give you a defective amplifier or introduce trouble into an audio amplifier which you have connected. You will service it, following the techniques discussed in this experiment. Prepare a Troubleshooting Report chart similar to the one shown (Fig. 25-3) and keep a step-by-step record of the checks made. Correct the trouble and notify your instructor, who will give you another amplifier to service. Service it and fill out a separate troubleshooting report. Service as many troubles as time permits.

Answers to Self-Test

1. supply voltage
2. open C_3
3. c
4. d
5. b
6. a

TROUBLESHOOTING REPORT

Date _____

Name _____ Amplifier Make _____

Teammates _____ Model # (or Fig. #) _____

1. Describe symptoms _____

_____

2. Preliminary inspection _____

_____

3. Supply-voltage(s) measurement _____ V

4. Describe procedure (list steps in order performed)

Test Point	Voltages, ac or dc	Other

5. Describe trouble found and list parts used for repair.

_____

_____

6. Was amplifier operation normal after repair? _____

If not, describe trouble. _____

_____

7. Time required to make repair: _____

8. Approved: _____

Instructor's Signature

Fig. 25-3.

JUNCTION FIELD-EFFECT TRANSISTOR (JFET) FAMILIARIZATION AND CHARACTERISTIC CURVES

OBJECTIVES

1. To determine the effect of drain-to-source voltage V_{DS} on drain current I_D
2. To determine the effect of reverse gate-to-source bias voltage V_{GS} on I_D
3. To determine and plot the family of drain characteristics of a JFET
4. To determine and plot a JFET transfer curve, I_D versus V_{GS}, for a specified value of V_{DS}

INTRODUCTORY INFORMATION

JFET Operation

The transistors you have studied to this point are called *bipolar transistors*. They are two-junction devices whose operation depends on the action of two types of charge carriers, holes and electrons. There is another class called *field-effect transistors* (FETs). These are *unipolar* devices because their action depends on only *one* type of charge carrier. FETs include the junction type (JFET) and the metal-oxide semiconductor type (MOSFET). This experiment will deal with JFETs, the first to be developed in the family of FETs.

Figure 26-1 illustrates an N-channel JFET. The three elements of the transistor are called the *source,* the *gate,* and the *drain.* The body or channel of the transistor is an N-type semiconductor. Terminal leads for the drain and source make ohmic contacts at the top and bottom of the channel. They are *not* semiconductor junctions. The P-type material, called the *gate,* which is embedded on both sides of the channel forms a semiconductor junction; hence the name "junction FET." Ohmic contacts to the P-type material serve as terminal leads for the gates. When gates 1 and 2 are internally connected in the manufacturing process, the device is a single-gate FET (Fig. 26-1a). When separate leads are brought out at each junction, a dual-gate FET results (Fig. 26-1b).

In the FET the drain corresponds to the collector of a bipolar transistor, the source to the emitter, and the gate to the base. However, the operation of a unipolar transistor is completely different from a bipolar transistor. The chief operational difference is that drain current (I_D) in the JFET is controlled by gate-to-source *voltage* (V_{GS}), whereas collector current in the bipolar transistor is controlled by base *current*.

To understand the operation of a JFET, consider the N-channel semiconductor with ohmic contacts at the top (drain) and bottom (source) of the channel in Fig. 26-2. If a battery V_{DD} is connected across the channel, with the polarity shown, the negative-charge carriers (electrons) in the N channel move toward the positive terminal of the battery, and electrons from the negative terminal of the battery move through the source into the N channel to replace those that left at the drain. Current in this circuit will continue as long as the circuit is complete. A limited *control* of current is possible by varying V_{DD}.

A much more effective way of controlling drain current is to add a gate and reverse-bias it with respect to the source, as

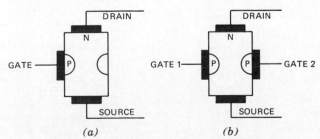

(a)   *(b)*

Fig. 26-1. N-channel JFET. *(a)* Single gate; *(b)* dual gate.

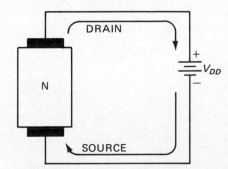

Fig. 26-2. A voltage source across a semiconductor channel will cause current to flow in the circuit.

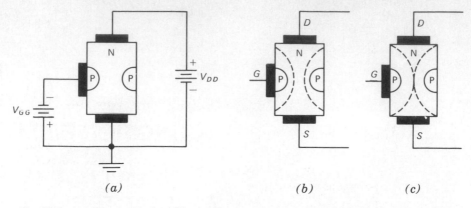

Fig. 26-3. *(a)* Reverse biasing of the gate to source has the effect of *(b)* widening the gate. *(c)* If the reverse-gate bias is made high enough, drain current is completely cut off.

in Fig. 26-3*a*. Observe that the first effect of adding the gate, even when unbiased, is to *narrow* the channel somewhat physically. This immediately restricts the current in the channel. By reverse-biasing the gate, the electric field at the junction has the effect of widening the gate, as in Fig. 26-3*b*, further reducing the channel width. If the negative bias is increased sufficiently, the gate becomes so wide (Fig. 26-3*c*) that the channel is blocked and *no* drain current flows.

Note that there is no gate current because the gate junction is reverse-biased. So gate control of drain current I_D results from varying the *negative voltage* on the gate, *not* from gate current.

The FET we have been discussing is an N-channel, P-gate device whose symbol is shown in Fig. 26-4. Observe that the gate arrow points toward the N channel (vertical line). The gate arrow may be centered on the channel (Fig. 26-4*a*), or the gate arrow may be drawn close to the source (Fig. 26-4*b*).

It is also possible to make a P-channel, N-gate JFET whose symbol is shown in Fig. 26-5*a* and *b*. For a P-channel device the arrow points away from the channel. For a P-channel JFET all battery polarities V_{GG} and V_{DD} must be reversed in connecting the FET in a circuit.

JFET Drain Characteristics

The effect of drain-to-source voltage (V_{DS}) on I_D in an N-channel JFET may be determined experimentally by the circuit of Fig. 26-6*a*. V_{GG} serves as the gate bias supply and V_{DD} as the drain-voltage source. The voltmeter M_2 measures V_{DS}. The milliammeter M_1 measures I_D as V_{DD} is varied.

The circuit of Fig. 26-6*a* will be used in this experiment to determine the effect of reverse-gate bias on I_D, when V_{DS} is held constant, and to plot the family of drain characteristic curves of an N-channel JFET. The drain characteristic curves will show how I_D varies with V_{DS} for constant values of gate bias voltage V_{GS}.

Figure 26-7 shows the variation of I_D with V_{DS} for $V_{GS} = 0$, the condition when the gate is effectively shorted to the source, Fig. 26-6*b*. As V_{DS} is increased from 0 V to V_P, called the *pinchoff voltage*, I_D increases from 0 to the maximum drain current that can be attained without destroying the JFET, the value I_{DSS}. As V_{DS} increases from V_P to V_{DS} max, drain current remains relatively constant at the value I_{DSS}. The JFET is normally operated in this interval V_P to V_{DS} max where no change in I_D occurs.

If V_{DS} is increased beyond the point V_{DS} max, there is an avalanche increase in I_D which quickly destroys the JFET. V_{DS} max, therefore, is the maximum drain-to-source voltage at which the FET may be operated safely, when $V_{GS} = 0$.

The value V_P (pinchoff voltage) is the start of the interval V_P to V_{DS} max during which I_D remains constant. It will be observed that when V_{GS} is equal to the value $-V_P$, that is, when the gate bias is $-V_P$, drain current is cut off.

Another drain characteristic curve may be determined by reverse-biasing the gate at some voltage, say -0.5 V. For this condition, I_D will vary as shown in Fig. 26-8. Note that there is again an interval during which I_D remains constant and that I_D is lower than I_{DSS}. Other curves may be determined by setting V_{GS} at -1.0 V, then at -1.5 V, and so on.

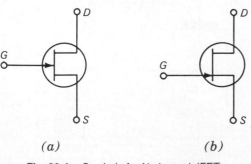

Fig. 26-4. Symbols for N-channel JFET.

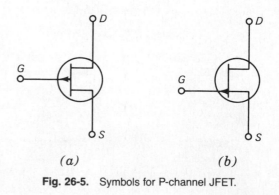

Fig. 26-5. Symbols for P-channel JFET.

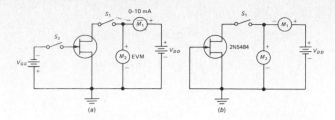

Fig. 26-6. Experimental circuits for determining the drain characteristics of an N-channel JFET.

V_{GS} may be made increasingly negative until I_D cutoff is reached, which is $V_{GS} = -2$ V (approx) in Fig. 26-8. This final value of V_{GS} at which drain current is cut off is designated $V_{GS\ (off)}$.

Transfer Characteristic

Another curve, the *transfer* or *transconductance characteristic,* is useful in evaluating the operating conditions of an FET. This curve is also plotted by using the experimental circuit in Fig. 26-6a. Now, however, V_{DS} is kept at some constant value while V_{GS} is varied, and I_D is measured. The results are graphed and resemble the curve in Fig. 26-9. It is also possible to transfer the appropriate values of I_D and V_{GS} to plot the transfer curve from the family of drain characteristics. A required value of V_{DS} is selected. The corresponding values of V_{GS} and I_D, read from the family of drain characteristics, serve as the coordinates of points on the I_D versus V_{GS} curve. Note in the transfer characteristic that the end points of the curve are (1) $V_{GS} = 0$, $I_D = I_{DSS}$, and (2) $V_{GS} = V_{GS\ (off)} = -2$, $I_D = 0$. Point 1 defines the maximum drain current, when $V_{GS} = 0$; point 2 the minimum drain current, when $V_{GS} = V_{GS\ (off)}$.

The transfer curve in Fig. 26-9 is called a *square-law curve,* because of the squared term in the equation from which it is determined

$$I_D = I_{DSS}\left(1 - \frac{V_{GS}}{V_{GS\ (off)}}\right)^2 \qquad (26\text{-}1)$$

Because of their square-law characteristic, FETs are useful in the tuners of radio and TV receivers.

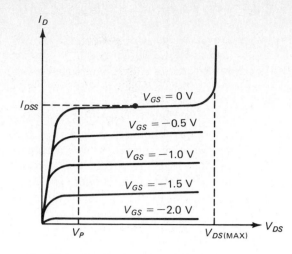

Fig. 26-8. Family of drain characteristics of a JFET.

FETs also have a very high input impedance, since the input circuit, the gate, draws no current. This characteristic is desirable in electronic voltmeters, many of which are designed with FETs in the input. These EVMs are usually called *FET VOMs* or simply *FET meters.*

SUMMARY

1. FETs are called *unipolar transistors* because they use only one charge carrier, that in the channel, for their operation.
2. The two types of FETs are the *junction* (JFET) and the *metal-oxide semiconductor* (MOSFET).
3. The elements of a JFET are the drain, the source, and the gate. These correspond, respectively, to collector, emitter, and base of a bipolar transistor.
4. An FET is a *voltage*-controlled device, whereas a bipolar transistor is a current-controlled semiconductor.
5. In the JFET the gate is biased negatively, relative to the source.
6. The family of drain characteristic curves of a JFET reveals that:
   (a) I_{DSS} is the maximum drain current that this transistor develops in normal operation.

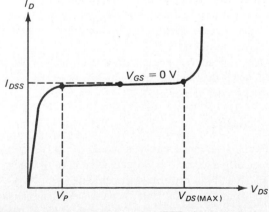

Fig. 26-7. Drain characteristics of a JFET when gate is shorted to source.

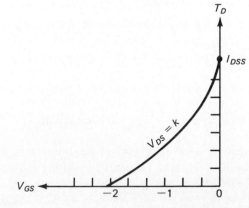

Fig. 26-9. Transfer of transconductance characteristic of an FET.

Junction Field-Effect Transistor (JFET) Familiarization and Characteristic Curves **153**

(b) As the reverse-gate-to-source bias increases, I_D decreases, until at $V_{GS\ (off)}$ I_D is cut off.

7. JFETs can be constructed so that the channel is either N semiconductor *or* P semiconductor. When the channel is N, the gate is P, and vice versa.

8. Battery polarities for normal operation of a P-channel JFET are the reverse of those for an N-channel JFET.

9. The input impedance (or resistance at low frequencies) of a JFET is very high. The use of FET amplifiers in electronic voltmeters results in meters with extremely high input impedance.

SELF-TEST

Check your understanding by answers these questions.

1. The channel of a JFET can be restricted until drain current is cut off by increasing the _____ (*forward, reverse*) bias on the gate to source.

2. In a P-channel JFET the semiconductor material of the gate is _____ (*P, N*).

3. The maximum safe drain current I_{DSS} is obtained when the _____ is shorted to the source.

4. The maximum drain voltage at which a JFET can be operated safely when $V_{GS} = 0$ is designated _____ .

5. The pinchoff voltage V_P is the voltage on the drain at which drain current is _____ (*stabilized, cut off*).

6. In the interval $V_P \rightarrow V_{DS}$ max, I_D remains _____ .

7. When $V_{GS} = -V_P$, I_D is _____ .

8. The drain characteristic of a unipolar transistor corresponds to the collector characteristic of a bipolar transistor. _____ (*true, false*)

9. The drain characteristic curve shows how _____ varies with _____ when _____ is held constant.

10. The input impedance of an FET amplifier is very _____ .

MATERIALS REQUIRED

- Power supply: Two independent variable low-voltage dc sources
- Equipment: EVM; 0–10-mA milliammeter
- Semiconductors: 2N5484 (N-channel JFET) or equivalent
- Miscellaneous: Two SPST switches

PROCEDURE

Gate Shorted to Source, $V_{GS} = 0$

1. Connect the circuit of Fig. 26-6b. S_1 *off.*

2. Set the output of V_{DD} as 0 V; S_1 *on;* **power on.** Measure, and record in Table 26-1, the drain current I_D for $V_{DS} = 0$, $V_{GS} = 0$.

3. Increase the output of V_{DD} to $V_{DS} = 0.5$ V. Measure and record in Table 26-1 the value of I_D for $V_{DS} = 0.5$ V, $V_{GS} = 0$ V.

4. Reset V_{DS} to each of the values listed in the table. For each value of V_{DS} measure and record in Table 26-1 the corresponding value of I_D.

5. *Open S_1.* **Power off.**

Gate is Reverse-Biased

6. Modify the gate circuit of Fig. 26-6b to conform with that in Fig. 26-6a. S_1 and S_2 are *open.*

7. Set the output of V_{DD} and V_{GG} to 0 V. *Close S_1 and S_2.* **Power on.**

8. Adjust V_{GG} so that V_{GS} measures -0.25 V. Maintain it at this level for steps 9 and 10.

9. Measure I_D and record its value in Table 26-1 for $V_{DS} = 0$ V.

10. Increase V_{DD} in turn to each of the values for V_{DS} listed in the table. For each value of V_{DS} measure and record I_D in Table 26-1.

11. Reduce V_{DD} to 0 V. Increase V_{GG} to -0.5 V. Maintain it at this level for step 12.

12. Repeat steps 9 and 10 for each value of V_{DS} listed in Table 26-1.

13. Repeat steps 11 and 12 for each value of V_{GS} and V_{DS} listed in Table 26-1, *until drain-current cutoff is reached.*

14. **Power off.**

Transfer Characteristic

15. S_1 and S_2 are *open.* Set V_{DD} at 15 V, $V_{GG} = -2.5$ V.

16. *Close S_1 and S_2.* Keeping V_{DS} constant at 15 V, measure and record in Table 26-2 the values of I_D for each value of V_{GS} listed. **Power off.**

17. On graph paper draw the family of drain characteristics using the data in Table 26-1. V_{DS} is the horizontal axis, I_D the vertical. Identify each characteristic curve by its V_{GS} value. Identify V_P.

18. On a separate graph paper draw the transfer characteristic using the data in Table 26-2. V_{GS} is the horizontal axis, I_D the vertical.

Extra Credit

19. Explain the procedure you would use to verify experimentally that there is no gate current in the range over which a JFET is normally operated.

20. Following the procedure, measure the gate current, if any, over the range of normal operation of the JFET assigned to you.

TABLE 26-1. Data for Drain Characteristics

V_{DS}, V	I_D, mA									
	V_{GS}, V									
	0	−0.25	−0.5	−0.75	−1.0	−1.25	−1.5	−1.75	−2.0	−2.5
0										
0.5										
1.0										
1.5										
2.0										
2.5										
3.0										
3.5										
4.0										
4.5										
5.0										
7.0										
9.0										
11.0										
13.0										
15.0										

TABLE 26-2. Data for Transfer Characteristic

V_{GS}, V	−2.5	−2.0	−1.75	−1.50	−1.25	−1.0	−0.75	−0.50	−0.25	0
I_D, mA										

QUESTIONS

1. (a) From your graphs, what is the value of V_P? (b) What factors helped you identify V_P?
2. (a) From your graphs, what is the value of I_{DSS}? (b) For what values of V_{GS} and V_{DS} is I_{DSS} defined?
3. From the data in Table 26-1 and the family of drain characteristics, compare the level of drain current for each value of V_{GS}, in the interval V_{DS} = 5 to 15 V. What conclusions can you draw?
4. Does your experiment indicate which is more effective in controlling drain current, V_{DS} or V_{GS}? Explain, referring to your data.
5. What is the value of V_{GS} which cuts off I_D in this experiment? How does this compare with V_P?
6. From the *drain characteristic curves* derived from Table 26-1, determine the value of I_D for each value of V_{GS} in Table 26-2, at V_{DS} = 15 V. How do these values compare with those obtained in step 16? Explain any differences.

Answers to Self-Test

1. reverse
2. N
3. gate
4. V_{DS} max
5. stabilized
6. constant
7. cut off
8. true
9. I_D; V_{DS}; V_{GS}
10. high

MOSFET COMMON-SOURCE AMPLIFIER

OBJECTIVES

1. To determine experimentally the drain characteristics of a metal-oxide semiconductor field-effect transistor (MOSFET)
2. To consider biasing arrangements for an FET amplifier
3. To measure the voltage gain of a MOSFET common-source amplifier

INTRODUCTORY INFORMATION

Like the JFET the MOSFET is a field-effect transistor whose drain current I_D is controlled by the voltage on the gate. The MOSFET and the JFET differ physically and in operation. The manner in which the MOSFET is constructed determines whether it is a depletion or an enhancement type. The MOSFET has a much higher input impedance than the JFET.

Enhancement-Type MOSFET

In the enhancement type, there is no channel between the drain and source, but separating the N-type drain and source is a P-type substrate, as in Fig. 27-1. Deposited over the substrate is a *very thin, fragile layer* of silicon dioxide (SiO_2), an insulator. Deposited on the SiO_2 is a metallic film which acts as the gate. Ohmic contacts are brought out for the gate, drain, source, and substrate. When the substrate is internally connected to the source, there is no substrate lead.

From Fig. 27-1 it is clear that the gate is insulated from the body of the FET, and for this reason the MOSFET is also called an *insulated gate FET (IGFET)*. Although there is no physical channel, for reasons which will become clear, the MOSFET in Fig. 27-1 is called an *N-channel enhancement type*.

The gate and the substrate act as the plates of a capacitor separated by the SiO_2 insulator. When the gate is made positive relative to the source to which the substrate is connected (Fig. 27-2a), the capacitor charges. Since the gate is positive, negative charges appear in the substrate between drain and source (Fig. 27-2b), in effect creating an N channel which will permit current in the source-to-drain circuit.

The N-channel enhancement-type MOSFET conducts only when the gate is positive relative to the source. It is cut off when there is zero or gate-to-source bias. For this reason it is designated a normally OFF MOSFET. Because the gate is insulated from the substrate, there is no dc gate current even though the gate is positive relative to the substrate. Therefore the MOSFET is a high-impedance transistor.

It should be noted that a P-channel normally OFF MOSFET may be made by using an N substrate and P-type drain and source. The symbols for N- and P-channel enhancement types are shown in Fig. 27-3a and b. Note the broken vertical channel line indicates a normally OFF MOSFET.

Depletion-Type MOSFET

The MOSFET in Fig. 27-4 is constructed like the one in Fig. 27-1 with one exception. Figure 27-4 *has* an N-type channel,

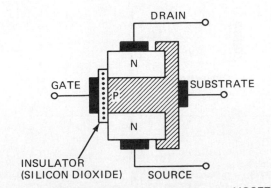

Fig. 27-1. Construction of an enhancement-type MOSFET.

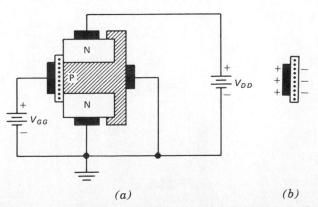

Fig. 27-2. (a) Biasing an enhancement-type MOSFET. (b) By capacitor action an N channel is induced between drain and source, providing a path for current from source to drain.

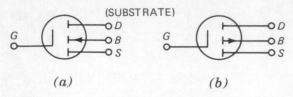

(a)   *(b)*

Fig. 27-3. Symbols for enhancement-type MOSFET: *(a)* N channel; *(b)* P channel.

while Fig. 27-1 does not. The FET in Fig. 27-4 can be operated with both a positive and a negative gate. When the gate is positive, the FET operates in the enhancement mode; when negative, in the depletion mode.

Consider the effect of a positive gate on the charge carriers in the N channel of Fig. 27-5a. The gate-to-channel capacitor charges, inducing negative-charge carriers in the N channel (Fig. 27-5b), thus increasing the conductivity of the channel and increasing the drain current. The more positive the gate is made, the more drain current flows. This is the enhancement mode.

When the gate is shorted to the source, that is, when there is zero gate voltage, drain current will flow in the channel, from source to drain, if a voltage source V_{DD} is connected as in Fig. 27-5a. The drain current with zero gate voltage is less than the drain current with positive gate voltage. Since there is drain current, even with 0-V gate bias, this is a normally ON MOSFET.

What happens when the gate is made negative relative to the source, as in Fig. 27-6a? The capacitor charge distribution on the gate and N-type channel is shown in Fig. 27-6b. Electrons on the gate repel the negative charge carriers in the N channel, in effect depleting the channel of negative carriers and reducing drain current. The more negative the gate, the less the drain current. When V_{GS} is made sufficiently negative, drain current is cut off.

The depletion MOSFET may be constructed with a P channel and N substrate. The symbols for N- and P-channel depletion MOSFETs are shown in Fig. 27-7. Note that the vertical channel line is not broken here because the device is normally ON.

The drain curves for a MOSFET are similar to those for a JFET (Experiment 26). Figure 27-8 shows a family of drain curves for an N-channel depletion MOSFET.

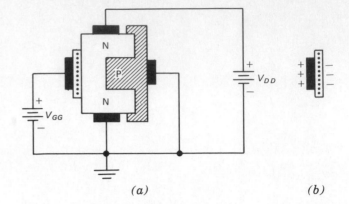

(a)   *(b)*

Fig. 27-5. *(a)* Enhancement mode; gate is positive relative to source. *(b)* Negative charge carriers are induced in the N channel.

Dual-Insulated-Gate MOSFET

For special applications MOSFETs may be constructed with two gates, G_1 and G_2, as in Fig. 27-9. This is the diagram of a four-terminal 3N187, a device which may be used as an amplifier for frequencies up to 300 MHz. Gate 1 is at terminal 3, gate 2 at terminal 2. The drain is terminal 1; the source, terminal 4. Note that the substrate is internally connected to the source. Two sets of back-to-back diodes are built into this device. One set of diodes is internally connected between gate 1 and the source, and the other set between gate 2 and the source. The manufacturer states that "the diodes bypass any voltage transients which exceed approximately ± 10 V," protecting the gates against damage in normal handling and use.

You will recall that the silicon dioxide insulator in a MOSFET is a thin glasslike deposit on the substrate. Since it is very fragile, it may be easily fractured by static voltages to which the MOSFET is subject in ordinary handling. The back-to-back diodes connected between gates and source protect the MOSFET against static charges and other voltage transients.

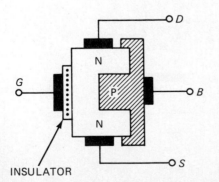

Fig. 27-4. Depletion-enhancement type MOSFET.

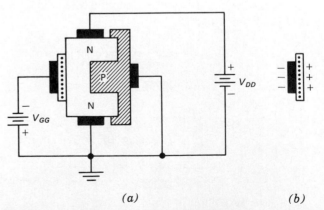

(a)   *(b)*

Fig. 27-6. *(a)* Depletion mode; gate is negative relative to source. *(b)* Negative charge carriers in the N channel are depleted.

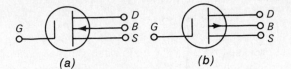

Fig. 27-7. Symbols for depletion-type MOSFET: (a) N channel; (b) P channel.

Biasing JFETs

The gate-biasing network of a JFET is similar to the base-biasing network of a bipolar transistor, except that the JFET gate must be reverse-biased, while the bipolar base is forward-biased.

Voltage-Divider Bias

Figure 27-10a shows an N-channel JFET gate-bias arrangement employing voltage-divider bias. The voltage at the gate to ground V_G is given by the equation

$$V_G = \frac{R_1}{R_1 + R_2} \times V_{DD} \qquad (27\text{-}1)$$

The voltage on the source to ground V_S is the IR drop across the source resistor R_S. It depends on the value of drain current I_D (since drain and source current are the same) and on R_S.

$$V_S = I_D \times R_S \qquad (27\text{-}2)$$

The gate bias V_{GS} is the difference between V_G and V_S. That is,

$$V_{GS} = V_G - V_S \qquad (27\text{-}3)$$

To ensure reverse bias of the gate, V_S must be larger than V_G, so that $V_G - V_S$ is a negative quantity. Parameters are chosen to ensure this and to provide the proper operating point for the circuit.

By injecting an ac signal (v_g) between gate and ground the circuit in Fig. 27-10a acts as an ac amplifier. However, the

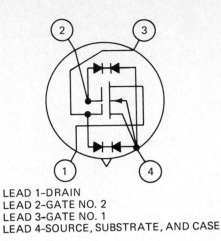

LEAD 1–DRAIN
LEAD 2–GATE NO. 2
LEAD 3–GATE NO. 1
LEAD 4–SOURCE, SUBSTRATE, AND CASE

Fig. 27-9. Terminal diagram of a 3N187, dual-insulated gate MOSFET. (RCA)

gain of the circuit is very low, because an ac voltage (v_s) is developed across R_S, and the gate-to-source signal (v_{gs}) which the amplifier "sees" is the difference between v_g and v_s. That is,

$$v_{gs} = v_{in} = v_g - v_s \qquad (27\text{-}4)$$

The degenerative signal voltage developed across R_S can be eliminated by providing a bypass capacitor C_1 across R_S, as in Fig. 27-10b. To approximate the value of C_1 we use the relationship

$$X_{C1} = \frac{R_S}{10} \qquad (27\text{-}5)$$

for the *lowest* frequency which the amplifier must handle. You will recall that this is the same arrangement used to bypass the emitter resistor in a bipolar transistor amplifier.

In Fig. 27-10, the output signal is developed across R_L and is taken from the drain.

Self-Bias

The circuit of Fig. 27-11 illustrates a self-bias arrangement for an N-channel JFET. Since the gate is returned to ground

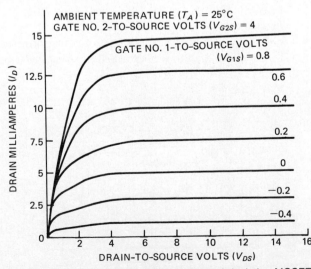

Fig. 27-8. Drain curves for 3N200 N-channel depletion MOSFET. (RCA)

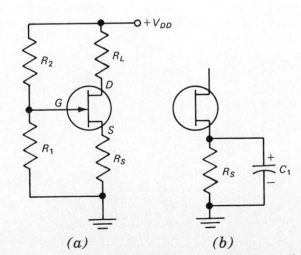

Fig. 27-10. (a) Voltage-divider bias for JFET. (b) Bypassing the source resistor prevents ac degeneration.

MOSFET Common-Source Amplifier **159**

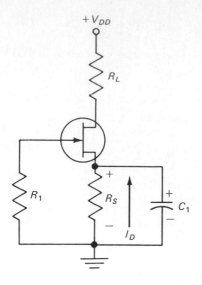

Fig. 27-11. Self-bias arrangement for an N-channel JFET.

through R_1 and there is no gate current, the voltage on the gate V_G is 0 V. Current (I_D), in the FET and in the external circuit (R_S and R_L), will develop a voltage drop V_S across R_S.

$$V_S = I_D \times R_S \qquad (27\text{-}6)$$

Since $V_G = 0$, the voltage difference between V_G and V_S is the gate bias V_{GS}. That is,

$$V_{GS} = 0 - V_S = -I_D \times R_S \qquad (27\text{-}7)$$

For example, if $I_D = 1$ mA and $R_S = 2.2$ kilohms (kΩ), the gate bias is

$$V_{GS} = -(1)10^{-3} \times 2.2 \times 10^3 = -2.2 \text{ V}$$

Source Bias

If two independent power supplies are available, source bias, as illustrated in Fig. 27-12, may be used. Here $-V_{SS}$ is the source supply and $+V_{DD}$ is the drain supply. The circuit parameters are chosen to supply the proper gate bias for the required drain current. If the voltage $-V_{SS}$ is high enough, drain current may be made independent of the variation in characteristics of the FET.

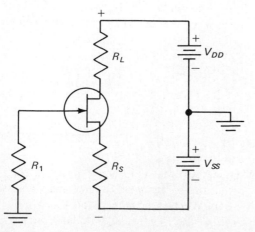

Fig. 27-12. Source bias requires two independent power supplies.

MOSFET Biasing

Voltage-divider gate bias, self-bias, and source bias may be used in biasing MOSFETs. Circuit arrangements are similar to those shown in Figs. 27-10 to 27-12. The polarity of the bias required will depend on the nature of the channel (N or P) and on the type of MOSFET (enhancement or depletion).

Depletion MOSFETs may be operated at zero bias, that is, $V_{GS} = 0$ V. An ac signal fed to the gate of an N-channel depletion MOSFET causes the FET to operate in the enhancement mode on the positive alternation, and in the depletion mode on the negative alternation. The opposite is true of P-channel depletion MOSFETs.

MOSFET Common-Source Amplifier Circuit and Operation

The circuit of Fig. 27-13 is an experimental common-source amplifier, using an N-channel depletion-type dual-gate MOSFET. Gates 1 and 2, externally connected together, receive the input signal. The amplifier is operated with self-bias. The output signal is developed in the drain across the 10-kΩ load resistor.

The voltage gain of the amplifier A_v may be determined experimentally by measuring the output and input signals and substituting the measured values in the equation

$$A_v = \frac{v_{\text{out}}}{v_{\text{in}}} \qquad (27\text{-}8)$$

The voltage gain of the MOSFET amplifier is relatively low. The phase relationship between the input and output signals may also be determined with an oscilloscope by the methods used in previous experiments.

SUMMARY

1. The MOSFET is a metal-oxide semiconductor field-effect transistor whose gate voltage controls drain current.
2. The two basic types of MOSFET are (*a*) depletion and (*b*) enhancement.
3. The gate of a MOSFET is insulated from the substrate. The device is therefore also called an *insulated gate field-effect transistor (IGFET)*.

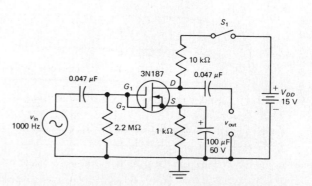

Fig. 27-13. Experimental MOSFET common-source amplifier.

4. The enhancement-type MOSFET has no physical channel between drain and source. The channel, consisting of charge carriers, is electrically induced in the substrate by biasing the gate.
5. Since there is no physical channel in an enhancement MOSFET, its symbol contains a broken (vertical) channel line (see Fig. 27-3).
6. The gate of an N-channel enhancement-type MOSFET must be biased positively. The gate of a P-channel enhancement-type MOSFET must be biased negatively.
7. Depletion-type MOSFETs (also called *depletion-enhancement type*) contain a channel (Fig. 27-5).
8. The gate of a depletion-enhancement-type MOSFET may be biased either positively, zero, or negatively.
9. When the gate of an N-channel depletion MOSFET is biased positively, it operates in the enhancement mode.
10. When the gate of an N-channel depletion MOSFET is biased negatively, it operates in the depletion mode.
11. MOSFETs constructed with two independent gates are called *dual-gate*.
12. MOSFETs are very fragile. To protect the MOSFET against destruction due to transient voltages, back-to-back diodes may be built into the device, as in Fig. 27-9.
13. FETs may use voltage-divider bias as in Fig. 27-10, self-bias as in Fig. 27-11, or source bias as in Fig. 27-12. For the latter arrangement two independent power sources are required.
14. FET amplifiers (Fig. 27-13) have a high input impedance and a relatively low voltage gain.

SELF-TEST

Check your understanding by answering these questions.

1. Figure 27-1 shows that there _____ (*is, is not*) a physical channel between the drain and the source in an enhancement-type MOSFET.
2. The gate of an enhancement-type MOSFET must be biased _____ relative to the source in order to permit current flow.
3. In an N-channel depletion-type MOSFET the N-type substrate acts as the channel. _____ (*true, false*)
4. If the gate of an N-channel depletion-type MOSFET is biased positive relative to the source, the transistor operates in the _____ mode.
5. The vertical channel line in the symbol for a depletion-type MOSFET (*is, is not*) a broken line.
6. The range of gate bias (relative to the source) for the 3N200 MOSFET is from _____ to _____ V (see Fig. 27-8).
7. In the circuit of Fig. 27-10a, the voltage on the gate to ground is $+3.2$ V; the voltage on the source to ground is $+2.5$ V. The bias on the gate relative to the source is therefore _____ V.
8. In the circuit of Fig. 27-11, $R_S = 3.3$ kΩ, $I_D = 1$ mA, and the gate draws no current. The bias on the gate relative to the source is _____ V.

MATERIALS REQUIRED

- Power supply: Two independent variable regulated low-voltage dc sources; EVM; 0–10-mA dc milliammeter or equivalent range on VOM
- Resistors: 1-kΩ, 10-kΩ; 2.2-MΩ ½-W
- Capacitors: Two 0.047-μF, 100-μF 50-V
- Semiconductors: 3N187 (MOSFET) or equivalent
- Miscellaneous: Two SPST switches; other resistors as required in extra-credit step 11, or a resistor decade box

PROCEDURE

CAUTION: Do *not* exceed the voltages listed in Table 27-1.

Drain Characteristics (Gate Control)

1. Figure 27-14 is a bottom view of the terminals of the 3N187. Connect the circuit of Fig. 27-15. S_1 and S_2 are *off*. Set V_{DD} at 0 V, V_{GG} at -0.8 V.
2. *Close* S_1 and S_2. Measure I_D for $V_{GS} = -0.8$ V, $V_{DS} = 0$, and record in Table 27-1.
3. Maintain V_{GS} at -0.8 V. Increase V_{DS} to $+1$ V, $+3$ V, etc., as in Table 27-1 and record the measured values of I_D.
4. Decrease the negative bias V_{GS} to -0.7 V. Measure and record I_D for each value of V_{DS} listed in Table 27-1.
5. Repeat step 4 for each negative value of V_{GS} and record I_D in Table 27-1.
6. S_1 and S_2 *off*. Reverse the polarity of V_{GG}, the gate bias supply, and repeat step 4 for $V_{GS} = 0$ V and for each positive value of V_{GS} and V_{DD}.

NOTE: If you have a calibrated curve tracer, steps 1 to 6 can be accomplished automatically, using the curve tracer to display I_D versus V_{DS} characteristic curves for the voltage values in Table 27-1. Photograph the curves or fill in the values of I_D in Table 27-1 from the characteristic-curve display.

Common-Source Amplifier

7. **Power off.** Connect the circuit of Fig. 27-13. Set V_{DD} at $+15$ V. Set the output of the sine-wave generator at 1000 Hz, *minimum* output.
8. **Power on.** With an oscilloscope, monitor the output signal v_{out}. Gradually increase the input signal for the maximum undistorted signal, v_{out}.
9. Measure and record in Table 27-2 the peak-to-peak voltages of v_{out} and v_{in}. Compute and record the gain. Measure also and record V_{GS} and V_{DS}.
10. From the data in Table 27-1 draw the drain characteristic curves of the 3N187. Identify each curve.

TABLE 27-1. Drain Characteristics

V_{DS}, V	I_D, mA								
	0	1	3	5	7	9	11	13	15
V_{GS}, V									
−0.8									
−0.7									
−0.6									
−0.5									
−0.4									
−0.3									
−0.2									
−0.1									
0									
+0.1									
+0.2									
+0.3									
+0.4									
+0.5									
+0.6									
+0.7									
+0.8									

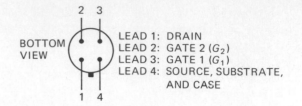

LEAD 1: DRAIN
LEAD 2: GATE 2 (G_2)
LEAD 3: GATE 1 (G_1)
LEAD 4: SOURCE, SUBSTRATE, AND CASE

Fig. 27-14. Bottom view of 3N187 showing terminal arrangement.

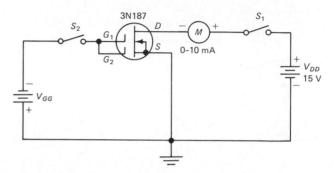

Fig. 27-15. Experimental circuit to determine gate control characteristics of a 3N187.

TABLE 27-2. Amplifier Gain

v_{in}, V p-p	v_{out}, V p-p	V_{GS}, V	V_{DS}, V	Gain

Extra Credit

11. Experimentally determine the effect of increasing and decreasing the drain load resistor **(a)** on input-signal level v_{in} which the amplifier can handle without distortion, **(b)** on amplifier output and gain.

QUESTIONS

1. From your data in Table 27-1, what is the value of V_{GS} at which drain current is cut off?
2. If your answer to question 1 indicates that I_D was not cut off in the experiment, how can you determine what value of V_{GS} will cut off I_D? Explain fully.
3. What is meant by a depletion-type MOSFET?
4. Do your data confirm that the 3N187 is a depletion-type transistor? Explain.
5. In which mode of operation of the 3N187 is the transistor likely to be cut off? Why?
6. What is the effect on current (charge) carriers in the channel of the 3N187 of biasing gate 1 positively? Why?
7. What is the effect on I_D of biasing the 3N187 positively? Negatively? Refer to your data.
8. Is it possible, using the drain characteristic curves, step 10, to predict the effect on amplifier gain of changing the drain load resistor? Explain how.

Answers to Self-Test

1. is not
2. positive
3. false
4. enhancement
5. is not
6. −0.4 to +0.8
7. +0.7
8. −3.3

THE DIFFERENTIAL AMPLIFIER (DA)

OBJECTIVES

1. To observe the output waveforms of a DA resulting from a single-ended input and note their phase relative to the input waveform
2. To observe the output waveforms of a DA resulting from two input signals of the opposite phase, differential mode, and note their phase relative to the input
3. To observe the output waveforms of a DA resulting from two input signals of the *same phase (common mode)*

INTRODUCTORY INFORMATION

Simple Differential Amplifier

The differential amplifier has its greatest application in direct-coupled linear integrated circuit (IC) amplifiers, such as the operational amplifier (op amp). Its design is therefore normally related to IC fabrication techniques, although it is also used in some circuits employing discrete components. We will begin our study of integrated circuits in Experiment 29. In this experiment, however, we will consider the DA as made up of individual components and our experimental procedure will deal with such discrete components.

Figure 28-1 is the schematic diagram of the basic differential amplifier. It consists of two transistors with two inputs and a single output. The circuit is symmetrical; that is, the two transistors Q_1 and Q_2 have identical characteristics. The emitter resistor R_E is common to both transistors. Collector load resistors $R_{L1} = R_{L2}$. The two input circuits are also identical; that is, $v_1 = v_2$ and $R_1 = R_2$.

The output signal is proportional to the difference between the two input signals, and the equation for v_{out} is

$$v_{out} = A(v_1 - v_2) \qquad (28\text{-}1)$$

where A is the gain of each transistor and v_1 and v_2 are the signal voltages at each base relative to ground. When the phase of the two input signals is the same, and their amplitudes are equal, *the condition for common-mode operation,*

$$v_1 - v_2 = 0$$

and

$$v_{out} = A\,(0) = 0 \qquad (28\text{-}2)$$

That is, in the common mode a DA rejects the common-mode signal, and its output for the common-mode signal is 0, ideally. In practice, the two halves of the DA amplifier are never completely balanced, and there is a very low output for the common-mode signal.

When the two input signals are equal in amplitude but 180° out of phase, *the condition for the differential mode* of operation,

$$v_1 = -v_2 \qquad \text{or} \qquad v_2 = -v_1$$

$$v_{out} = A\,[v_1 - (-v_1)] = A\,(2v_1) \qquad (28\text{-}3)$$

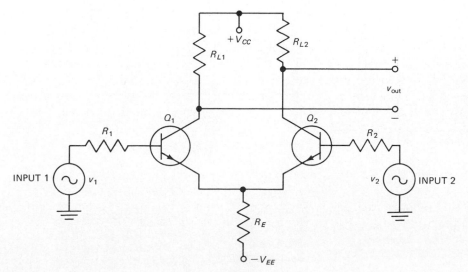

Fig. 28-1. Differential amplifier.

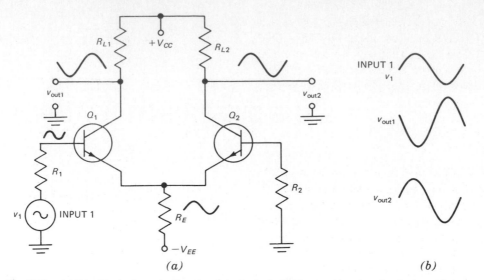

Fig. 28-2. *(a)* DA with single-ended input and dual output. *(b)* Phase of input and output waveforms.

Therefore in the differential or *non*-common mode of operation, the DA does amplify the input signals, and the output signal is equal to twice the gain times the input signal.

Single-Ended Input

The DA is normally operated with a double-ended input as in Fig. 28-1. However, it may be used with a single-ended input as in Fig. 28-2. Moreover, the output signal can be taken between the two collectors, as in Fig. 28-1, or there can be two separate outputs, v_{out1} and v_{out2}, from each collector to *ground,* as in Fig. 28-2. It will be simpler to understand the operation of the double-ended input DA if we first consider the single-ended input.

Consider a sine wave v_1 with the polarity shown in Fig. 28-2*b* applied to input 1. The amplified signal v_{out1} appearing at the collector of Q_1 to ground will be 180° out of phase with the input. Moreover, because emitter resistor R_E is unbypassed, a sine wave will develop across R_E which is *in*

phase with the input v_1. Since the base of Q_2 is returned to ground, there will be no signal voltage from the base of Q_2 to ground. However, an oscilloscope connected from base 2 to the emitter will show a sine wave v_2 which is 180° out of phase with input 1. Moreover, the amplitude of v_2 will equal that of v_1. The sine wave v_2 between base and emitter of Q_2 is effectively an input signal to Q_2, which is amplified by Q_2 and appears as a *positive-going* sine wave v_{out2} between the collector of Q_2 and ground (see Fig. 28-2*b*). Since the amplitudes of v_1 and v_2 are equal, the output waveforms are equal in amplitude but 180° out of phase.

A DA with a single-ended input and dual outputs produces two output waveforms, equal in amplitude but 180° out of phase.

If a single output from the DA in Fig. 28-2 is taken between the collector of Q_2 and the collector of Q_1, as in Fig. 28-3, the output waveform v_{out} is in phase with the input v_1. Input 1 to Q_1 is therefore called a *noninverting* input.

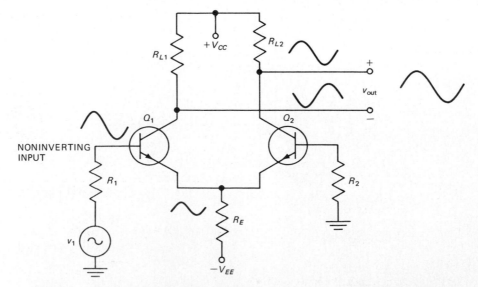

Fig. 28-3. Single-ended input to Q_1 and output taken between collectors of Q_2 and Q_1.

If the base resistor of Q_1 is returned to ground and the input signal is applied to the base of Q_2, as in Fig. 28-4, and a single output waveform is taken between the collector of Q_2 and the collector of Q_1, the output waveform, v_{out}, is 180° out of phase with the input. Input 2 (to Q_2) is called an *inverting* input.

Differential-Mode Input (Non-Common-Mode Operation)

The operation of a DA with differential-mode input (Fig. 28-5) will now be considered. The transistors Q_1 and Q_2, connected as a differential amplifier, receive input sine waves from opposite ends of the secondary of center-tapped transformer T. The input signals to the bases of Q_1 and Q_2 are equal in amplitude but 180° out of phase, the condition for differential-mode operation. Assume an instant in time when the sine wave on base 1 is positive-going, that on base 2 negative-going, as in Fig. 28-5. Now consider the action of transistor Q_1, as though Q_2 were not in the circuit. With a positive-going signal on the base of Q_1, an amplified negative-going waveform develops at collector 1. Moreover, a positive-going sine wave is developed across R_F, the unbypassed emitter resistor, because of the emitter-follower action of Q_1.

Now consider Q_2 in the circuit and Q_1 out. A negative-going signal on the base of Q_2 will result in an amplified positive-going signal at collector 2. Moreover, a negative-going sine wave is developed across R_E because of the emitter-follower action of Q_2. The signal voltages developed across R_E because of the opposite actions of Q_1 and Q_2 are equal in amplitude but 180° out of phase. Therefore, when we consider the effect of both Q_1 and Q_2 acting *together*, it is evident that the signal voltages across the emitter resistor cancel each other, and *no* signal is developed across the emitter resistor. As a result no signal current flows in the emitter resistor, which acts as though it were bypassed for ac. Note that R_E, in this case, does *not* introduce degeneration.

Ideally, then, there is no coupling between Q_1 and Q_2, and the output at each collector is an amplified sine wave 180° out

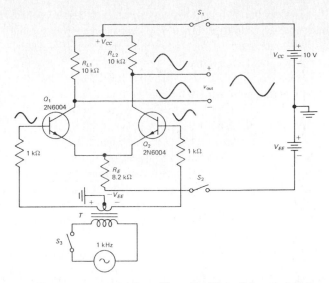

Fig. 28-5. Experimental amplifier with differential-mode input.

of phase with its input on the base. If v_{out} is taken as the signal between the collectors of Q_2 and Q_1, v_{out} is a positive-going wave, whose amplitude is *twice* the amplitude of the signal voltage from either collector to ground.

It is possible to take two outputs from the DA; from the collector of Q_1 and from the collector of Q_2, with respect to ground. These signal voltages are equal in amplitude but 180° out of phase.

Common-Mode Input (Common-Mode Operation)

The circuit of Fig. 28-6 shows the DA connected for common-mode input. Note that the secondary of transformer T supplies the same amplitude and the same phase signal voltages v_1 and v_2 to the bases of Q_1 and Q_2. The action of the circuit is now dramatically changed. The in-phase signals at the bases of Q_1 and Q_2 cause in-phase signal voltages to appear across R_E, which *add* together. Moreover, the phase of the combined signal across R_E is the same as the phase of the signals at the bases of Q_1 and Q_2.

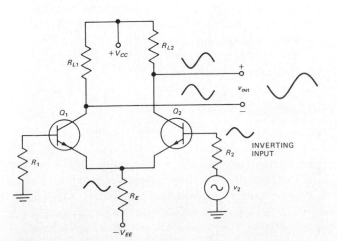

Fig. 28-4. Single-ended input to Q_2 and output taken between collectors of Q_2 and Q_1.

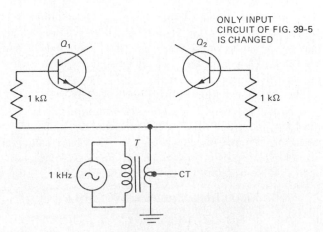

Fig. 28-6. Experimental amplifier with common-mode input.

The Differential Amplifier (DA)   **165**

It is clear that R_E provides degenerative or negative feedback to each amplifier. Since R_E is *common* to both Q_1 and Q_2, its degenerative effect is twice as great (approximately) as it would be if each transistor were acting alone. Therefore, the effective signal (base-to-emitter) which each amplifier "sees" is negligibly small, and the output at each collector is negligibly small. Ideally, the output would be 0. In practice this is never realized.

The *superposition theorem* offers another means for understanding the action of common-mode operation. This theorem states that the voltage across any circuit component is the algebraic sum of the voltages across that component, due to each individual source acting alone. To apply this theorem, we find the response of *one* source across a specific component in a circuit, while replacing all other sources with their internal resistances. In the same way we find the response of a second source across the selected component. The response of each source, in turn, is found, and then all responses are added algebraically for the total or overall response.

Let us apply this theorem to the circuit of Fig. 28-1. Assume both ac sources supply Q_1 and Q_2 with signals v_1 and v_2 equal in amplitude and having the same phase, the condition for common-mode operation. Now applying the superposition theorem, replace source v_2 with its internal resistance, assumed to be 0. The result is the single-ended DA of Fig. 28-3. The outputs at the collectors of Q_1 and Q_2, respectively, due to source 1, are negative- and positive-going waveforms equal in amplitude, say 1 V p-p. Next, replace source v_1 with its internal resistance, also assumed to be 0. The result is the single-ended circuit of Fig. 28-4. Now the output at the collector of Q_1 due to source 2 is 1 V p-p but positive-going, while the output at the collector of Q_2 is 1 V p-p but negative-going. Combining the signal voltages at the collector of Q_1 due to sources v_1 and v_2, we find the result is 0. Similarly the combined response at the collector of Q_2 due to sources v_1 and v_2 is 0.

The circuit of Fig. 28-6 will be used in this experiment to demonstrate *common-mode operation*. However, it should be understood that the basic DA circuit of Fig. 28-1 can operate in both the differential mode *and* the common mode. What ensures differential-mode operation are out-of-phase, equal-amplitude signals fed to the bases of Q_1 and Q_2. Common-mode operation results when in-phase, equal-amplitude signals appear on the bases of Q_1 and Q_2. Both types of signals may appear at the same time on the bases of Q_1 and Q_2, and the DA will *reject* the common-mode signal while amplifying the differential-mode signal.

The operation of a DA with differential input is very useful in amplifying desired signals (non-common-mode) and rejecting undesired signals (common-mode). It should be noted that common-mode signals may result from induction into the DA inputs of stray magnetic fields or from power-supply ripple or hum.

Common-Mode Rejection Ratio (CMRR)

A differential amplifier should have high gain for differential-mode input signals and very low gain for common-mode signals. Let us specify the gain of the amplifier for differential-mode signals by A_{DM} and common-mode signals by A_{CM}. The ratio A_{DM}/A_{CM} is designated the *common-mode rejection ratio*, or *CMRR*. That is,

$$\text{CMRR} = \frac{A_{DM}}{A_{CM}} \qquad (28\text{-}4)$$

The CMRR value is therefore an index of the effectiveness of a DA; the higher the ratio, the better the DA.

Effect of Unbypassed Emitter Resistor R_E on DA Operation

It was noted that for common-mode signals the unbypassed emitter resistor R_E provides degenerative or negative feedback. Therefore the larger the resistance of R_E, the higher the negative feedback voltage in the common mode, and the more effective the DA is in *rejecting* common-mode signals.

In the differential mode R_E acts as though it were completely bypassed and does not reduce the gain of the amplifier for differential-mode signals.

It would appear therefore that a large-valued resistor R_E is desirable for a DA, because it ensures the least common-mode gain (A_{CM}) without affecting the differential-mode gain (A_{DM}). That is, the higher the value of R_E, the higher the value of CMRR.

There is a practical limit to the size of R_E, because R_E determines the value of emitter current I_E. You will recall the approximate equation for I_E,

$$I_E = \frac{V_{EE}}{R_E} \qquad (28\text{-}5)$$

Moreover, the emitter supplies equal currents to both collectors in the DA. By increasing R_E, both emitter and collector currents are reduced.

It is possible to replace R_E with a constant current transistor Q_3, as in Fig. 28-7. The collector-to-emitter resistance of Q_3 is high without unduly limiting the collector current of Q_3, and therefore without limiting the emitter current I_E which supplies collector currents for Q_1 and Q_2. The emitter resistor R_E in Fig. 28-7 can be relatively small in value. DC bias for Q_3 is determined by the voltage divider R_3 and R_4. The addition of Q_3 to the circuit therefore ensures a high effective emitter resistance R_E without the restriction on collector currents for Q_1 and Q_2.

DA Circuit Symbol

DA circuitry may be more complex than the circuit in Fig. 28-7. It is simpler to represent the DA as a block (Fig. 28-8) rather than by conventional circuit symbols. This is particularly true of IC differential amplifiers where terminals of the IC represent input connections to and outputs from the DA. Figure 28-8a represents a DA with two inputs and a single output. Figure 28-8b is the block symbol for a two-input DA with two separate outputs. Note that the single output in (a) is in fact the output voltage from outputs 3 to 4 in (b).

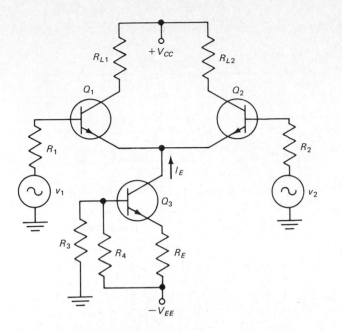

Fig. 28-7. Improved DA using constant-current transistor Q_3 to replace the emitter resistor R_E in Fig. 28-1.

SUMMARY

1. The differential amplifier is a two-input terminal device utilizing at least two transistors.
2. The DA transistors Q_1 and Q_2 in Fig. 28-1 are matched so that their characteristics are the same or almost the same. The resistors $R_1 = R_2$ and $R_{L1} = R_{L2}$. The equality of the matched circuit components makes the DA circuit arrangement completely symmetrical.
3. The output v_{out} from the DA (Fig. 28-1) is given by the equation $v_{out} = A (v_1 - v_2)$, where A is the gain of the transistor and v_1 and v_2 are the input signals.
4. There may be *one* output from a DA, taken between the collectors of Q_2 and Q_1 (Fig. 28-1).
5. There may be *two* outputs from a DA, as in Fig. 28-2. The outputs here are taken from the collector of each transistor with respect to ground.
6. One of the two inputs to a DA amplifier may be grounded, as in Fig. 28-2. This arrangement is called a *single-ended* input.
7. The output signal in the DA of Fig. 28-3 is in phase with the single-ended input signal on the base of Q_1. The Q_1 *input* is therefore called the *noninverting* input.

8. The output signal in the DA of Fig. 28-4 is 180° out of phase with the input signal on the base of Q_2. The Q_2 *input* is therefore called the *inverting* input.
9. The signals to a DA are defined as common mode or differential mode. Common-mode signals are equal in amplitude and have the same phase. The circuit in Fig. 28-6 operates in the common mode. Differential-mode signals are equal in amplitude but 180° out of phase, as in Fig. 28-5.
10. A DA may receive both common-mode and differential-mode signals. However, the DA *rejects* common-mode signals but amplifies differential-mode signals.
11. If the gain of a DA for differential-mode signals is designated A_{DM}, and the gain for common-mode signals is A_{CM}, then the ratio A_{DM}/A_{CM} is called *CMRR*, or the *common-mode rejection ratio*.
12. The larger the CMRR ratio, the more effective the DA.

SELF-TEST

Check your understanding by answering these questions.

1. A differential amplifier is a symmetrical circuit using at least two transistors and a common _____ resistor.
2. The output of a DA is proportional to the _____ (*difference, sum*) between (of) the two input signals.
3. A DA can operate in one or both of two modes, the _____ mode and the _____ mode.
4. When the two input signals to a DA are equal in amplitude and in the same phase, the output is _____ .
5. A DA will _____ (*amplify, reject*) differential-mode signals.
6. A DA with a single-ended input (Fig. 28-2) will result in _____ (*equal, unequal*) amplitude outputs.
7. The input to Fig. 28-4 is called an _____ (*inverting, noninverting*) input.
8. The outputs in Fig. 28-2 are equal in _____ but opposite in _____ .
9. In the circuit of Fig. 28-5, the unbypassed emitter resistor R_E introduces negative feedback, reducing the gain of the amplifier. _____ (*true, false*)
10. Consistent with good design, the larger the resistance of R_E in Fig. 28-5, the higher the common-mode rejection ratio. _____ (*true, false*)

MATERIALS REQUIRED

- Power supply: Regulated variable *dual* dc source
- Equipment: Oscilloscope; EVM or VOM
- Resistors: Two 1000-, 8200-, two 10,000-Ω ½-W
- Transistors: Two matched (if possible) 2N6004 or equivalent
- Miscellaneous: Two SPST switches; transformer, 120-V primary, 25-V center-tapped secondary

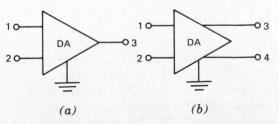

(a)   *(b)*

Fig. 28-8. DA circuit symbols: *(a)* 1 and 2 are inputs, 3 is output; *(b)* 1 and 2 are inputs, 3 and 4 are dual outputs taken with respect to ground.

The Differential Amplifier (DA)   **167**

Single-Ended Input

1. Connect the circuit of Fig. 28-9a. A 1-kilohertz (kHz) sine wave is applied across the primary of T and acts as the source, v_1, for the DA. S_1, S_2, and S_3 are *open*.
2. Set V_{CC}, the collector supply, at 10 V. Set V_{EE}, the emitter supply, at 9 V. *Close S_1 and S_2.*
3. Measure the dc voltage from the collector of Q_1 to ground (V_{C1}) and from the collector of Q_2 to ground (V_{C2}). These voltages should be approximately equal to 5 V. If the collector voltages are not 5 V, adjust V_{EE} until they are. Record in Table 28-1 V_{C1}, V_{C2}, and the measured voltage from the common emitter junction to ground (V_E). Also measure and record V_{EE}. *Do not change* the setting of V_{EE} for the remainder of the experiment. Measure and record the bias voltage from the base to the emitter of Q_1 (V_{B1}) and from the base to the emitter of Q_2 (V_{B2}).
4. *Close S_3.* With an oscilloscope measure v_1, the input 1-kHz signal voltage. Adjust the output of the signal generator until $v_1 = 50$ mV.
5. Externally trigger/sync the oscilloscope with the generator signal.
6. Observe the waveform at the base of Q_1. Adjust the oscilloscope controls until one or two waveforms appear on the scope. Center the trace until the reference waveform v_1 appears as in Table 28-2. Now observe, measure, record, and draw in proper phase with the reference waveforms v_{out1}, v_{out2}, and v_E (the signal from emitter to ground).
7. *Open S_1, S_2, S_3.* Reconnect the circuit so that the input v_2 is on the base of Q_2, as in Fig. 28-9b.
8. *Close S_1, S_2, and S_3,* and repeat steps 4 and 6.

Differential Mode

9. *Open S_1, S_2, and S_3.* Do *not* change settings of V_{CC} and V_{EE}.
10. Modify the circuit for differential-mode input, as in Fig. 28-5.

TABLE 28-1. DC Voltages in Differential Amplifier

V_{C1}	V_{C2}	V_E	V_{EE}	V_{B1}	V_{B2}

TABLE 28-2. Waveforms in DA—Single-ended Input

Test Point	Waveform	V p-p
Reference v_1		50 mV
v_{out1}		
v_{out2}		
v_E		
Reference v_2		50 mV
v_{out1}		
v_{out2}		
v_E		

11. *Close S_1, S_2, and S_3.* Set the output of the 1-kHz generator for a 100-millivolt (mV) p-p sine wave across the secondary of T. The input-signal voltages v_1 and v_2 from the base to the ground of Q_1 and Q_2, respectively, should each measure 50 mV p-p.
12. With the oscilloscope still externally triggered/synchronized by the generator waveform, observe v_1, the waveform on the base to ground of Q_1. Adjust oscilloscope controls as in step 6. Center the reference waveform v_1 so that it appears as in Table 28-3. Measure and

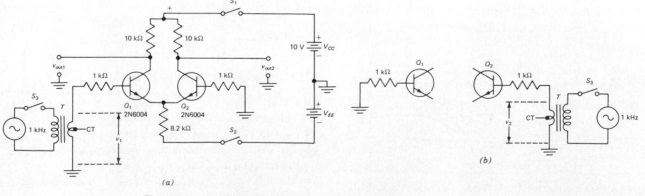

Fig. 28-9. Experimental DA with single-ended input to (a) Q_1, (b) Q_2.

adjust, if necessary, the output of the sine-wave generator so that $v_1 = 50$ mV p-p.

13. Observe and draw in Table 28-3 in proper phase with the reference the waveforms at the collector of Q_1 to ground (v_{c1}), base of Q_2 to ground (v_2), collector of Q_2 to ground (v_{c2}), the output waveform from collector of Q_2 to collector of Q_1 (v_{out}), and the emitter-to-ground (v_E).

NOTE: *The measurement of v_{out} will be possible only if the oscilloscope has a differential input. Otherwise do not attempt it.*

Measure also and record the peak-to-peak amplitude of these waveforms.

14. Compute the gain of Q_1 and Q_2 (v_{out1}/v_1 and v_{out2}/v_2, respectively) and record in Table 28-3.

Common Mode

15. *Open S_1, S_2, and S_3. Do not change settings of V_{CC} and V_{EE}. Modify the experimental circuit to conform with Fig. 28-6. Change only the input circuit.*

TABLE 28-3. Waveforms in DA—Differential Mode

Test Point	Waveform	V p-p
Reference v_1		50 mV
v_{c1}		
v_2		
v_{c2}		
v_{out}		
v_E		
Gain of Q_1		
Gain of Q_2		

16. *Close S_1, S_2, and S_3.* Set the output of the 1-kHz generator for a 50-mV p-p sine wave across the secondary of T.

17. With the oscilloscope externally triggered/synchronized by the generator signal (across the primary of T), observe the waveform v_1, base of Q_1 to ground. Adjust the oscilloscope controls and center the reference waveform so that it appears as in Table 28-4.

18. Observe and draw in Table 28-4 in proper phase with the reference the waveforms v_{c1}, v_2, v_{c2}, v_{out}, and v_E. Measure also and record the peak-to-peak voltages of each waveform. Remember, the oscilloscope must have a differential input to measure v_{out}.

19. Compute the gains of Q_1 and Q_2 and record in Table 28-4.

TABLE 28-4. Waveforms in DA—Common Mode

Test Point	Waveform	V p-p
Reference v_1		50 mV
v_{c1}		
v_2		
v_{c2}		
v_{out}		
v_E		
Gain of Q_1		
Gain of Q_2		

QUESTIONS

1. In the experimental single-ended DA (Fig. 28-9), what is the purpose of the unbypassed emitter resistor? Confirm your answer by referring specifically to the data in Table 28-2.

2. Refer to Table 28-2. Compare and explain the phase of v_{out1} and v_{out2}, with reference to the input waveform.

3. From the data in Table 28-2, what is the value of the output waveform taken from the collector of Q_2 to the collector of Q_1? Show your computations.

4. In the experimental circuit of Fig. 28-5, what do the waveforms collector to ground (v_{c1} and v_{c2}) of Q_1 and Q_2 prove? Refer specifically to the data in Table 28-3.

5. In the experimental circuit of Fig. 28-5, does the common emitter resistor provide degenerative feedback to the cir-

cuit? Which of the waveform(s) in Table 28-3 confirm(s) your answer?

6. Does the amplifier in the differential mode exhibit gain? Refer to the experimental data to substantiate your answer.

7. In the experimental circuit of Fig. 28-6, does the common emitter resistor provide degenerative feedback to the circuit? Which of the waveform(s) in Table 28-4 confirm(s) your answer?

8. How does the gain of the DA in Fig. 28-6 differ from the gain of the DA in Fig. 28-5? Why?

Answers to Self-Test

1. emitter
2. difference
3. common; differential
4. 0
5. amplify
6. equal
7. inverting
8. amplitude; phase
9. false
10. true

INTEGRATED CIRCUITS: THE LINEAR AMPLIFIER

OBJECTIVES

1. To become familiar with IC construction techniques
2. To connect an IC medium-power audio-frequency amplifier and check its operation

INTRODUCTORY INFORMATION

Integrated circuits (ICs) have radically altered the world of electronics. The profound effect of these microelectronic "giants" was first felt in the computer and digital field. First-generation electronic computers used vacuum tubes. Vacuum tubes were replaced by transistors in second-generation computers. Third-generation computers used digital ICs, greatly reducing computer size and increasing computer speed and reliability.

As the state of the art advanced, specially designed ICs were employed in *linear* circuits in communications, military, and industrial applications. ICs and large-scale ICs (LSICs) have changed the design of electronic devices from the use of only discrete components to hybrid solid-state devices which mix discrete components with ICs. Modular electronics owes its growth to ICs.

The tempo at which electronics is moving and changing is breathtaking. The vacuum tube, which ushered in the age of electronics, was discovered early in the twentieth century. The body of knowledge and applications expanded gradually, compared with today's pace. The impetus of the Second World War speeded up the process. The transistor, discovered in 1948, at the start of solid-state technology, foreshadowed the coming obsolescence of vacuum tubes. It was also the forerunner of IC technology. The transistor made possible reductions in product weight, size, power requirements, and cost, and improved reliability. Integrated circuits have further reduced product dimensions and cost, while ensuring even greater reliability.

Physical and Electrical Characteristics of an IC

What is an integrated circuit? It is not just a more efficient transistor or several transistors inside a small case. It is a complete "circuit," consisting of active and passive ele-

ments connected in a unique circuit configuration housed in a container no larger than a small transistor. The "active" elements are transistors and diodes. The passive components are resistors and capacitors. The elements in an IC are *not* discrete components wired together in a miniature circuit. Rather the IC is a complete circuit formed by an intricate microphotolithographic process, on a silicon chip no larger than the head of a pin. The IC "roadmap" of circuits can be seen only under high magnification. Visualize a ½-W audio amplifier, complete except for a few external, discrete components, mounted on a pin head!

Figure 29-1 shows a magnified closeup of an individual chip of a Motorola MC 1533 operational amplifier, a high-performance unit with a voltage gain of 60,000. Figure 29-2 is a digital integrated circuit mounted on a metal header. The header measures approximately ⅜ in in diameter. The relative size of an IC can be appreciated by noting Fig. 29-3, which shows an RCA IC 3502 next to a dime. It has been estimated that a thimblefull of ICs contains enough circuitry to build thousands of radios.

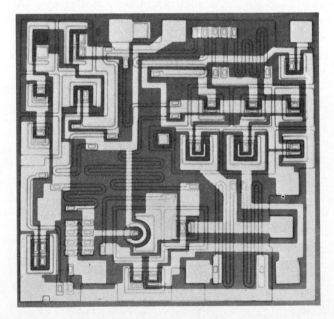

Fig. 29-1. Magnified view of an IC chip MC1533 operational amplifier. *(Motorola Semiconductor Products, Inc.)*

Fig. 29-2. Digital IC mounted on a metal header. *(Motorola Semiconductor Products, Inc.)*

Construction

IC fabrication processes are based on the production technology of silicon-diffused, or "planar," transistors. The student is referred to standard transistor texts which describe in detail the process for making planar transistors.

Briefly, silicon ICs are fabricated on a silicon disk, approximately 1 in in diameter, about the thickness of this paper. The disk is altered in a series of individual steps. The top of the disk is first oxidized, then covered with a photosensitive lacquer, called a *resist*. Circuit patterns are etched into the oxides by a microphotolithographic process. After heating, minute quantities of "impurities" or dopants, such as boron or arsenic, are diffused into the silicon to form the P and N islands on the disk. The process is repeated many times until all the circuit elements, transistors, diodes, resistors, and capacitors are made on the disk. The various elements are connected by depositing vaporized aluminum in the desired pattern to form the circuits (see Fig. 29-4). The completed disk may have as many as 200 ICs on it. These are then diced. The chips are separated, tested, and mounted on ceramic or metal bases. Then aluminum wires, about one-third the thickness of a human hair, are bonded between the IC contacts and the header leads. Then the package is sealed. The familiar transistor TO-5 case, flatpack, or dual-in-line packages (Fig. 29-5) are used.

Fig. 29-3. IC 3502 and dime for comparison. *(RCA)*

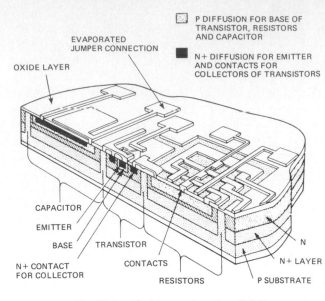

Fig. 29-4. IC chip cross section. *(RCA)*

IC Circuit Configurations

The concept of integrated circuits was an outgrowth of printed-circuit techniques and the diffused planar-transistor-fabricating technology. It was natural, at the start of IC technology, to attempt to develop ICs in the image of conventional transistor circuits, and so early IC circuit design closely resembled transistor circuits using discrete components. However, as knowledge and experience grew, IC circuits took on a completely unique character. An engineer or

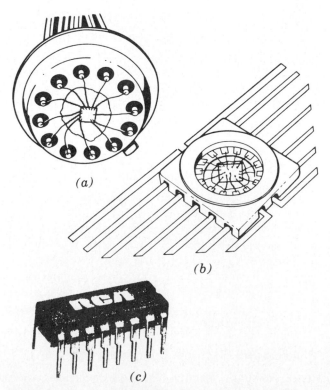

Fig. 29-5. ICs housed in *(a)* transistor-type TO-5 case, *(b)* flatpack, and *(c)* dual in-line package. *(RCA)*

technician, knowledgeable in conventional transistor circuitry, seeing an IC circuit arrangement for the first time might be rather startled and find it hard to relate the circuit design to the specific function.

The reason IC circuitry is unique is closely related to the methods and economics of the fabrication process and to the size of the chip. In conventional circuit design employing discrete components, the "active" elements (transistors and diodes) are more costly than resistors and capacitors. In IC technology transistors and diodes are less costly than resistors and capacitors, in that order, since transistors and diodes take up very little space on the chip, whereas resistors and capacitors require a relatively large amount of room. At this time, practical values for IC resistors are limited to 100 to 25,000 $\Omega \pm 30$ percent for P-type diffused resistors; 10 to 500 $\Omega -50$ to $+100$ percent for *pinch* resistors; and 10 to 500 $\Omega \pm 50$ percent for epitaxial resistors. Capacitors are limited to a range of approximately 3 to 30 picofarads (pF). Inductors are not fabricated. For this reason IC designers make extensive use of *active* elements and sparing use of resistors and capacitors. Thus direct coupling is favored over capacitive coupling. Integrated circuit *RC* rather than *LC* "tuning" is used. Coupling transformers are avoided if possible. Where not possible, external transformers are employed. As the state of the art advances, even greater changes in circuitry are anticipated.

An important consideration in IC economics is flexibility. Chips are designed to be used in more than one circuit configuration. Their arrangements involve various discrete external components. For example, the same chip in different external circuit arrangements may be used as a high-frequency wideband amplifier, a low-frequency amplifier, an oscillator and amplifier combined, or even a high-frequency amplifier-detector audio amplifier. This flexibility is facilitated by the fact that circuit leads on the body of the IC make it possible to "break in" on the circuit arrangement at critical points within the IC configuration. By way of illustration, the RCA CA 3018 IC contains an array of four transistors with 12 external connections. Figure 29-6 shows the schematic diagram of a CA 3018 broadband amplifier. The IC transistors shown inside the triangle are labeled Q_1, Q_2, Q_3, and Q_4. The amplifier uses the four transistors in this IC as discrete components. External connections of conventional resistors and capacitors are brought to the 12 IC terminals, numbered 1 through 12. Power is supplied from a $+6$-V source. The arrangement in Fig. 29-7 uses this same IC as a 15-megahertz (MHz) RF amplifier, and in Fig. 29-8 the IC is used as a class B audio amplifier. The CA 3018 presents four transistors in one case. Circuit design employing this IC is traditional.

Figure 29-9 shows the circuit diagram of a 10-terminal integrated circuit, the Motorola MC 1533, incorporating the *complete* circuit concept. Included on the IC are 14 transistors, 3 diodes, and 15 resistors, a total of 32 parts, interconnected as shown. The manufacturer's application notes for this IC give four arrangements in which this circuit can be used—source follower, filter and oscillator, voltage regulator,

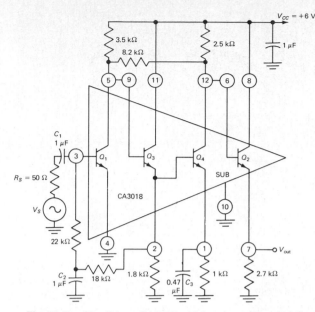

Fig. 29-6. Circuit for a CA3018 wideband amplifier. *(RCA)*

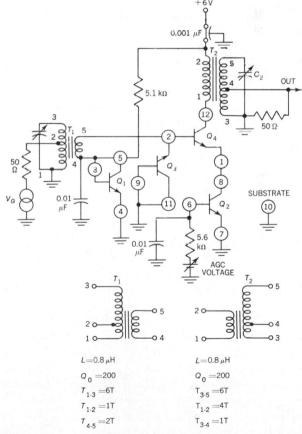

#22 WIRE ON Q·2 MATERIAL, CF107 TORROID FROM INDIANA GENERAL.

C_1, C_2 = ARCO 425 OR EQUIV.

Fig. 29-7. CA3018 15-MHz amplifier. *(RCA)*

and high-input-impedance voltmeter. In the case of the MC 1533 a complete circuit description is given in the application notes. It is evident that this IC, like all ICs, exhibits certain characteristics which must be considered in individual applications. But circuit use of the IC requires a "systems"

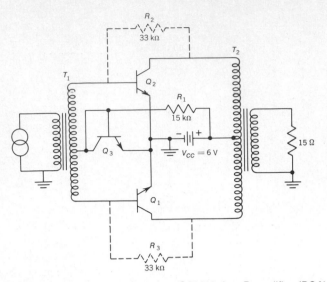

Fig. 29-8. Schematic diagram for a CA3018 class B amplifier. *(RCA)*

design orientation as opposed to use of discrete components in conventional circuit arrangements.

Figure 29-9 is typical of IC design of linear amplifiers in that direct coupling is used throughout. Note also the use of differential amplifier pairs, for example, Q_2–Q_3 and Q_8–Q_9.

Linear IC Audio Power Amplifier

The manufacturer describes the CA 3020 as a "monolithic integrated circuit, a multipurpose, multifunction power

amplifier designed for use in portable and fixed audio communication equipment and servo control systems." The schematic diagram of the CA 3020 is shown in Fig. 29-10; the functional block diagram in Fig. 29-11. This direct-coupled amplifier performs preamplifier, phase-inverter, driver, and power-output functions without transformers. The circuit can operate from a +3- to +9-V power source. The power output is determined by the supply voltage. The range of direct output is from 65 mW at +3 V to 550 mW at +9 V. A temperature-compensating voltage regulator permits operation over the temperature range −55 to +125°C.

Series-connected diodes D_1, D_2, and D_3 together with resistors R_{11} and R_{10} constitute the voltage regulator. The external power-supply voltage is connected between terminals 9 and 12 and supplies relatively constant voltages of 1.4 V (junction of D_1 and D_2) for base supply and 2.1 V (junction of D_1 and R_{11}) for the collectors.

Transistors Q_2 and Q_3, together with collector resistors R_1 and R_3, emitter resistor R_2, and base-biasing resistors R_4, R_5, R_6, and R_7, constitute the (differential) amplifier and phase inverter. The input ac signal may be capacitor-coupled into terminal 3 or terminal 10. If applied to terminal 10, Q_1 is operated as a buffer amplifier, an emitter follower whose output is then coupled to terminal 3, the input of Q_2 and Q_3. Coupling from Q_2 and Q_3 is achieved through the common unbypassed emitter resistor R_2. Two equal-amplitude signals, 180° out of phase, are developed at the collectors of Q_2 and Q_3. These signals are direct-coupled to the bases of emitter-follower amplifiers Q_4 and Q_5. Negative feedback to Q_2 is provided by R_5 and R_7 for dc and ac stability of Q_2 and Q_3. The signal is direct-coupled from the emitters of Q_4 and

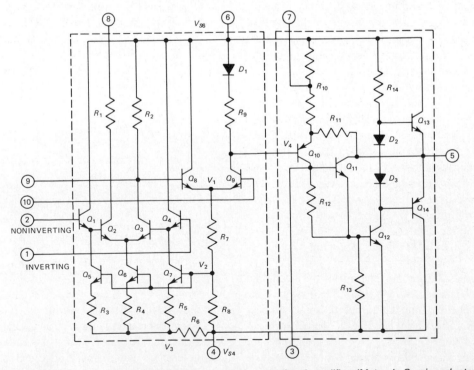

Fig. 29-9. Schematic diagram of MC1533, a high-voltage, monolithic operational amplifier. *(Motorola Semiconductor Products, Inc.)*

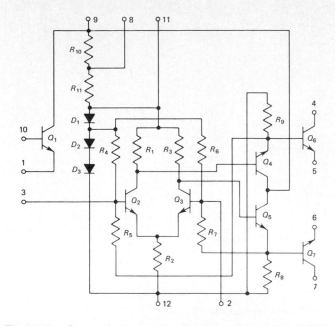

Fig. 29-10. Schematic diagram of a CA3020 audio amplifier. *(RCA)*

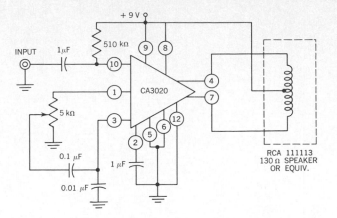

Fig. 29-12. CA3020 audio amplifier without transformer. *(RCA)*

Q_5 to the bases of Q_6 and Q_7, the power amplifiers. Q_6 and Q_7 operate in a class B push-pull mode to deliver power to the load, which may be a high-impedance center-tapped speaker fed directly as in Fig. 29-12 or a center-tapped output transformer feeding a low-impedance speaker, Fig. 29-13.

In this experiment you will have the opportunity to try out this versatile amplifier with a variety of audio sources.

SUMMARY

1. An integrated circuit (IC) is a complete circuit microphotographically and chemically etched into a silicon chip (Fig. 29-1).
2. An IC consists of active elements (transistors and diodes) and passive elements (resistors and capacitors) connected in a desired circuit arrangement and housed in a small case. The case may be the familiar transistor TO-5 case, flatpack, or dual-in-line package (Fig. 29-5).
3. The components in an IC are not discrete, and if it were possible to see them (which it is not) they would *not* resemble the familiar individual resistors, transistors, diodes, and capacitors.

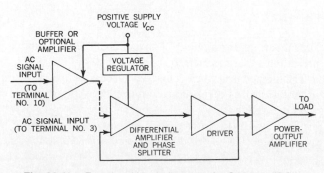

Fig. 29-11. Functional block diagram of a CA3020. *(RCA)*

4. ICs are economically feasible because they are manufactured simultaneously in large quantities.
5. There are many IC configurations or circuits. However, they may be broadly classified as linear ICs and digital ICs. Linear ICs are used mainly in signal-processing circuits (amplifiers, oscillators, etc.) in communications electronics, for example in TV, radio receivers, and transmitters, and in hi-fi equipment. Digital ICs are found in computers, calculators, and other electronic "counting" devices.
6. Because of economic considerations and the manufacturing process, IC circuit arrangements differ markedly from circuits using discrete components.
7. ICs make extensive use of direct coupling. Moreover, because of their many advantages in circuit design, differential amplifiers are used extensively in linear ICs.
8. IC chips are either highly specialized (for example, the chips used in a calculator) or designed with the greatest flexibility in mind. In the latter case, the addition of external discrete components can modify an IC circuit for a variety of purposes.
9. The CA 3020 housed in a TO-5 case (Fig. 29-14a) is an example of a linear IC.

SELF-TEST

Check your understanding by answering these questions.

1. An IC is a complete circuit consisting of discrete components. _____ (*true, false*)
2. ICs usually come in one of three different cases. These are: (*a*) _____ , (*b*) _____ , and (*c*) _____ .
3. The most economical component to fabricate in an IC is a _____ .
4. The active components in an IC are _____ and _____ .
5. ICs may be classified as _____ or _____ devices.
6. An _____ _____ is an example of a linear IC.

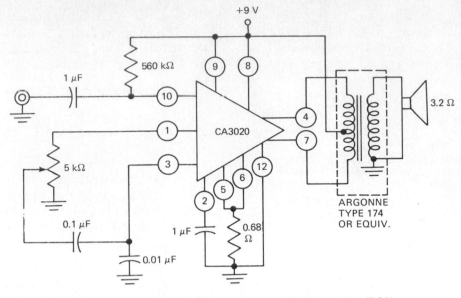

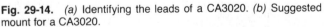

Fig. 29-13. IC amplifier driving a low-impedance speaker. *(RCA)*

7. In Fig. 29-10, the inputs to the IC amplifier are either at terminal _____ or _____ .
8. In Fig. 29-10, the circuit consisting of Q_2, Q_3, and their associated resistors is called a _____
_____ .

MATERIALS REQUIRED

- Power supply: Variable regulated low-voltage dc source; isolation transformer
- Equipment: Oscilloscope; EVM, ac generator
- Resistors: 0.68-Ω, 560-kΩ ½-W
- Capacitors: miniature 0.01-μF 12-V; miniature 0.1-μF 12-V, two 1.0-μF 12-V
- Semiconductors: RCA CA 3020 integrated circuit with mounting, or equivalent
- Miscellaneous: Push-pull output transformer (Kelvin 149-18 or equivalent); low-impedance speaker; 5000-Ω 2-W potentiometer; audio source such as high-impedance microphone, phonograph, or AM/FM tuner; as required for extra-credit steps

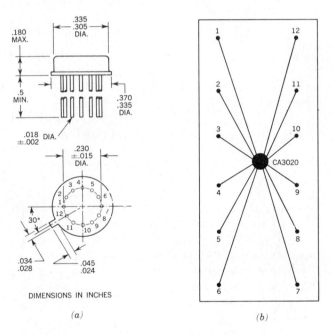

DIMENSIONS IN INCHES

(a)

(b)

Fig. 29-14. *(a)* Identifying the leads of a CA3020. *(b)* Suggested mount for a CA3020.

PROCEDURE

NOTE: Figure 29-14*a* is a side and bottom view of the CA 3020. The leads are identified by reading clockwise from the tab. The tab is adjacent to contact 12 on the IC.

The CA 3020 is a high-gain audio amplifier. In wiring the external circuit extreme care must be used to prevent oscillation. There is no problem in its commercial use, where lead lengths can be kept short and external components uniquely oriented to minimize undesired regenerative feedback.

Breadboarding the circuit by conventional means used in the preceding experiments leads to instability.

The method described below was successful in eliminating instability and is therefore recommended. A special mount was prepared for the CA 3020 from an etched board $3 \times 1\frac{1}{2}$ in. See Fig. 29-14*b*. (The mounting process is described in the *Instructor's Guide for Basic Electronics*.) The terminal pins on this board, numbered 1 to 12 in a counterclockwise direction, correspond to the terminal leads of the CA 3020.

Miniature capacitors and resistors were placed on the board in accordance with the particular circuit configuration required in this experiment. The capacitor and resistor pigtails were wound around the terminal pins, making good mechanical and electrical connections. Placement of parts was arranged to keep the shortest lead lengths. Where connection to external components was required, such as to the transformer, lead length was also kept short. Connection between terminals of the CA 3020, where required, was achieved by wrapping bare flexible solid wire jumpers around the terminals.

CAUTION: In the following procedure, be sure that any audio source that is not line isolated is plugged into an isolation transformer.

1. Connect the circuit of Fig. 29-13. In this arrangement the CA 3020 serves as a ½-W amplifier driving a low-impedance speaker.
2. (a) To the amplifier input connect a high-impedance microphone. Try out this PA system and comment on its effectiveness.
   Sound volume: _____ (*good or poor*).
   Quality of sound: _____ (*good or poor*).
   (b) With an oscilloscope connected across the primary of output transformer, observe the patterns (waveform) of each of the vowel sounds.
3. To the amplifier input connect the output of a phonograph pickup. Comment on the
   (a) Volume of sound _____ .
   (b) Quality of sound _____ .

4. To the amplifier input connect the output of an AM or FM tuner. Comment on the
   (a) Volume of sound _____ .
   (b) Quality of sound _____ .

Extra Credit—Frequency Response

5. Describe in detail a method you would use to determine the frequency response of the amplifier connected as in Fig. 29-15.
6. Following your procedure, measure the frequency response, and list the data in a specially prepared table. Use an input signal of 10 mV (approx).
   Plot a graph of your results, and show the bandpass of the amplifier 30 percent down from reference (at 1 kHz).

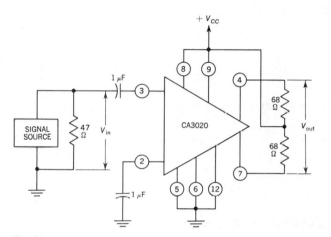

Fig. 29-15. Test setup for measuring frequency response. *(RCA)*

QUESTIONS

1. How does an integrated circuit differ from a conventional transistorized circuit?
2. What advantages does an IC offer over conventional transistorized circuitry?
3. What is the relationship, if any, between amplifier power output and V_{CC}?
4. Explain in detail how signal coupling is accomplished from Q_2 to Q_3.
5. Why would you expect that a CA 3020 amplifier would be wideband?
6. (*a*) What advantages does the IC audio amplifier used in this experiment have over discrete-component AF amplifiers? (*b*) What disadvantages does it have?
7. Suggest a method for using the CA 3020 in conjunction with discrete components to provide more audio power.

Answers to Self-Test

1. false
2. TO-5; flatpack; dual-in-line
3. transistor
4. transistors; diodes
5. linear; digital
6. audio amplifier
7. 10; 3
8. differential amplifier

OPERATIONAL AMPLIFIER (OP AMP)

OBJECTIVES

1. To verify experimentally that the gain of an op amp can be made dependent on only the external negative feedback loop from output to input
2. To operate an op amp as a noninverting amplifier
3. To operate an op amp as an inverting *summer*

INTRODUCTORY INFORMATION

IC Operational Amplifiers

An op amp is a high-gain, direct-coupled *differential* linear *amplifier* whose response characteristics are externally controlled by negative feedback from the output to the input. OP amps, widely used in computers, can perform mathematical operations such as summing, integration, and differentiation. OP amps are also used as video and audio amplifiers, oscillators, etc., in communication electronics. Because of their versatility op amps are widely used in all branches of electronics both in digital and linear circuits.

OP amps lend themselves readily to IC manufacturing techniques. Improved IC manufacturing techniques, the op amp's adaptability, and extensive use in the design of new equipment have brought the price of IC op amps from very high to very reasonable levels. These facts ensure a very substantial role for the IC op amp in electronics.

Figure 30-1 shows the symbol for an op amp. Note that the operational amplifier, like the differential amplifier studied in Experiment 28, has *two* inputs, marked (−) and (+). The minus input is the *inverting* input. A signal applied to the minus terminal will be shifted in phase 180° at the output. The plus input is the *noninverting* input. A signal applied to the plus terminal will appear in the same phase at the output as at the input. Because of the complexity of the internal circuitry of an op amp, the op amp symbol is used exclusively in circuit diagrams.

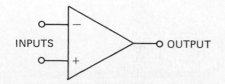

Fig. 30-1. Symbol for operational amplifier.

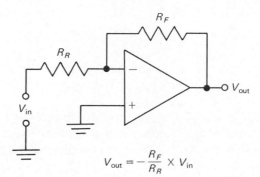

$$V_{out} = -\frac{R_F}{R_R} \times V_{in}$$

Fig. 30-2. Circuit of op amp showing negative feedback loop.

Negative Feedback Control

Figure 30-2 shows the basic circuit, including the negative feedback loop of an op amp. Note that the output is fed back to the *inverting* input terminal in order to provide *negative* feedback for the amplifier. The input signal is applied to the inverting input in Fig. 30-2. As a result, the output will be inverted. It is possible to operate the op amp as a noninverting amplifier by applying the signal to the *plus* input, as in Fig. 30-3. In this circuit the feedback network is still connected to the inverting input.

Analysis of the gain of op amps is outside the scope of this book. The student is referred to standard texts for such analysis. Several of the gain formulas are given here, however; they can be applied when specified conditions for the op amp are met.

For the circuit of Fig. 30-2, the output of the amplifier is defined by Eq. (30-1).

$$V_{out} = -\frac{R_F}{R_R} V_{in} \qquad (30\text{-}1)$$

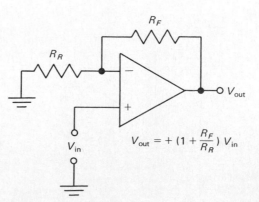

$$V_{out} = +\left(1 + \frac{R_F}{R_R}\right) V_{in}$$

Fig. 30-3. Op amp operated as noninverting amplifier.

The minus sign indicates that the sign of the output is inverted as compared to the input. The equation for the gain of this amplifier is

$$\text{Gain} = \frac{R_F}{R_R} \qquad (30\text{-}2)$$

For the noninverting amplifier in Fig. 30-3:

$$V_{\text{out}} = \left(1 + \frac{R_F}{R_R}\right) V_{\text{in}} \qquad (30\text{-}3)$$

and its gain is

$$1 + \frac{R_F}{R_R} \qquad (30\text{-}4)$$

Equations (30-1) to (30-4) indicate that the output voltage is dependent only on the ratio of the feedback resistors R_F and R_R and that the gain of the op amp is dependent only on these resistors.

Several examples are given here to illustrate the use of the gain and voltage output formulas. In Fig. 30-2, if R_F = 10,000 Ω and R_R = 2500 Ω, the overall gain of the amplifier is

$$\text{Gain} = \frac{10,000}{2500} = 4$$

If $R_R = R_F$ = 10,000 Ω, the gain of the circuit is 1, and the amplifier becomes an inverter with unity gain. That is, the voltage of the output is the same as the input, but the output is 180° out of phase with the input.

In Fig. 30-3, if $R_R = R_F$ = 10,000 Ω, then

$$V_{\text{out}} = \left(1 + \frac{10,000}{10,000}\right) V_{\text{in}} = 2 \times V_{\text{in}}$$

Here the output is twice the input voltage and in phase with the input. The gain of the amplifier is 2.

It is evident from Eq. (30-2) that if $R_R > R_F$, the output voltage is smaller than the input. Thus, if $R_R = 2R_F$,

$$V_{\text{out}} = -\tfrac{1}{2}V_1$$

From Eq. (30-3) it can be seen that whether R_F is smaller or larger in value than R_R, the output voltage of Fig. 30-3 always is greater than the input voltage. In the special case where $R_R = \infty$, that is, where R_R doesn't exist, $V_{\text{out}} = V_{\text{in}}$.

The circuit of an op amp connected as a noninverting voltage follower is shown in Fig. 30-4.

The preceding discussion has emphasized that by the proper choice of feedback resistors, op amps can become constant-multiplier or constant-divider amplifiers. The gain ratio of this type of op amp is called the *scaling factor* of the amplifier.

Before leaving this discussion, note that slight imbalances in the internal circuitry of an op amp will lead to drift and dc imbalances in the output. To eliminate these imbalances, where necessary, op amps frequently provide two offset null terminals. The null circuitry and instructions for the balancing procedure are supplied by the manufacturer.

Another possible source of input error in op amp circuitry may result from the bias current. This can be eliminated by connecting a resistor R in the noninverting input whose value is the resistance of the parallel-combination of R_F and R_R (Fig. 30-5). The input resistances of the inverting and noninverting inputs are thus equalized, eliminating this bias-current error.

Op Amp Connected as a Summer

In the circuit of Fig. 30-6, the op amp is connected as an algebraic *summer*. It can be seen that if $R_F = R_1 = R_2$, the output voltage

$$V_0 = -(V_1 + V_2) \qquad (30\text{-}5)$$

That is, the output voltage is the sum of the input voltages with the sign inverted. If the input voltages are of opposite sign, the circuit of Fig. 30-6 acts to invert and subtract the input voltages. For example if $V_1 = +3$ V and $V_2 = -2.5$ V, then

$$V_0 = -(+3 - 2.5) = -0.5 \text{ V}$$

The output voltage can be made equal to the sum of a desired ratio of input voltages, depending on the values of R_F, R_1, and R_2. For example if $R_F = 2R_1 = 3R_2$, then since

$$V_0 = -\left(\frac{R_F}{R_1} V_1 + \frac{R_F}{R_2} V_2\right)$$

$$V_0 = -(2V_1 + 3V_1) \qquad (30\text{-}6)$$

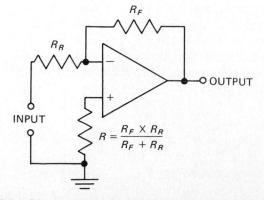

Fig. 30-5. *R* is connected to the noninverting input to compensate for possible input-bias error. The resistance of both inverting and noninverting inputs is equalized.

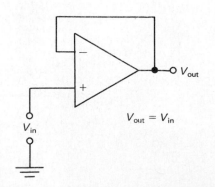

Fig. 30-4. Op amp operated as a voltage follower.

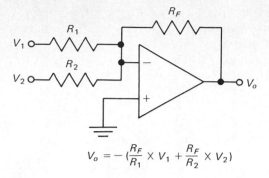

$$V_o = -\left(\frac{R_F}{R_1} \times V_1 + \frac{R_F}{R_2} \times V_2\right)$$

Fig. 30-6. Operational amplifier connected as a summer.

The summer of Fig. 30-6 can be modified to provide three or more inputs.

Op Amp Specifications

The manufacturer provides a circuit diagram, a basic diagram, and specifications for each op amp type, including performance graphs. A circuit diagram for the 741C op amp which you will use in this experiment is shown in Fig. 30-7a. Figure 30-7b is a basing diagram showing all terminal connections.

The 741 is available in two grades, the military (M) and commercial (C). It also comes in different packages. The commercial grade, can type IC, designated 741HC, will be used in this experiment.

The manufacturer lists the following *maximum* ratings for the 741HC:

Supply voltage	± 18 V
Internal power dissipation	500 mW
Differential input voltage	± 30 V
Operating temperature	0 to 70° C

The following characteristics are also of interest to the technician and designer:

Input bias current	800 nA
Input resistance	0.3 to 2 MΩ
Input voltage range	± 13 V, typical
Common-mode rejection ratio for $R_s \leqslant 10$ kΩ	90 dB, typical
Output resistance	75 Ω
Output short-circuit current	25 mA
Supply current	2.8 mA, typical
Power consumption	85 mW, typical
Large-signal voltage gain, $R_L \geqslant 2$ kΩ and $V_{out} = \pm 10$ V	15,000
Output voltage swing, $R_L \geqslant 2$ kΩ	± 13 V, typical

Typical performance curves for the 741C are shown in Fig. 30-8.

SUMMARY

1. An operational amplifier is a high-gain, direct-coupled differential amplifier whose gain is controlled by external negative feedback circuitry.
2. An op amp has two inputs, one inverting ($-$) and one noninverting ($+$) (Fig. 30-1).
3. In the inverting amplifier (Fig. 30-2) the feedback network is connected from the output to the inverting ($-$) input. The signal is also applied to the inverting input.
4. The output voltage of the inverting amplifier, Fig. 30-2 is

$$V_{out} = -\frac{R_F}{R_R} V_{in}$$

The minus sign represents the 180° phase shift which occurs from input to output.
5. The gain of the inverting amplifier is R_F/R_R.

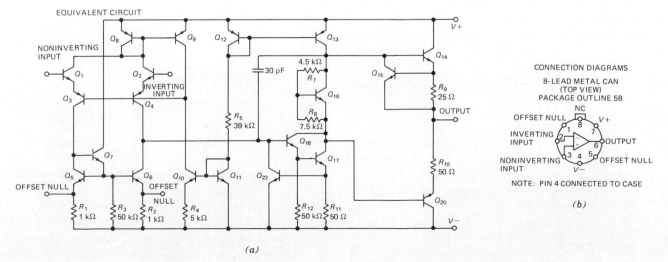

Fig. 30-7. (a) Circuit diagram; (b) terminal diagram of 741C. (Fairchild)

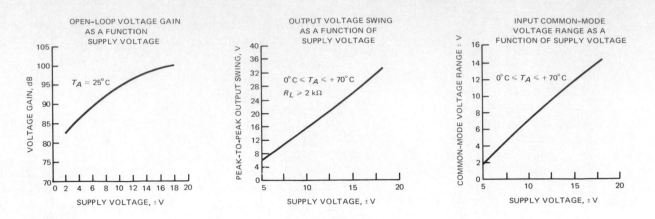

NOTES:
1. Rating applies to ambient temperatures up to 70°C. Above 70°C ambient derate linearly at 6.3 mW/°C for the Metal Can, 8.3 mW/°C for the DIP, 5.6 mW/°C for the Mini DIP, and 7.1 mW/°C for the Flatpak.
2. For supply voltages less than ± 15 V, the absolute maximum input voltage is equal to the supply voltage.
3. Short circuit may be to ground or either supply. Rating applies to +125°C case temperature or 75°C ambient temperature.

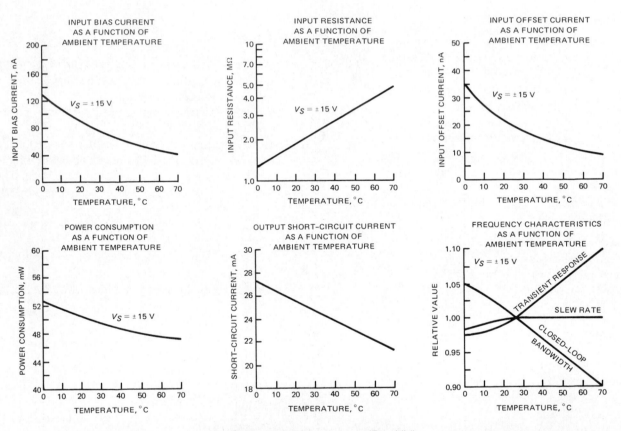

Fig. 30-8. Performance curves. *(Fairchild)*

6. In the noninverting amplifier (Fig. 30-3) the signal is applied to the noninverting (+) input, while the feedback network is connected to the inverting input.

7. The output voltage of the noninverting amplifier of Fig. 30-3 is

$$V_{\text{out}} = \left(1 + \frac{R_F}{R_R} \right) V_{\text{in}}$$

and the gain of this amplifier is $1 + (R_F/R_R)$.

8. The gain of the inverting amplifier can be made equal to, greater than, or less than 1.

9. The gain of the noninverting amplifier can be made equal to or greater than 1.

10. Op amps can be connected as summers (Fig. 30-6).

11. Frequently op amps contain *offset null* terminals to compensate for imbalance and drift in the internal circuits.

12. Op amp circuit diagrams, basing diagrams, and specifications are provided by the manufacturer.

SELF-TEST

Check your understanding by answering these questions.

1. In an op amp it is possible to change the internal (open-loop) gain of the amplifier by adding _____ _____ .

2. In the circuit of Fig. 30-2, $R_F = 5000 \ \Omega$, $R_R = 1250 \ \Omega$. The gain of the circuit is _____ .

3. If a $+0.5$-V dc voltage were applied to the input of the amplifier in question 2, the output would be _____ V.

4. In the circuit of Fig. 30-3, $R_F = 10,000 \ \Omega$, $R_R = 10,000 \ \Omega$, and the input ac voltage is 1 V p-p. The output voltage is _____ V p-p.

5. In the circuit of Fig. 30-3 a positive-going signal on the input will develop a _____ -going signal on the output.

6. In the circuit of Fig. 30-4, a $+1.5$-V dc voltage on the input will result in _____ V on the ouput.

7. In Fig. 30-6, $R_F = 10,000 \ \Omega = R_1 = R_2$. If $V_1 = V_2 = -1.5$ V, the output voltage is _____ V.

8. In Fig. 30-6, $R_F = 10,000 \ \Omega$, $R_1 = 5000 \ \Omega$, $R_2 = 2500 \ \Omega$. If $V_1 = +1.5$ V and $V_2 = -1.5$ V, the output voltage is _____ V.

MATERIALS REQUIRED

- Power supply: Dual variable regulated low-voltage dc source
- Equipment: Oscilloscope; AF sine-wave generator; EVM; resistor decade box (5 decades, 0 to 99,999)
- Resistors: Two 10,000-Ω ½-W
- Semiconductor: 741C
- Miscellaneous: Four SPST switches (SPDT may be used to replace SPST switches)

PROCEDURE

Gain of an Op Amp

1. Connect the circuit of Fig. 30-9. $R_F = R_R = 10,000 \ \Omega$. S_1 and S_2 are *open*. Set each of the two supplies to 9 V. Set the sine-wave generator at 1000 Hz, zero output. Connect the oscilloscope to the output of the op amp. *Externally* trigger/sync the oscilloscope with the output from the generator.

2. *Close* S_1 and S_2, applying power to the circuit.

3. Gradually increase the output from the signal generator just *below* the point where the waveform distorts. Measure and record in Table 30-1 the peak-to-peak output-signal voltage. This is the maximum undistorted output signal for the feedback resistors in the circuit.

4. With the oscilloscope measure and record in Table 30-1 the input signal v_{in} to the amplifier (output of signal generator).

5. Compute and record the gain of the amplifier (v_{out}/v_{in}).

6. Compare the phase of the input and output signals and indicate in Table 30-1 whether they are in phase or 180° out of phase.

7. Reduce the output of the generator to 0.

TABLE 30-1. Gain of Inverting Op Amp

R_F, Ω	R_R, Ω	V p-p		Gain	Phase
		Output	Input	(v_{out}/v_{in})	
10,000	10,000				
	5,000				
	3,333				
	2,500				
	20,000				
	30,000				

8. Repeat steps 3 through 7 for each value of R_R shown in Table 30-1. Use a resistor decade box to set up the required value of R_R.

Noninverting Amplifier

9. **Power off.**

10. Modify the circuit to conform to that in Fig. 30-10. The power supplies remain connected as in Fig. 30-9, each set at 9 V. The generator output is at 1000 Hz, 0 V.

11. For each value of R_F and R_R shown, complete and record the data required in Table 30-2, following the same procedure as in Table 30-1.

Op Amp As an Inverting Summer

12. **Power off.** Keep the power supplies connected as in Fig. 30-9, each at 9 V. Modify the experimental circuit as in Fig. 30-11. $R_F = R_1 = R_2 = 10,000 \ \Omega$. Use the resistor decade box for R_2. V_1 and V_2 are 1.5-V batteries.

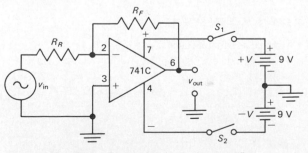

Fig. 30-9. Experimental inverting amplifier.

Operational Amplifier (OP AMP) **183**

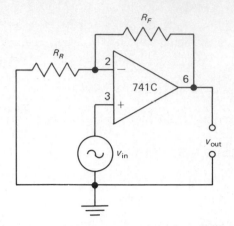

Fig. 30-10. Experimental noninverting amplifier.

TABLE 30-2. Noninverting Op Amp

R_F, Ω	R_R, Ω	V p-p Output	V p-p Input	Gain (v_{out}/v_{in})	Phase
	10,000				
	5,000				
10,000	3,333				
	20,000				
	30,000				

13. **Power on.** *Close S_3. S_4 is open.* Measure and record in Table 30-3 V_1 and V_{out}.
14. *Open S_3. Close S_4.* Repeat step 13.
15. *Close S_2.* Both S_3 and S_4 are now closed. Measure and record V_{out} and V_{in}.
16. Reverse the polarity of V_1. Measure and record V_{out} and V_{in} with S_3 and S_4 closed.

Extra Credit

17. Modify the summer circuit so that with the two 1.5-V inputs in Fig. 30-11, $V_{out} = -4.5$ V (approx). Show the values of all resistors and V_1 and V_2 polarities. Measure the output voltage and record it in a specially prepared table.
18. Modify the summer circuit so that with two 1.5-V inputs, the output voltage = +1.5-V (approx). Show and record the values of all resistors and V_1 and V_2 polarities.
19. Experimentally determine if there is a practical limit on the undistorted gain of the inverting op amp. Record the maximum voltage and gain.

TABLE 30-3. Op Amp as a Summer

Condition S_3	Condition S_4	Input Polarity V_1	Input Polarity V_2	V_{in}, V V_1	V_{in}, V V_2	V_{out}, V
ON	OFF	+	X		X	
OFF	ON	X	+	X		
ON	ON	+	+			
ON	ON	−	+			

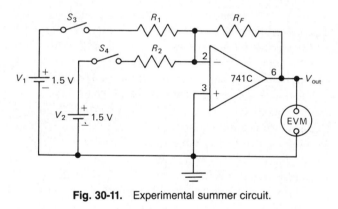

Fig. 30-11. Experimental summer circuit.

QUESTIONS

1. Does the maximum undistorted sine-wave output of the op amp vary with amplifier gain? Refer to the experimental data to confirm your answer.
2. What is the relationship, if any, between the polarity of the output and input voltages in your experimental op amp? Refer to your data.
3. What is the relationship, if any, between experimental inverting amplifier gain and R_F and R_R? Refer to your data to support your answer.
4. Do the data in Table 30-1 bear out the gain formula (R_F/R_R) in each case? Show your computations.

5. Do the data in Table 30-2 bear out the gain formula for a noninverting amplifier in each case? Show your computations.
6. In which procedural step was the *summer* operated as a *subtractor*?

Answers to Self-Test

1. external feedback resistors
2. 4
3. −2
4. 2
5. positive
6. +1.5
7. +3
8. +3

OP AMP CHARACTERISTICS

OBJECTIVES

1. To get data for input bias current
2. To measure and null the output offset voltage
3. To calculate the slew rate of a 741
4. To observe the effect of power bandwidth

INTRODUCTORY INFORMATION

Figure 31-1 shows a simplified schematic diagram for a 741 and many later-generation op amps. To pin down some important ideas, we will analyze how this circuit works.

Input Stage

Q_1 and Q_2 are a *diff amp* (differential amplifier). This diff amp is biased by Q_{14}, which acts like a constant current source (discussed in Experiment 28). The diff amp drives an active load consisting of Q_3 and Q_4. An input signal V_{in} produces an amplifier current which goes into the base of Q_5.

Second and Third Stages

The second stage is an emitter follower Q_5. It steps up the input impedance of the third stage Q_6 by a factor of β. Transistor Q_6 acts like a driver for the output stage. The plus sign on the collector of Q_5 means it is internally connected to the positive supply pin; similarly, the minus sign at the bottoms of R_2 and R_3 mean these are connected to the negative supply pin.

Output Stage

The last stage is a class B push-pull emitter follower (Q_9 and Q_{10}). Because of the *split supply* (equal positive and negative voltages), the quiescent (no-signal) output is ideally 0 V. Q_{11} is part of an output biasing network that produces a small idling current and eliminates crossover distortion.

The output stage uses *diode bias*, found mainly in integrated circuits. Diodes Q_7 and Q_8 are known as *compensating diodes*; they have *IV* curves that match the base-emitter diode curves of Q_9 and Q_{10}. Diode bias compensates for changes in temperature that otherwise would increase the idling current to dangerous levels. This type of bias is possible with integrated circuits because all components are on the same chip.

Compensating Capacitor

C_C is called a *compensating capacitor*. Typically 30 pF, this capacitor has a pronounced effect on the frequency response and prevents *oscillations* (unwanted signals producd by the amplifier). A later experiment tells you more about the compensating capacitor and oscillations.

Active Loading

All CE stages discussed up to now have used a passive load (resistor, transformer primary, etc.). In Fig. 31-1, we have an example of a CE stage (Q_6) driving an *active load* (Q_{11}). Ideally, Q_{11} acts like a constant current source; therefore, its impedance approaches infinity. Because of this, any signal current out of Q_6 is forced into the final output transistors (Q_9 or Q_{10}), whichever is conducting.

Active loading (using transistors for loads instead of resistors) is very popular in integrated circuits because it is easier and less expensive to fabricate transistors on a chip than resistors. Metal-oxide semiconductor (MOS) digital integrated circuits use active loading almost exclusively; in

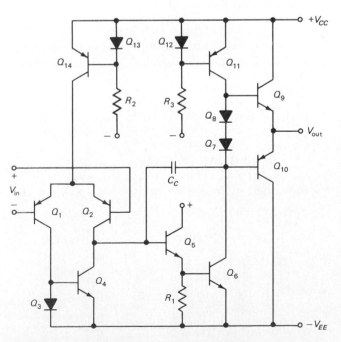

Fig. 31-1. Simplified schematic for 741 and other typical op amps.

these ICs, one metal-oxide semiconductor field-effect transistor (MOSFET) is the active load for another MOSFET.

Input Bias Current

The op amp of Fig. 31-1 is equivalent to what is inside the IC package. For the circuit to work, you need to connect V_{CC} and V_{EE} supplies. But that is not all. You also have to connect external dc returns for the floating input bases. In other words, the Q_1 and Q_2 base currents have to flow to ground to complete the circuit, because the other ends of the power supplies are grounded.

The two base currents of the input diff amp must flow through external resistances. These base currents are close in value, but not necessarily equal. When slightly unequal base currents flow through external resistances, they produce a small differential input voltage or unbalance; this represents a false input signal. When amplified, this small input unbalance produces an offset in the output voltage. The smaller the base currents are, the better, because the output offset is minimized.

The *input bias current* shown on data sheets is the average of the two input base currents. It tells you approximately what each input current is. As a guide, the smaller the input bias current, the smaller the possible unbalance. The 741 has a worst-case input bias current of 500 nA, which is acceptable in many applications. But in critical applications, a later-generation op amp may be preferred.

Input Offset Current

The *input offset current* is the difference between the two input currents; it tells you how much larger one current is than the other. The 741 has a worst-case input offset current of 20 nA. This means we may find up to 20 nA more current in one base than the other. Again, the general guide is this: The smaller the input offset current, the better the op amp.

Input Offset Voltage

Ideally, the ouput voltage should be 0 when the voltage between the inverting and noninverting inputs is 0. In reality, the output voltage may still have a slight offset or unbalance. This output offset is caused by internal mismatches, tolerances, etc. In other words, even if you short the inverting and noninverting inputs together to eliminate the effect of input bias current, the output may still have a slight offset from 0.

The *input offset voltage* is the input voltage needed to null or zero the quiescent output voltage. For example, a 741 has a worst-case input offset voltage of 5 mV. Under no-signal conditions, therefore, we may have to apply an input of 5 mV to produce an output voltage of exactly 0.

CMRR was defined for a diff amp. As discussed in Experiment 28,

$$CMRR = \frac{A_{DM}}{A_{CM}} \tag{31-1}$$

The CMRR of a 741 is approximately 30,000. Given equal common-mode and differential-mode input signals, the common-mode signal will be 30,000 times smaller than the differential-mode signal at the output of the 741.

Most forms of interference like static, ripple, induced noise voltages, etc., drive an op amp as common-mode signals. As a result, they receive very little amplification compared to the desired signal which drives the op amp in the differential mode.

Slew Rate

Among all specifications affecting the ac operation of an op amp, *slew rate* is the most important because it places a severe limit on large-signal operation. Slew rate is defined as the maximum rate at which the output voltage can change. The 741, for instance, has a typical slew rate of 0.5 volts per microsecond (V/μs). This is the ultimate speed of a typical 741; its output voltage can change no faster than 0.5 V/μs.

Figures 31-2a and b illustrate the idea. If we overdrive a 741 with large step input (Fig. 31-2a), the output slews (rises linearly) as shown in Fig. 31-2b. It takes 20 μs for the output voltage to change from 0 to 10 V (nominal output swing). It is impossible for the output of a typical 741 to change faster than this. Without getting into details, compensating capacitor C_C in Fig. 31-1 is the cause of slew rate; it must be charged before the voltage can increase. Because of the high op amp gain, all we see at the output is the earliest part of an exponential charge (Fig. 31-2b).

We can also get slew-rate limiting with a sinusoidal signal. Figure 31-3a shows a sine wave with a peak value of 10 V. As long as the initial slope of the sine wave is less than or equal to the slew rate S_R, there is no slew-rate limiting. But when the initial slope of the sine wave is greater than S_R, we get the slew-rate distortion shown in Fig. 31-3b. The output begins to look triangular; the higher the frequency, the smaller the swing and the more triangular the waveform.

Power Bandwidth

Slew-rate distortion of a sine wave starts at the point where the initial slope of the sine wave equals the slew rate of the op amp. With advanced mathematics, it is possible to derive this useful formula:

$$f_{\max} = \frac{S_R}{2\pi V_P} \tag{31-2}$$

where $f_{\max}$ = highest undistorted frequency
S_R = slew rate of op amp
V_P = peak voltage of output sine wave
As an example, if the output sine wave has a peak voltage of

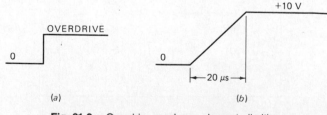

Fig. 31-2. Overdrive produces slew-rate limiting.

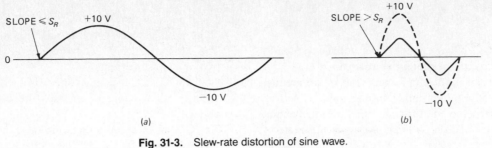

Fig. 31-3. Slew-rate distortion of sine wave.

10 V and the op amp a slew rate of 0.5 V/μs, the maximum frequency for large-signal operation is

$$f_{max} = \frac{0.5 \text{ V/μs}}{2\pi \times 10 \text{ V}} = 7.96 \text{ kHz}$$

Frequency f_{max} is called the *power bandwidth* of an op amp. We have just found the 10-V power bandwidth of a 741 is approximately 8 kHz. This means the undistorted bandwidth for large-signal operation is 8 kHz. Try to amplify higher frequencies of the same peak value and you will get slew-rate distortion.

Tradeoff

One way to increase the power bandwidth is to accept less peak voltage. Figure 31-4 is a graph of Eq. (31-2) for three different slew rates. By trading off amplitude for frequency, we can improve the power bandwidth. For instance, if peak amplitudes of 1 V are acceptable in an application, the power bandwidth of a 741 increases to 80 kHz (bottom curve).

If an op amp has a slew rate of 50 V/μs (top curve), its 10-V power bandwidth is 800 kHz, and its 1-V power bandwidth is 8 MHz.

SUMMARY

1. In a 741, the first stage is a *diff amp*.
2. The output stage of a 741 is a class B push-pull emitter follower. A small idling current eliminates crossover distortion.
3. The 741 uses *active loading* in some stages. This means using transistors for loads instead of resistors.
4. The compensating capacitor inside an op amp controls the frequency response and prevents oscillations.
5. Active loading is very popular in integrated circuits because it is easier to fabricate transistors than resistors.
6. The *input bias current* is the average of the two base currents in the input stage of an op amp under no-signal conditions.

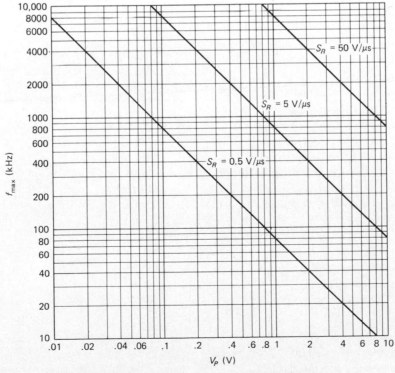

Fig. 31-4. Trading off amplitude for power bandwidth.

7. The *input offset current* is the difference of the two base currents.
8. The *input offset voltage* is the input voltage needed to null or zero the quiescent output voltage.
9. The CMRR of an op amp is the ratio of differential voltage gain to common-mode voltage gain.
10. *Slew rate* is the maximum rate at which the output voltage can change. A 741 has a typical slew rate of 0.5 V/μs.
11. *Power bandwidth* is the highest undistorted frequency an op amp can deliver; it is directly proportional to slew rate and inversely proportional to amplitude.
12. A typical 741 has a 10-V power bandwidth of 8 kHz. One way to increase the power bandwidth is to have less amplitude in the output signal. Another way is to get an op amp with a higher slew rate.

SELF-TEST

Check your understanding by answering these questions.

1. The input stage of a 741 is a _____ amp.
2. The output stage of a 741 is a class B _____ emitter follower.
3. In some stages, the 741 uses _____ loading, which means _____ loads instead of resistor loads.

4. The input bias current of an op amp is the _____ of the two input base currents under no-signal conditions.
5. The input _____ current is the difference of the two input base currents.
6. The input _____ voltage is the differential input voltage needed to null or zero the quiescent output voltage.
7. The CMRR of an op amp is the ratio of _____ voltage gain to _____ voltage gain.
8. A 741 has a _____ rate of 0.5 V/μs. This is the fastest rate at which the _____ can change.
9. Power bandwidth is the _____ undistorted frequency out of an op amp. It depends on the _____ rate of the op amp and the _____ of the output signal.

MATERIALS REQUIRED

- Two power supplies: 15-V
- Equipment: AC generator, oscilloscope
- Resistors: Two 100-Ω; 1-k, 10-k, 100-k, two 200-kΩ; three 1-MΩ ½-W
- Potentiometer: 5-kΩ (or nearest available value)
- Op amps: Three 741C
- Capacitors: Two 1-μ, 10-μF

PROCEDURE

Input Bias Current

1. Connect the circuit of Fig. 31-5a.
2. The dc voltages you are about to measure are in the millivolt region. One convenient way to measure these voltages is with an oscilloscope using the dc input (Fig. 31-5b).
3. Connect point A (scope input) to the inverting input of the op amp. (If the signal has excessive ripple and noise, insert the filter shown in Fig. 31-5c. This filter will transmit dc but stop ac.) Record the dc voltage in Table 31-1.
4. Connect point A to the noninverting input. Record the dc voltage in Table 31-1.
5. Replace the op amp by another 741C. Then repeat steps 3 and 4.

TABLE 31-1. DC Return Voltages

	Inverting	*Noninverting*
First 741C		
Second 741C		
Third 741C		

6. Replace the op amp by a third 741C. Then repeat steps 3 and 4.
7. Using Ohm's law and Table 31-1, calculate the input currents for the first op amp. Average these two currents and enter the result in Table 31-2. (This is what the manufacturer calls the "input bias current.")
8. Repeat step 7 for the second and third op amps.

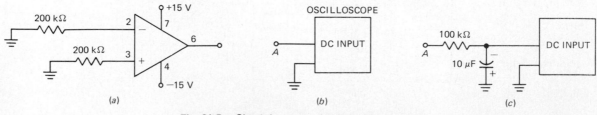

Fig. 31-5. Circuit for measuring input bias current.

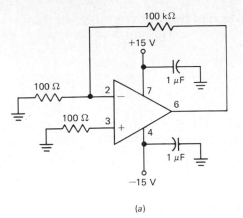

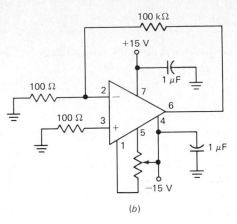

(a) (b)

Fig. 31-6. Circuit for measuring input offset voltage.

TABLE 31-2. Calculated Bias Currents

	Input bias current
First 741C	
Second 741C	
Third 741C	

TABLE 31-3. Offset Voltages

	V_{out}	V_{in}
First 741C		
Second 741C		
Third 741C		

Output Offset Voltage

9. Connect the circuit of Fig. 31-6a. Measure the dc output voltage (pin 6) and record the result in Table 31-3.
10. Repeat step 9 for the second and third op amps.
11. As discussed in Experiment 30, the voltage gain is approximately equal to the ratio of the feedback resistor to the input resistor. In Fig. 31-6a, this means the voltage gain is approximately 1000. With the output voltages of Table 31-3, calculate the input offset voltages using

$$V_{in} = \frac{V_{out}}{1000} \qquad (31-3)$$

Record the input offset voltages in Table 31-3.

12. Add a 5-kΩ potentiometer to the circuit as shown in Fig. 31-6b. Look at the output voltage (pin 6) with an oscilloscope. Adjust the potentiometer until the output offset voltage is 0. (This is how you eliminate output offset.)

Slew Rate

13. Connect the circuit of Fig. 31-7.
14. Use the oscilloscope to look at the output of the op amp (pin 6). Set the ac generator to 10 kHz. Adjust the signal level to overdrive the op amp. Then adjust the oscilloscope timing to get a couple of cycles like Fig. 31-7b.
15. Measure the voltage change ΔV and the time change ΔT of the waveform (Fig. 31-7b). Record the results in Table 31-4.

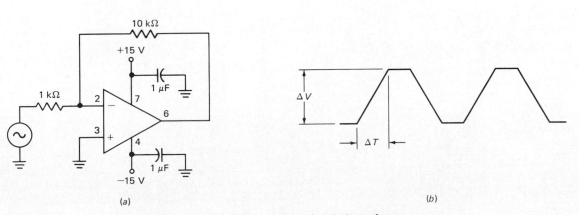

(a) (b)

Fig. 31-7. (a) Measuring slew rate; (b) typical waveform.

TABLE 31-4. Slew Rates

	ΔV	ΔT	S_R
First 741C			
Second 741C			
Third 741C			

16. Repeat step 15 for the other 741Cs.
17. Calculate the slew rate using

$$S_R = \frac{\Delta V}{\Delta T} \qquad (31\text{-}4)$$

Record the slew rates in Table 31-4.

Power Bandwidth

18. Using the circuit of Fig. 31-7a, set the ac generator at 1 kHz. Adjust the signal level to get 20 V p-p out of the op amp.
19. Increase the frequency and watch the waveform. Somewhere above 10 kHz, slew-rate distortion will become evident because the waveform will appear triangular and the amplitude will decrease.

QUESTIONS

1. An op amp has a slew rate of 2 V/μs. If a large input signal overdrives the op amp, how long will it take for output to change from −10 V to +10 V?
2. What is the 10-V power bandwidth of an op amp with a slew rate of 5 V/μs?
3. Use the highest input bias current in Table 31-2. Approximately how much voltage does this current produce across the 100-Ω resistors of Fig. 31-6?
4. An op amp has a differential voltage gain of 100,000 and a common-mode voltage gain of 3. What is the CMRR?
5. Use the largest input bias current in Table 31-2 to work out the following: If the 200-kΩ resistors in Fig. 31-5a are replaced by 1-MΩ resistors, approximately how much dc voltage will there be across these resistors?
6. Calculate the 10-V power bandwidth of the three 741Cs used in this experiment.
7. What is the 0.5-V power bandwidth of the three 741Cs used in this experiment?

Answers to Self-Test

1. diff
2. push-pull
3. active; transistor
4. average
5. offset
6. offset
7. differential; common-mode
8. slew; output
9. maximum; slew; amplitude

NEGATIVE FEEDBACK

OBJECTIVES

1. To measure closed-loop voltage gain
2. To get data for calculating the gain-bandwidth product
3. To trade off gain for bandwidth

INTRODUCTORY INFORMATION

In a feedback control system, the output is sampled and a fraction of it is sent back to the input. The returning signal combines with the original input, producing unusual changes in system performance. *Negative feedback* means the returning signal has a phase that opposes the input signal. The advantages of negative feedback are: stabilizing the gain, improving input and output impedances, and increasing bandwidth.

Basic Idea

Figure 32-1 illustrates the general idea behind negative feedback. The input to the amplifier (block A) is called the *error voltage*. This error voltage is amplified by a factor of A. Part of the output voltage is then fed back to the input through block B. The quantity B is a less than or equal to unity; it represents the fraction of output fed back to the input.

The error voltage is the difference between the input voltage and the feedback voltage. In symbols,

$$v_{error} = v_{in} - Bv_{out} \qquad (32\text{-}1)$$

The feedback is negative because the feedback voltage opposes the input voltage.

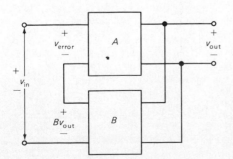

Fig. 32-1. Noninverting amplifier with negative feedback.

Here is why feedback stabilizes gain. Suppose the voltage gain A increases because of temperature change or some other reason. The output voltage will rise. This means more voltage is fed back to the input. Since the feedback voltage subtracts from the input voltage, the error voltage decreases. The reduced error voltage to the amplifier almost completely offsets the original increase in voltage gain A. The result is that v_{out} hardly increases at all.

A similar argument applies to a decrease in voltage gain A. If A decreases for any reason, the output voltage decreases. In turn, the feedback voltage decreases, causing v_{error} to increase. This increase in error voltage almost completely offsets the original decrease in voltage gain A. As a result, the output voltage shows only the slightest decrease.

Voltage Gain with Negative Feedback

A mathematical analysis of Fig. 32-1 leads to this formula for the voltage gain

$$\frac{v_{out}}{v_{in}} = \frac{A}{1 + AB} \qquad (32\text{-}2)$$

For the negative feedback to be effective, the product AB must be much greater than 1. When this condition is satisfied, Eq. (32-2) reduces to

$$\frac{v_{out}}{v_{in}} = \frac{1}{B} \qquad (32\text{-}3)$$

This result is important. Why? Because it says the overall voltage gain no longer depends on the internal gain A, which is temperature- and transistor-dependent. Instead, the overall gain depends only on the value of B. The feedback circuit is usually a voltage divider with precision resistors. This means B is an accurate and stable value. Because of this, the voltage gain of a negative-feedback circuit becomes a rock-solid value equal to $1/B$. For instance, if $B = 0.1$, then the gain is 10. If $B = 0.01$, then the gain is 100.

Open-Loop and Closed-Loop Gain

The internal gain A is called the *open-loop gain* because it is gain we would get if the feedback path were opened. On the other hand, the overall gain with feedback is called the *closed-loop gain* because it is the gain we get when there is a closed loop or signal path all the way around the circuit. This is why Eqs. (32-2) and (32-3) often appear as

$$A_{CL} = \frac{A_{OL}}{1 + A_{OL}B} \qquad (32\text{-}4)$$

and

$$A_{CL} = \frac{1}{B} \qquad (32\text{-}5)$$

where A_{CL} is the closed-loop gain and A_{OL} is the open-loop gain.

Input and Output Impedances

Negative feedback also affects the input and output impedances. It can be shown that the input impedance of a noninverting amplifier like Fig. 32-1 is

$$z_{in(CL)} = \frac{A_{OL}}{A_{CL}} z_{in(OL)} \qquad (32\text{-}6)$$

This means the input impedance increases by a factor of A_{OL}/A_{CL}. For instance, suppose the amplifier has an input impedance of 1 kΩ and A_{OL}/A_{CL} is 100, then

$$z_{in(CL)} = 100 \times 1\ \text{k}\Omega = 100\ \text{k}\Omega$$

Negative feedback has a different effect on the output impedance of a noninverting amplifier. An advanced mathematical derivation shows that

$$z_{out(CL)} = \frac{z_{out(OL)}}{A_{OL}/A_{CL}} \qquad (32\text{-}7)$$

This says that negative feedback reduces the output impedance by a factor of A_{OL}/A_{CL}. As an example, if the internal amplifier has an output impedance of 100 Ω and A_{OL}/A_{CL} is 100, then

$$z_{out(CL)} = \frac{100\ \Omega}{100} = 1\Omega$$

The Op Amp as a Noninverting Amplifier

Figure 32-2 shows an op amp used as a noninverting amplifier with negative feedback. The input signal drives the noninverting input of the op amp. The op amp provides open-loop gain A_{OL}. The external resistors R_1 and R_2 form the feedback voltage divider. Since the returning feedback volt-

age drives the inverting input, it opposes the input voltage. In other words, the feedback is negative.

The fraction of output voltage fed back to the input is

$$B = \frac{R_1}{R_1 + R_2}$$

Therefore, the approximate closed-loop gain is

$$A_{CL} = \frac{1}{B} = \frac{R_1 + R_2}{R_1}$$

which is usually written as

$$A_{CL} = \frac{R_2}{R_1} + 1 \qquad (32\text{-}8)$$

This says the closed-loop gain depends only on the ratio of feedback resistors. As already indicated, these can be precision resistors, which means we get a precise value of closed-loop gain. Despite temperature change or op amp replacement, the closed-loop gain has a rock-solid value.

For Eq. (32-8) to be accurate to within 1 percent, the open-loop gain must be at least 100 times greater than the closed-loop gain. (In symbols, $A_{OL}/A_{CL} > 100$.) For instance, the 741C has a typical open-loop gain of 100,000. As long as the closed-loop gain is less than 1000, Eq. (32-8) is accurate to within 1 percent.

Upper Cutoff Frequency

The op amp of Fig. 32-2 has no lower cutoff frequency because it is direct-coupled. But it does have an open-loop upper cutoff frequency, designated f_{OL}. The upper cutoff frequency f_{CL} of the overall amplifier is greater than f_{OL} because of the negative feedback.

Here is why. When the input frequency increases, we eventually reach the internal cutoff frequency f_{OL}. At this frequency the open-loop gain is down to 0.707 of its maximum value. Because less voltage is fed back to the input, the error voltage increases. As a result, the closed-loop gain shows almost no decrease at all.

As the input frequency keeps increasing, the open-loop gain keeps decreasing until it approaches the value of closed-loop gain. At this point, the closed-loop gain starts dropping noticeably. The frequency where the closed-loop gain decreases to 0.707 of its maximum value is called the *closed-loop cutoff frequency*.

An advanced derivation results in this formula

$$f_{CL} = \frac{A_{OL}}{A_{CL}} f_{OL} \qquad (32\text{-}9)$$

For instance, if f_{OL} is 10 Hz and A_{OL}/A_{CL} is 1000, then

$$f_{CL} = 1000 \times 10\ \text{Hz} = 10\ \text{kHz}$$

Constant Gain-Bandwidth Product

Equation (32-9) can be rearranged as

$$A_{CL}f_{CL} = A_{OL}f_{OL} \qquad (32\text{-}10)$$

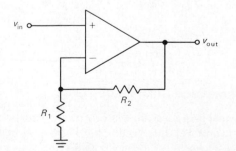

Fig. 32-2. Noninverting op amp with negative feedback.

The right-hand side of this equation is the product of the internal gain and cutoff frequency. For instance, the 741C has a typical A_{OL} of 100,000 and an f_{OL} of 10 Hz. Therefore, the product of its internal gain and cutoff frequency is

$$A_{OL}f_{OL} = 100,000 \times 10 \text{ Hz} = 1 \text{ MHz}$$

This says the internal or open-loop gain-bandwidth product of a 741C equals 1 MHz.

The left-hand side of Eq. (32-10) is the product of closed-loop gain and closed-loop cutoff frequency. No matter what the values of R_1 and R_2 in Fig. 32-2, the product of $A_{CL}f_{CL}$ must equal the product $A_{OL}f_{OL}$. In other words, the closed-loop gain-bandwidth product equals the open-loop gain-bandwidth product. Given a typical 741C, the product $A_{CL}f_{CL}$ always equals 1 MHz, regardless of the values of R_1 and R_2.

Equation (32-10) is often summarized by saying the gain-bandwidth product is a constant. For this reason, even though A_{CL} and f_{CL} change when we change external resistors, the product of these two quantities remains constant for a particular op amp.

Unity-Gain Frequency

Figure 32-3 shows the open-loop response for a typical 741C. The open-loop gain has a maximum value of 100,000. When the operating frequency increases to 10 Hz, the open-loop gain decreases to 0.707 of its maximum value. With increasing frequency, the gain keeps dropping off. Well above f_{OL}, the gain decreases by a factor of 10 for each decade increase in frequency. (This is equivalent to 20 dB/decade.)

The *unity-gain frequency* is the frequency where the open-loop gain has decreased to unity. In Fig. 32-3, f_{unity} equals 1 MHz. Data sheets sometimes list the value of f_{unity} because it represents the upper limit on the useful gain of an op amp. For instance, the data sheet of a 741C gives an f_{unity} of 1 MHz, whereas the data sheet of a 318 lists an f_{unity} of 15 MHz. Although it costs more, the 318 gives us usable gain to much higher frequencies than a 741C.

A final point: Since the voltage gain is 1 for a signal frequency of f_{unity}, the open-loop gain-bandwidth product of an op amp equals f_{unity}. Therefore, the closed-loop gain-bandwidth product is

$$A_{CL}f_{CL} = f_{\text{unity}} \qquad (32\text{-}11)$$

Given the f_{unity} of any op amp, you immediately know its closed-loop gain-bandwidth product. In other words, no matter what the values of R_1 and R_2 in Fig. 32-2, the product of A_{CL} and f_{CL} must equal f_{unity}.

A Visual Summary

Figure 32-4 summarizes all the key ideas discussed so far. The open-loop gain of a 741C is down to 0.707 of its maximum value when the signal frequency is f_{OL}. A_{OL} continues decreasing until it approaches the value of closed-loop gain. Then, A_{CL} starts to decrease. When the signal frequency equals f_{CL}, the closed-loop gain is down to 0.707 of its maximum value. Thereafter, the curves for A_{OL} and A_{CL} superimpose and decrease to unity at f_{unity}.

If the feedback resistors are changed, the closed-loop gain will change to a new value and so too will the closed-loop cutoff frequency. Figure 32-5 illustrates the curves for closed-loop gains of 10 and 1000. Notice the tradeoff between gain and bandwidth. When the A_{CL} is 1000, f_{CL} is 1 kHz. When A_{CL} is 10, f_{CL} is 100 kHz. In other words, by decreasing the closed-loop gain, we can increase the bandwidth.

SUMMARY

1. In a feedback circuit the output is sampled and a fraction of it is sent back to the input.
2. *Negative feedback* means the returning signal has a phase that opposes the input signal.
3. The internal gain is called the *open-loop gain*. The gain of the overall circuit is the *closed-loop gain*.

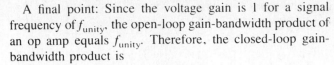

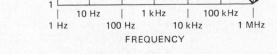

Fig. 32-3. Voltage gain versus frequency for typical 741C.

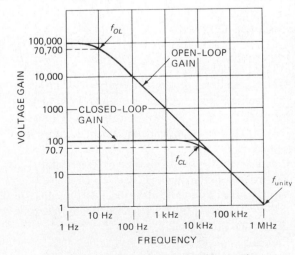

Fig. 32-4. Open-loop and closed-loop gains.

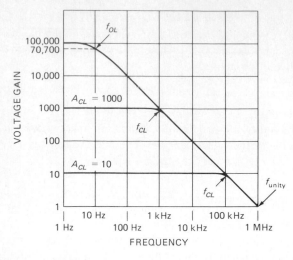

Fig. 32-5. Another example of open-loop and closed-loop gain.

4. For a noninverting amplifier with negative feedback, the closed-loop voltage gain depends only on the values of external resistors.
5. Negative feedback increases the input impedance and decreases the output impedance of a noninverting amplifier.
6. Negative feedback also increases the upper cutoff frequency of a noninverting amplifier.
7. The product of gain and upper cutoff frequency is called the *gain-bandwidth product.*
8. The closed-loop gain-bandwidth product always equals the open-loop gain-bandwidth product.
9. Given a particular op amp, the closed-loop gain-bandwidth product is a constant.
10. The frequency where the voltage gain equals unity is called the *unity-gain frequency,* designated f_{unity}.
11. The gain-bandwidth product, either open-loop or closed-loop, equals f_{unity}.
12. Because the gain-bandwidth product is a constant, we can trade off gain for bandwidth. This means we can decrease the gain to get a higher cutoff frequency.

SELF-TEST

Check your understanding by answering these questions.

1. The error voltage is the difference between the input voltage and the _____ voltage.
2. If the open-loop gain increases for any reason at all, the returning feedback signal increases and the error voltage _____ .
3. Negative feedback increases the _____ impedance of a noninverting amplifier.
4. Negative feedback _____ the output impedance of a noninverting amplifier.
5. Since the closed-loop gain is smaller than the open-loop gain in a negative-feedback circuit, the closed-loop cutoff frequency is _____ than the open-loop cutoff frequency.
6. The product of gain and bandwidth, whether open-loop or closed-loop, equals a _____ for a particular op amp.
7. If the f_{unity} of an op amp equals 2 MHz, the gain-bandwidth product equals _____ .
8. A negative-feedback circuit has a gain-bandwidth product of 2 MHz. If $A_{CL} = 100$, what does f_{CL} equal?
9. Suppose $A_{CL} = 1000$ and $f_{CL} = 5$ kHz. If the feedback resistors are changed to get $A_{CL} = 200$, what is the new value of f_{CL}?

MATERIALS REQUIRED

- Power supplies: Two 15-V
- Equipment: AC generator, oscilloscope
- Resistors: Two 1-kΩ, one 10-kΩ, one 47-kΩ, 100-kΩ ½-W
- Op amp: 741C

PROCEDURE

Closed-Loop Gain

1. Connect the circuit shown in Fig. 32-6 using an R_2 of 10 kΩ.
2. With an oscilloscope across the output 1-kΩ resistor, set the frequency to 1 kHz and the signal level to 3 V p-p.
3. Move the oscilloscope leads to the input (pin 3) and measure the peak-to-peak input voltage. Record this voltage in Table 32-1.
4. Calculate the closed-loop voltage gain using v_{out}/v_{in}. Record this gain in Table 32-1.
5. Move the oscilloscope leads to the inverting input (pin 2) and measure the feedback voltage. Record the peak-to-peak value in Table 32-1.
6. Repeat steps 2 to 5 for the other values of R_2 shown in Table 32-1.

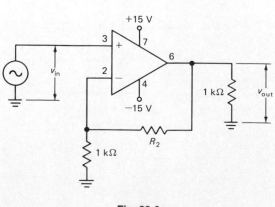

Fig. 32-6.

TABLE 32-1. Closed-loop Gain

R_2	v_{out}	v_{in}	A_{CL}	Bv_{out}
10 kΩ	3 V p-p			
47 kΩ	3 V p-p			
100 kΩ	3 V p-p			

TABLE 32-2. Gain-bandwidth Product

R_2	f_{CL}	$A_{CL}f_{CL}$
10 kΩ		
47 kΩ		
100 kΩ		

Gain-Bandwidth Product

7. Increase the frequency until the output voltage decreases to 2.1 V p-p. (This is the cutoff frequency, assuming the input voltage has not changed significantly.) Record the closed-loop cutoff frequency in Table 32-2. Calculate gain-bandwidth product and record.

8. Repeat step 7 for the other values of R_2. Remember to start at 1 kHz with 3-V p-p output.

Tradeoff

9. Because of measurement errors, tolerances, etc., the values of gain-bandwidth product in Table 32-2 are only approximations. Average the three products and record the average gain-bandwidth product here:

$$A_{CL}f_{CL} = \underline{\hspace{2cm}}$$

10. We want an amplifier with a bandwidth of 20 kHz. Calculate the necessary closed-loop voltage gain and record the answer here:

$$A_{CL} = \underline{\hspace{2cm}}$$

11. If $R_1 = 1$ kΩ, what is the value of R_2 needed to produce the closed-loop voltage gain of step 10? Record the value here:

$$R_2 = \underline{\hspace{2cm}}$$

12. Connect the circuit of Fig. 32-6 with the value of R_2 in step 11. (Use the nearest available value.)

13. For an input frequency of 1 kHz, measure the input voltage that produces an output of 3 V p-p. Record the input voltage here:

$$v_{in} = \underline{\hspace{2cm}}$$

14. Calculate the closed-loop voltage gain for step 13 and record the answer here:

$$A_{CL} = \underline{\hspace{2cm}}$$

This should be approximately equal to the A_{CL} of step 10.

QUESTIONS

1. Compare the input voltage and the feedback voltage in Table 32-1. What conclusion do you arrive at? What can you say about the error voltage?
2. Use Eq. (32-8) to calculate the theoretical A_{CL} values for the resistances listed in Table 32-1.
3. Look at Table 32-2 and comment on the statement: "The closed-loop gain-bandwidth product is a constant for a given op amp."
4. What is the approximate f_{unity} for the op amp used in this experiment?
5. What is the approximate value of $A_{OL}f_{OL}$ for the op amp used in this experiment?

6. Suppose you change R_2 to 240 kΩ in Fig. 32-6. Using the gain-bandwidth product recorded in step 9, what is the bandwidth?
7. Using the $A_{CL}f_{CL}$ of step 9, what is the f_{OL} for an A_{OL} of 100,000?

Answers to Self-Test

1. feedback
2. decreases
3. input
4. decreases
5. greater
6. constant
7. 2 MHz
8. 20 kHz
9. 25 kHz

BASIC OP AMP CIRCUITS

OBJECTIVES

1. To test a voltage-to-current converter
2. To test a current-to-voltage converter
3. To test a current amplifier

INTRODUCTORY INFORMATION

The noninverting op amp with negative feedback is one of four basic negative-feedback circuits. These four basic circuits have different effects on input and output impedances.

Voltage Amplifier

Figure 33-1 shows a noninverting op amp with negative feedback. This is the circuit discussed in Experiment 32. As you recall, it has stable voltage gain, high input impedance, and low output impedance. Ideally, this circuit approaches a perfect *voltage amplifier* described by these equations:

$$\frac{v_{\text{out}}}{v_{\text{in}}} = \frac{R_2}{R_1} + 1 \tag{33-1}$$

$$z_{\text{in}} = \infty \tag{33-2}$$

$$z_{\text{out}} = 0 \tag{33-3}$$

In an ideal voltage amplifier, the gain is constant. Furthermore, the input impedance is infinite, which means the amplifier will not load the circuit driving it. Also, the zero output impedance means the amplifier can drive small resistive loads without a decrease in voltage gain.

With a 741, we can build a voltage amplifier that approaches the ideal. If A_{OL}/A_{CL} is greater than 1000, the input impedance is in megohms and the output impedance is in fractions of an ohm.

Voltage-to-Current Converter

Figure 33-2 shows another feedback circuit. Notice that the feedback is negative because the returning voltage opposes the input voltage. A mathematical analysis of this circuit shows that it acts approximately like a perfect *voltage-to-current converter*, a circuit with these equations:

$$i_{\text{out}} = \frac{v_{\text{in}}}{R} \tag{33-4}$$

$$z_{\text{in}} = \infty \tag{33-5}$$

$$z_{\text{out}} = \infty \tag{33-6}$$

In a perfect voltage-to-current converter, the output current depends only on the input voltage and the value of R. For instance, if $v_{\text{in}} = 2$ V and $R = 1$ kΩ, then

$$i_{\text{out}} = \frac{2 \text{ V}}{1 \text{ k}\Omega} = 2 \text{ mA}$$

The infinite input impedance means the voltage-to-current converter will not load down the circuit driving it. Also, the infinite output impedance implies the circuit acts like a current source; it will force exactly 2 mA through any load resistance.

One application of the voltage-to-current converter is in building an electronic voltmeter. The outstanding advantage of this type of voltmeter is its extremely high input impedance. The high input impedance means it will not disturb the circuit whose voltage is being measured.

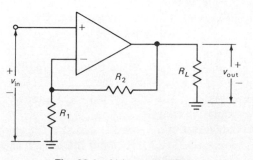

Fig. 33-1. Voltage amplifier.

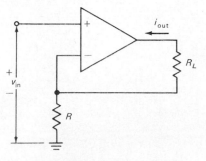

Fig. 33-2. Voltage-to-current converter.

Current-to-Voltage Converter

Figure 33-3 illustrates a third type of negative feedback. A mathematical analysis shows that this circuit acts approximately like a perfect *current-to-voltage converter*, one described by these equations:

$$v_{out} = R \times i_{in} \qquad (33\text{-}7)$$

$$z_{in} = 0 \qquad (33\text{-}8)$$

$$z_{out} = 0 \qquad (33\text{-}9)$$

In a perfect current-to-voltage converter, the output voltage depends only on the input current and the value of R. For example, if $i_{in} = 5$ mA and $R = 1$ kΩ, then

$$v_{out} = 1 \text{ k}\Omega \times 5 \text{ mA} = 5 \text{ V}$$

The zero input impedance means the converter looks like a perfect current sink (ground). The zero output impedance means the circuit will produce 5 V, no matter how small the load resistance.

One application of the current-to-voltage converter is building an electronic ammeter. The outstanding advantage of this type of ammeter is its extremely low input impedance. Since ammeters are connected in series, the almost zero input impedance will not disturb the circuit whose current is being measured.

Current Amplifier

Figure 33-4 illustrates the fourth type of negative feedback. Analysis of this circuit shows it acts approximately like a perfect *current amplifier* with these equations:

$$\frac{i_{out}}{i_{in}} = \frac{R_2}{R_1} + 1 \qquad (33\text{-}10)$$

$$z_{in} = 0 \qquad (33\text{-}11)$$

$$z_{out} = \infty \qquad (33\text{-}12)$$

A perfect current amplifier provides current gain rather than voltage gain. Furthermore, it has zero input impedance, which means it will not disturb the circuit driving it. Also, the infinite output impedance means it can force a fixed value of current through any size load resistance.

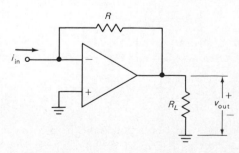

Fig. 33-3. Current-to-voltage converter.

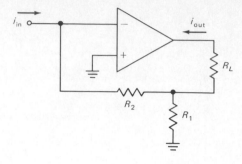

Fig. 33-4. Current amplifier.

SUMMARY

1. A *voltage amplifier* provides stable voltage gain, high input impedance, and low output impedance.
2. A *voltage-to-current converter* produces an output current that depends only on the input voltage and the value of resistor R.
3. A voltage-to-current converter has a high input impedance and a high output impedance.
4. The perfect voltage-to-current converter will not load down the circuit driving it. Furthermore, it can force a fixed current through any size load resistance.
5. A *current-to-voltage converter* produces an output current that depends only on the input current and resistor R.
6. A current-to-voltage converter has a low input impedance and a low output impedance.
7. A perfect current-to-voltage converter will not load down the circuit driving it. Also, its output voltage is unaffected by small load resistances.
8. The *current amplifier* produces current gain rather than voltage gain. It has a low input impedance and a high output impedance.

SELF-TEST

1. The circuit of Fig. 33-1 is a _____ amplifier.
2. A voltage amplifier has a high input impedance and a _____ output impedance.
3. A perfect voltage amplifier will have a stable voltage gain even for very _____ load resistors.
4. The circuit of Fig. 33-2 is a _____ converter. It has a very high input impedance and a very _____ output impedance.
5. If R_L changes in Fig. 33-2, i_{out} will not _____. This is because the output impedance approaches infinity.
6. The circuit of Fig. 33-3 is a _____ converter. It has a very _____ input impedance and a very low output impedance.
7. The circuit of Fig. 33-4 is a _____ amplifier. It has a very _____ input impedance and a very high output impedance.

MATERIALS REQUIRED

- Power supplies: Two 15-V
- Equipment: Two VOMs (or digital multimeters to measure voltage and current)
- Resistors: Two 1-kΩ, two 10-kΩ ½-W
- Potentiometer: 1-kΩ (or nearest available value)
- Op amp: 741C

PROCEDURE

Voltage-to-Current Converter

1. Connect the circuit of Fig. 33-5.
2. Adjust the potentiometer to get an input voltage of 1 V.
3. Read the output current and record the value in Table 33-1.
4. Repeat steps 2 and 3 for the remaining input voltages listed in Table 33-1.

Current-to-Voltage Converter

5. Connect the circuit of Fig. 33-6.
6. Adjust the potentiometer to get an input current of 1 mA.
7. Read the output voltage and record the value in Table 33-2.
8. Repeat steps 6 and 7 for the other input currents shown in Table 33-2.

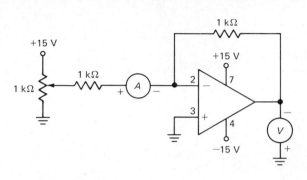

Fig. 33-6. Electronic ammeter.

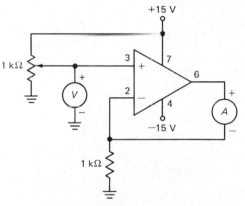

Fig. 33-5. Electronic voltmeter.

TABLE 33-1. Voltage-to-Current Converter

v_{in}, V	i_{out}, mA
1	
2	
3	
4	
6	
8	
10	

TABLE 33-2. Current-to-Voltage Converter

i_{in}, mA	v_{out}, V
1	
2	
3	
4	
6	
8	
10	

Current Amplifier

9. Connect the circuit of Fig. 33-7.
10. Adjust the potentiometer to get an input current of 0.1 mA.
11. Record the output current in Table 33-3.
12. Repeat steps 10 and 11 for the remaining input currents of Table 33-3.

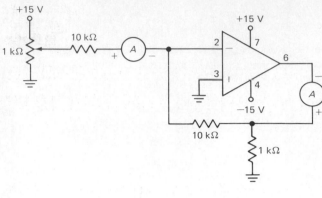

Fig. 33-7. Measuring current gain.

TABLE 33-3. Current Amplifier

i_{in}, mA	i_{out}, mA
0.1	
0.2	
0.3	
0.4	
0.6	
0.8	
1	

QUESTIONS

1. Ideally, the currents in Table 33-1 should be 1, 2, 3, 4, 6, 8, and 10 mA. Give three reasons why the measured currents were different.
2. After removing the potentiometer and voltmeter at the input of Fig. 33-5, the circuit that remains can be used as an electronic voltmeter. Why is this true?
3. Suppose we want 1 V to produce 2 mA of current in Fig. 33-5. What change can we make?
4. Based on theory, what is the approximate input imped-ance at the inverting input of Fig. 33-6?
5. In Fig. 33-6, what is the ideal value of output voltage when the input current is 5 mA?
6. After removing the potentiometer, 1-kΩ resistor, and ammeter from the input of Fig. 33-6, the converter acts like an electronic ammeter. Why is this true?
7. What change can we make in Fig. 33-6 to get an output voltage of 2 V when the input current is 1 mA?
8. What is the theoretical current gain in Fig. 33-7?
9. What is the current gain in Table 33-3 for an input current of 4 mA?

Answers to Self-Test

1. voltage
2. low
3. small
4. voltage-to-current; high
5. change
6. current-to-voltage; low
7. current; low

NONLINEAR OP AMP CIRCUITS

OBJECTIVES

1. To build a go–no go detector
2. To look at the output of an active half-wave rectifier
3. To measure the output of an active peak detector
4. To limit low-level signals

INTRODUCTORY INFORMATION

Comparator

The simplest way to use an op amp is open loop (no feedback resistors), as shown in Fig. 34-1a. Because of the high gain of the op amp, the slightest error voltage [typically in microvolts (μV)] produces maximum output swing. For instance, when V_1 is greater than V_2, the error voltage is positive and the output voltage goes to its maximum positive value, typically 1 to 2 V less than the supply voltage. On the other hand, if V_1 is less than V_2, the output voltage swings to its maximum negative value.

Figure 34-1b summarizes the action. A positive error voltage drives the output to $+V_{SAT}$, the maximum positive value of output voltage. A negative error voltage produces $-V_{SAT}$. When an op amp is used like this, it is called a *comparator* because all it can do is compare V_1 to V_2, producing a saturated positive or negative output, depending on whether V_1 is greater or less than V_2.

Active Half-Wave Rectifier

Op amps can enhance the performance of diode circuits. For one thing, an op amp can eliminate the effect of diode offset voltage, allowing us to rectify, peak-detect, clip, and clamp

low-level signals (those with amplitudes smaller than the offset voltage). And because of their buffering action, op amps can eliminate the effects of source and load on diode circuits. Circuits that combine op amps and diodes are called *active* diode circuits.

Figure 34-2 shows an active half-wave rectifier. When the input signal goes positive, the output goes positive and turns on the diode. The circuit then acts like a voltage follower, and the positive half-cycle appears across the load resistor. On the other hand, when the input goes negative, the op amp output goes negative and turns off the diode. Since the diode is open, no voltage appears across the load resistor. This is why the final output is almost a perfect half-wave signal.

The high gain of the op amp virtually eliminates the effect of offset voltage. For instance, if the offset voltage ϕ equals 0.7 V and open-loop gain is 100,000, the input that just turns on the diode is

$$V_{in} = \frac{0.7 \text{ V}}{100,000} = 7 \text{ } \mu\text{V} \qquad (34\text{-}1)$$

When the input voltage is greater than 7 μV, the diode turns on and the circuit acts like a voltage follower. The effect is equivalent to reducing the offset voltage by a factor of A.

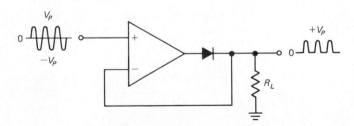

Fig. 34-2. Active half-wave rectifier.

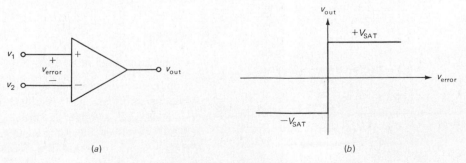

(a) (b)

Fig. 34-1. (a) Comparator; (b) input-output characteristic.

The active half-wave rectifier is useful with low-level signals. For instance, if we want to measure sinusoidal voltages in the millivolt region, we can add a milliammeter in series with the R_L of Fig. 34-2. With the proper value of R_L, we can calibrate the meter to indicate rms millivolts.

Active Peak Detector

A *peak detector* is a circuit whose output is a dc voltage equal to the peak value of the input waveform. For example, put a sine wave with a peak value of 10 V into an ideal peak detector and out comes a dc voltage equal to 10 V.

Figure 34-3a shows a passive peak detector. It conducts only on the positive half-cycles. The capacitor charges to the peak of the input voltage less the offset voltage. The passive peak detector is useful only when the peak input voltage is much greater than the offset voltage.

To peak-detect small signals, we can use an active peak detector like Fig. 34-3b. Notice how similar this is to the active half-wave rectifier. All we have done is connect a capacitor across the load resistor. The capacitor will charge to the peak input voltage. Because of the op amp, the effective offset voltage of the diode is reduced from 0.7 V to microvolts. As a result, we can peak-detect millivolt signals.

Active Positive Limiter

Figure 34-4 is an active positive limiter (also called a "clipper"). With the wiper all the way to the left, V_{REF} is 0 and the noninverting input is grounded. When V_{in} goes positive, the error voltage drives the op amp output negative and turns on the diode. This means the final output V_{out} is 0 (same as V_{REF}) for any positive value of V_{in}.

When V_{in} goes negative, the op amp output is positive, which turns off the diode and opens the loop. When this happens, the final output V_{out} is free to follow the negative half-cycle of input voltage. This is why the negative half-cycle appears at the output.

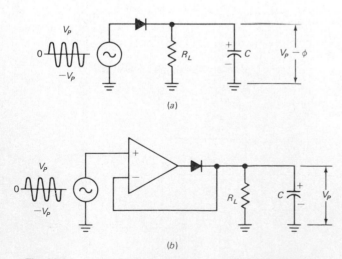

(a)

(b)

Fig. 34-3. *(a)* Passive peak detector; *(b)* active peak detector.

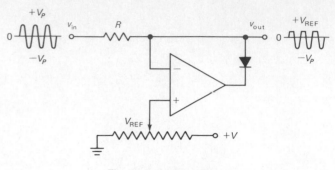

Fig. 34-4. Active limiter.

To change the limiting level, all we do is adjust V_{REF} as needed. In this case, clipping occurs at V_{REF}, as shown in Fig. 34-4.

As usual, the op amp effectively reduces the offset voltage to the microvolt region. Because of this, the active limiter of Fig. 34-4 can limit low-level signals.

Active Clamper

Figure 34-5 is an active positive clamper. The first negative half-cycle produces a positive op amp output which turns on the diode. This allows the capacitor to charge to the peak value of the input with the polarity shown. Just beyond the negative peak, the diode turns off. This clamps the output waveform at the 0 V.

SUMMARY

1. The simplest way to use an op amp is open loop. When an op amp is used like this, it is called a *comparator*.
2. Op amps can greatly reduce the effect of diode offset voltage. This allows active diode circuits to work in the millivolt region.
3. The active half-wave rectifier no longer has an offset voltage of 0.7 V. Instead, it has an effective offset voltage somewhere in the microvolt region.
4. An ideal *peak detector* produces a dc output voltage that equals the peak of the input waveform.
5. An active peak detector can peak-detect signals in the millivolt region.
6. The active limiter uses an op amp and a limiter circuit to clip low-level signals.
7. An active clamper can clamp low-level signals.

SELF-TEST

Check your understanding by answering these questions.

1. An op amp used open loop is called a _____ .
2. When op amps are used in diode circuits, the effect of _____ voltage is almost eliminated.
3. If the diode offset voltage is 0.7 V and an op amp has a voltage gain of 100,000, the effective offset voltage is _____ μV.

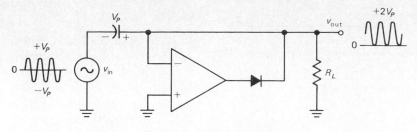

Fig. 34-5. Active clamper.

4. An active _____ rectifier produces a half-wave output whose peak voltage equals the peak voltage of the _____ .
5. A peak detector produces a _____ voltage equal to the peak value of the input waveform.
6. An _____ peak detector can peak-detect signals in the millivolt region.
7. An active _____ clips the signal at the level V_{REF}.

MATERIALS REQUIRED

- Power supplies: Two 15-V
- Equipment: AC generator, oscilloscope
- Resistors: 1-kΩ, 2.2-kΩ, 10-kΩ ½-W
- Potentiometer: 1-kΩ (or nearest available value)
- Diode: 1N914 (or any small-signal diode)
- LEDs: TIL221 and TIL222 (or equivalent red and green LEDs)
- Op amp: 741C
- Capacitor: 100-μF (at least 15 V)

PROCEDURE

Comparator

1. Connect the go–no go circuit of Fig. 34-6.
2. Vary the potentiometer and notice what the LEDs do.
3. Use the dc-coupled input of the oscilloscope to look at the input voltage to pin 3. Adjust the potentiometer to get +100-mV input. Record the color of the ON LED (Table 34-1).
4. Adjust the potentiometer to get an input of −100 mV. Record the color of the ON LED.

TABLE 34-1. Go–no go Detector

Color (step 3):	
Color (step 4):	

Half-Wave Rectifier

5. Build the circuit of Fig. 34-7a.
6. Connect the oscilloscope (dc input) across the 10-kΩ load resistor. Set the generator to 100 Hz and adjust the level to get a peak output of 1 V on the oscilloscope. (This should be a half-wave signal.)
7. Connect the oscilloscope to the input (pin 3). Record the peak value of the input sine wave (Table 34-2).
8. Adjust the signal level to get a half-wave output with a peak value of 100 mV. Then measure and record the input peak voltage (Table 34-2).

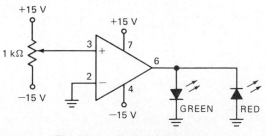

Fig. 34-6. Go–no go detector.

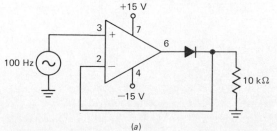

(a)

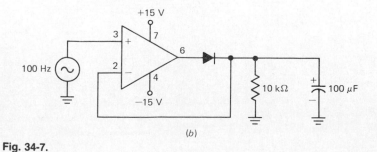

(b)

Fig. 34-7.

TABLE 34-2. Active Half-Wave Rectifier

V_P (step 7): _____

V_P (step 8): _____

Peak Detector

9. Connect a 100-μF capacitor across the load to get the circuit of Fig. 34-7b.
10. Adjust the generator to get an input peak value of 1 V. Measure and record the dc output value (Table 34-3).
11. Readjust the generator to get an input peak value of 100 mV. Measure and record the dc output (Table 34-3).

TABLE 34-3. Active Peak Detector

V_{dc} (step 10): _____

V_{dc} (step 11): _____

Limiter

12. Build the circuit of Fig. 34-8.
13. Adjust the generator to produce a peak value of 1 V at the left-hand end of the 2.2-kΩ resistor.
14. Look at the output signal while turning the potentiometer through its entire range.
15. Adjust the generator to produce a peak output of 100 mV at the left-hand end of the 2.2-kΩ resistor. Then repeat step 14.

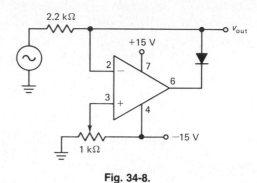

Fig. 34-8.

QUESTIONS

1. Explain why the circuit of Fig. 34-6 is called a "go–no go detector."
2. Suppose pin 2 is returned to $+1.5$ V instead of being grounded in Fig. 34-6. Describe what happens when you vary the potentiometer.
3. If the diode is reversed in Fig. 34-7a, what would the output be for an input sine wave with a peak of 100 mV?
4. If the input signal has a peak of 250 mV in Fig. 34-7a, what would the peak output voltage be?
5. Based on the data in Table 34-3, how much output would there be if the input waveform has a peak value of 375 mV?

6. Describe the output of Fig. 34-7b if the diode is reversed.
7. Is the circuit of Fig. 34-8 a positive or negative limiter?
8. If the diode were reversed in Fig. 34-8, what would the output be like?

Answers to Self-Test

1. comparator
2. offset
3. 7
4. half-wave; input

5. dc
6. active
7. clipper

35

ACTIVE FILTERS

OBJECTIVES

1. To test a first-order low-pass filter
2. To measure the cutoff frequency of a second-order low-pass filter
3. To experiment with a second-order high-pass filter

INTRODUCTORY INFORMATION

Passive Filters

A *low-pass filter* transmits low frequencies but stops high ones. Figure 35-1a shows one way to build a low-pass filter. At very low frequencies the inductive reactances approach 0 and the capacitive reactances approach infinity. This is equivalent to saying the inductors appear shorted and the capacitors appear open. Therefore, the output voltage equals the input voltage at very low frequencies.

As the frequency increases, the inductive reactances increase and the capacitive reactances decrease. At some point, the output voltage starts to decrease. For very high frequencies the inductors appear open and the capacitors appear shorted; therefore, the output voltage approaches 0.

Figure 35-1b illustrates how the voltage gain of a low-pass filter varies with frequency. Ideally, the voltage gain equals unity at lower frequencies. As the frequency increases, the voltage gain eventually starts to drop off. The *cutoff frequency* is where the voltage gain equals 0.707 (equivalent to the half-power point).

Figure 35-1c is an example of *high-pass filter*. In this case, the low frequencies are blocked and the high frequencies are transmitted. Figure 35-1d shows the graph of voltage gain versus frequency. Again notice the cutoff frequency; this is where the voltage gain drops to 0.707.

Decibels

Voltage gain is defined as the ratio of output voltage to input voltage:

$$A = \frac{v_{out}}{v_{in}}$$

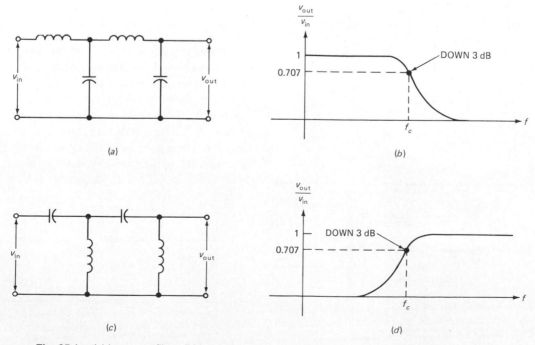

Fig. 35-1. (a) Low-pass filter; (b) low-pass response; (c) high-pass filter; (d) high-pass response.

In the low-pass filter of Fig. 35-1a, A equals unity at low frequencies. At the cutoff frequency, $A = 0.707$.

Decibels are units commonly used with filters. The decibel voltage gain is defined as

$$A_{dB} = 20 \log A \qquad (35\text{-}1)$$

where the logarithm is to the base 10. The abbreviation dB stands for "decibel" (one-tenth of a bel). Here is an example of calculating decibel voltage gain. If the voltage gain $A = 100$, then the decibel voltage gain is

$$A_{dB} = 20 \log 100 = 20(2) = 40 \text{ dB}$$

As another example, if the voltage gain equals 0.707, then

$$A_{dB} = 20 \log 0.707 = 20(-0.15) = -3 \text{ dB}$$

Table 35-1 lists some voltage gains and their decibel equivalents. As you see, a gain of 60 dB is equivalent to an ordinary voltage gain of 1000. A gain of 40 dB is equivalent to a gain of 100, and so on. Also, notice that -3 dB is equivalent to 0.707. This is why Fig. 35-1b and d show the gain down 3 dB at the cutoff frequency.

Active Low-Pass Filter

By using op amps and reactive elements, we can build *active filters*. Active filters have several advantages over passive filters. To begin with, we can eliminate the inductors, which are bulky and expensive at low cutoff frequencies. Active

TABLE 35-1. Decibel Equivalents

A	A_{dB}, dB
1000	60
100	40
10	20
8	18
4	12
2	6
1	0
0.707	-3
0.5	-6
0.25	-12
0.125	-18
0.1	-20
0.01	-40
0.001	-60

filters can also have variable voltage gain, allow easy tuning of cutoff frequency, etc.

Figure 35-2a shows one way to build an active low-pass filter. Here is what it does. At low frequencies the capacitor appears open, and the circuit acts like an inverting amplifier with a voltage gain of $-R_2/R_1$. As the frequency increases, the capacitive reactance decreases, causing the voltage gain to drop off. As the frequency approaches infinity, the capacitor appears shorted and the voltage gain approaches 0.

Figure 35-2b illustrates the output response. The output signal is maximum at low frequencies. When the frequency reaches the cutoff frequency, the output is down 3 dB. Well beyond this frequency, the gain decreases at an ideal rate of 6 dB/octave (a factor of 2 in frequency). For instance, if the cutoff frequency is 1 kHz, then the gain decreases approximately 6 dB when the frequency increases from 2 to 4 kHz. It will decrease another 6 dB when the frequency increases from 4 to 8 kHz, and so on.

A decrease of 6 db/octave is equivalent to 20 dB/decade. If the cutoff frequency is 1 kHz, then the gain decreases 20 dB when the frequency changes from 10 to 100 kHz. It changes another 20 dB when the frequency increases from 100 kHz to 1 megahertz (MHz).

A mathematical analysis leads to this formula for the cutoff frequency:

$$f_c = \frac{1}{2\pi R_2 C} \qquad (35\text{-}2)$$

The adjustable C of Fig. 35-2a allows us to vary the cutoff frequency, and the adjustable R_1 lets us control the gain. If a fixed response is desired, we can eliminate the adjustments and use a fixed R_1 and C. Because of the negative feedback, the output impedance approaches 0, which means the active filter can drive low-impedance loads.

Second-Order Low-Pass Filter

The filter of Fig. 35-2a is called a *first-order* filter because the gain decreases 6 dB/octave beyond the cutoff frequency. A *second-order* low-pass filter is one that decreases 12 dB/octave beyond the cutoff frequency.

Figure 35-3a shows a second-order active low-pass filter. At low frequencies both capacitors appear open, and the circuit becomes a voltage follower. As the frequency increases, the gain eventually starts to decrease until it is down 3 dB at the cutoff frequency. Because of the two

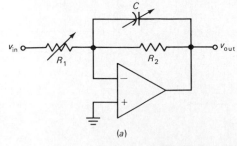

(a)

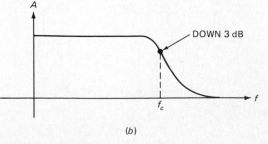

(b)

Fig. 35-2. *(a)* Active low-pass filter; *(b)* response.

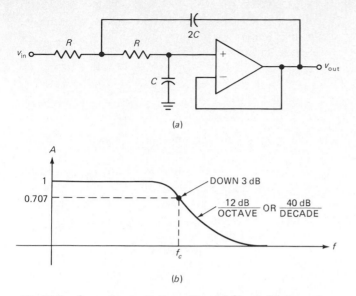

(a)

(b)

Fig. 35-3. Second-order low-pass filter. *(a)* Circuit; *(b)* response.

capacitors, the rate of decrease in gain is twice as fast as before. As a result, the gain drops off at a rate of 12 dB/octave or 40 dB/decade.

Figure 35-3*b* illustrates the gain versus frequency. First, notice the gain is down 3 dB at the cutoff frequency; this means the ordinary voltage gain equals 0.707 times the low-frequency value. Second, notice that the gain rolls off (decreases) at a rate of 12 dB/octave. For instance, suppose the cutoff frequency is 1 kHz. Then the gain ideally decreases 12 dB when the frequency changes from 2 to 4 kHz, decreases another 12 dB when the frequency changes from 4 to 8 kHz, and so on. Stated another way, the gain decreases 40 dB when the frequency changes from 2 to 20 kHz, another 40 dB when the frequency changes from 20 to 200 kHz, and so forth.

An advanced mathematical analysis shows that the cutoff frequency is given by

$$f_c = \frac{0.707}{2\pi RC} \tag{35-3}$$

Second-Order High-Pass Filter

Figure 35-4*a* is a second-order high-pass filter. At low frequencies the capacitors appear open, and the voltage gain approaches 0. At high frequencies the capacitors appear

shorted, and the circuit becomes a voltage follower. Figure 35-4*b* shows the response. The cutoff frequency is given by Eq. (35-3).

SUMMARY

1. A *low-pass filter* transmits low frequencies but stops high ones.
2. A *high-pass filter* blocks the low frequencies and passes the high frequencies.
3. *Active filters* use op amps and reactive elements.
4. One advantage of active filters is that they eliminate inductors which are bulky and expensive at low frequencies.
5. Well above the cutoff frequency, a *first-order* active low-pass filter has a voltage gain that decreases 6 dB/octave. This means the ordinary voltage gain decreases by a factor of 2 for each doubling of frequency.
6. A decrease of 6 dB/octave is equivalent to 20 dB/decade.
7. Well above cutoff, a *second-order* active low-pass filter has voltage gain that decreases 12 dB/octave, equivalent to 40 dB/decade.
8. At the cutoff frequency of a first- or second-order filter, the *decibel* voltage gain is down 3 dB. This means the voltage gain equals 0.707 of the maximum value.

SELF-TEST

Check your understanding by answering these questions.

1. At the cutoff frequency the voltage gain equals _____ of the maximum voltage gain.
2. In terms of decibels, the gain is down _____ dB at the cutoff frequency.
3. If $A = 8$, the decibel voltage gain equals _____ dB.
4. If $A_{dB} = -12$ dB, the voltage gain equals _____ .
5. If $R_1 = 1$ kΩ and $R_2 = 20$ kΩ in Fig. 35-2*a*, the voltage gain is _____ and the decibel voltage gain is _____ dB.
6. If $R_2 = 47$ kΩ and $C = 500$ pF in Fig. 35-2*a*, the cutoff frequency equals _____ kHz.
7. If the cutoff frequency equals 1 kHz in Fig. 35-3*a*, the decibel voltage gain decreases by _____ dB when the frequency changes from 10 to 20 kHz.

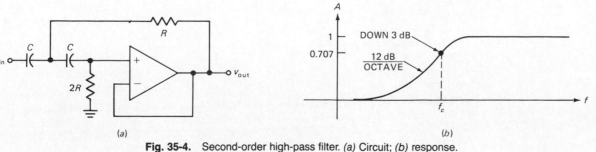

(a)

(b)

Fig. 35-4. Second-order high-pass filter. *(a)* Circuit; *(b)* response.

MATERIALS REQUIRED

- Power supplies: Two 15-V
- Equipment: AC generator, oscilloscope

- Resistors: 1-kΩ, two 10-kΩ, 20-kΩ ½-W
- Op amp: 741C
- Capacitors: Three 0.01-μF

PROCEDURE

First-Order Low-Pass Filter

1. Connect the circuit shown in Fig. 35-5.
2. Set the ac generator at 100 Hz. Adjust the signal level to get 1 V p-p at the output of the filter. Measure and record the peak-to-peak input voltage (Table 35-2).
3. Change the frequency to 200 Hz. Measure the input and output voltages. Record the data in Table 35-2.

4. Repeat step 3 for the remaining frequencies listed in Table 35-2.
5. Work out the voltage gain for each frequency in Table 35-2. Also calculate and record the equivalent decibel gain. (NOTE: Most pocket scientific and technical calculators include the logarithm function to allow easy calculation of decibels.)
6. Measure and record the cutoff frequency here:

$$f_c = \underline{\hspace{3cm}}$$

Second-Order Low-Pass Filter

7. Connect the circuit of Fig. 35-6. (Use two 0.01-μF capacitors in parallel for the 0.02-μF capacitor specified.)
8. Set the generator at 100 Hz. Adjust the signal level to get 1 V p-p at the output of the filter. Measure and record the peak-to-peak input voltage (Table 35-3).
9. Measure the input and output voltages for the other frequencies in Table 35-3.

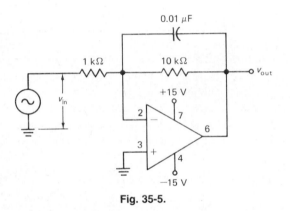

Fig. 35-5.

TABLE 35-2.   First-Order Low-Pass Filter

f	v_{in}, V	v_{out}, V	A	A_{dB}, dB
100 Hz		1		
200 Hz				
500 Hz				
1 kHz				
2 kHz				
5 kHz				
10 kHz				

TABLE 35-3.   Second-Order Low-Pass Filter

f	v_{in}, V	v_{out}, V	A	A_{dB}, dB
100 Hz		1		
200 Hz				
500 Hz				
1 kHz				
2 kHz				
5 kHz				
10 kHz				

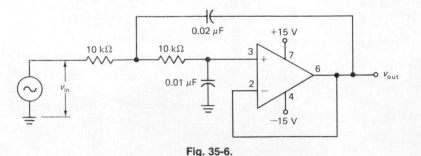

Fig. 35-6.

10. Calculate the voltage gain for each frequency in Table 35-3. Also work out the equivalent decibel gain.
11. Measure and record the cutoff frequency here:

$$f_c = \underline{\hspace{3cm}}$$

Second-Order High-Pass Filter

12. Connect the circuit of Fig. 35-7.
13. Set the generator to 10 kHz. Adjust the signal level to get 1 V p-p at the output of the filter. Measure and record the input voltage (Table 35-4).

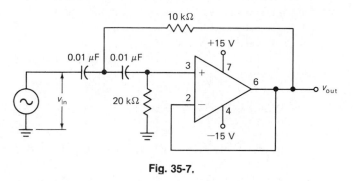

Fig. 35-7.

14. Set the generator to 5 kHz. Measure and record the input voltage.
15. Repeat step 14 for the other frequencies in Table 35-4.
16. Calculate the voltage gain for each frequency in Table 35-4. Also work out the equivalent decibel gain.
17. Measure and record the cutoff frequency here:

$$f_c = \underline{\hspace{3cm}}$$

TABLE 35-4. Second-order High-pass Filter

f	v_{in}, V	v_{out}, V	A	A_{dB}, dB
100 Hz				
200 Hz				
500 Hz				
1 kHz				
2 kHz				
5 kHz				
10 kHz		1		

QUESTIONS

1. What is the theoretical voltage gain at 100 Hz in Fig. 35-5? Explain why this may differ from the voltage gain in Table 35-2.
2. What is the theoretical cutoff frequency in Fig. 35-5? Explain why this may differ from the cutoff frequency measured in step 6.
3. Well above cutoff in Fig. 35-5, how fast should the voltage gain decrease? How much decrease is there between 5 and 10 kHz in Table 35-2?
4. In Fig. 35-6, what is the decibel voltage gain at 100 Hz?
5. What is the theoretical cutoff frequency for Fig. 35-7? What did you actually measure in step 11?

6. Well above the cutoff frequency of Fig. 35-7, how fast should the voltage gain decrease? Compare this to the data in Table 35-4.
7. What is the theoretical cutoff frequency in Fig. 35-7? Well below this cutoff frequency, how fast should the voltage gain decrease?

Answers to Self-Test

1. 0.707   5. 20; 26 dB
2. 3   6. 6.77
3. 18 dB   7. 12
4. 0.25

OBJECTIVES

1. To build a transistor voltage regulator
2. To observe current limiting
3. To measure the effects of line and load changes

INTRODUCTORY INFORMATION

Experiment 3 introduced the zener diode, a device used for voltage regulation. For voltage regulators capable of handling large currents, we need to combine the zener diode with negative-feedback amplifiers. In this experiment you will build a transistorized voltage regulator.

Zener-Diode Regulator

Figure 36-1 shows a bridge rectifier driving a zener-diode regulator. The dc voltage and ripple across the filter capacitor depend on the source resistance, the filter capacitance, and the load resistance. But as long as V_{in} is greater than V_Z, the zener diode operates in the breakdown region. Ideally, this means the final output voltage is constant. To a second approximation, the zener impedance causes the final output to change slightly with changes in line voltage and load current.

The limitation on a zener-diode regulator is this. Changes in load current produce equal and opposite changes in zener current. The changes in zener current flowing through the zener impedance produce changes in the final output voltage. The larger the changes in zener current, the larger the changes in output voltage. If the changes in zener current are only a few milliamperes, the changes in load voltage may be acceptable. But when the changes are tens of milliamperes or more, the changes in load voltage become too large for most applications.

Zener Diode and Emitter Follower

The simplest way to increase the current-handling ability of a zener-diode regulator is add an emitter follower as shown in Fig. 36-2. The load voltage still equals the zener voltage (less the V_{BE} drop of the transistor), but the changes in zener current are reduced by a factor of β. Because of this, the regulator can handle larger load currents and still maintain an almost constant load voltage.

This circuit is an example of a *series voltage regulator*. The collector-emitter terminals are in series with the load. As a result, the load current must pass through the transistor, and this is the reason the transistor is often called a *pass transistor*. The voltage across the pass transistor equals

$$V_{CE} = V_{in} - V_{out} \qquad (36\text{-}1)$$

and its power dissipation is

$$P_D = (V_{in} - V_{out})\, I_{out} \qquad (36\text{-}2)$$

Negative Feedback

In critical applications, zener voltages near 6 V are used because the temperature effects approach zero in this vicinity. The highly stable zener voltage, sometimes called a

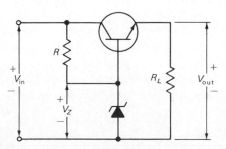

Fig. 36-2. Emitter follower increases current capacity.

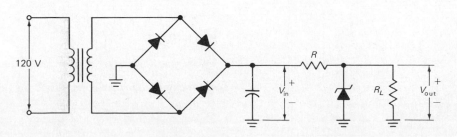

Fig. 36-1. Bridge supply drives zener-diode regulator.

reference voltage, can be amplified with a negative-feedback circuit to get higher voltages with essentially the same temperature stability as the reference voltage.

Figure 36-3 is an example of a negative-feedback regulator. Transistor Q_2 acts like an emitter follower as before. Transistor Q_1 provides voltage gain in a negative-feedback loop. Here is how the circuit operates. Suppose the load voltage tries to increase. The feedback voltage V_F will increase. Since the emitter voltage of Q_1 is held constant by the zener diode, more collector current flows through Q_1 and through R_3. This reduces the base voltage of Q_2. In response, the emitter voltage of Q_2 decreases, offsetting almost all the original increase in load voltage.

Similarly, if the load voltage tries to decrease, the feedback voltage V_F decreases. This reduces the current through Q_1 and R_3. The higher voltage at the base of Q_2 increases the emitter voltage of Q_2, and this almost completely offsets the original decrease in load voltage.

Therefore, any attempted change in load voltage is compensated for by negative feedback. The overall effect is to produce an almost rock-solid load voltage, despite changes in load resistance.

A mathematical analysis leads to this expression for the output voltage:

$$V_{out} = A_{CL} (V_Z + V_{BE}) \qquad (36\text{-}3)$$

where V_Z is the zener voltage and V_{BE} is the base-emitter drop of Q_1. Also, the closed-loop gain is

$$A_{CL} = \frac{R_2}{R_1} + 1 \qquad (36\text{-}4)$$

This means we can use a zener voltage around 6 V where the temperature stability of the zener diode is optimum. By adjusting the ratio of R_2 to R_1, we can produce a regulated output voltage with essentially the same stability as the zener voltage. The potentiometer of Fig. 36-3 allows us to adjust the output voltage to the exact value required in a particular application. In this way, we can adjust for the tolerance in zener voltages, V_{BE} drops, and feedback resistors.

A final point: For the regulator of Fig. 36-3 to work properly, the input voltage must be greater than the output voltage. As long as V_{in} is at least a volt or two greater than the V_{out}, Q_2 will continue to operate as an emitter follower.

Current Limiting

The voltage regulator of Fig. 36-3 is a *series regulator.* As it now stands, it has no short-circuit protection. If we accidentally place a short across the load terminals, we get an enormous current through Q_2. Either Q_2 will be destroyed or a diode in the power supply will burn out, or both. To avoid these possibilities, regulated supplies usually include *current limiting.*

Figure 36-4 shows one way to limit the load current to safe values even though the output terminals are accidentally shorted. For normal currents, the voltage drop across R_4 is small and Q_3 is OFF; under this condition, the regulator works as previously described. If excessive load current flows, however, the voltage across R_4 becomes large enough to turn on Q_3. The collector current of Q_3 flows through R_3; this decreases the base voltage of Q_2 and reduces the output voltage to prevent damage.

In Fig. 36-4 current limiting starts when the voltage across R_4 is around 0.6 to 0.7 V. At this point, Q_3 turns on and decreases the base drive to Q_2. Since R_4 is 1 Ω, current limiting begins when load current is in the vicinity of 600 to 700 mA. By selecting other values of R_4, we can change the level of current limiting.

The current limiting of Fig. 36-4 is a simple example of how it is done. In more advanced circuits, Q_3 is replaced by an op amp to increase the sharpness of the current limiting.

SUMMARY

1. A zener diode can be used to regulate voltage.
2. The limitation on a zener-diode regulator is the amount of current it can handle, typically in the milliamperes.
3. One way to increase the current capacity of a regulator is to add an emitter follower. This increases the current capability by a factor of β.
4. Temperature effects in zener diodes are minimum in the vicinity of 6 V.
5. By using a negative-feedback amplifier and a zener diode, we can build a voltage regulator with excellent voltage stability.
6. Current limiting is needed to avoid the damage that results from accidentally shorting the regulator output terminals.

SELF-TEST

Check your understanding by answering these questions.

1. The limitation on a zener-diode regulator like Fig. 36-1 is its _____ capacity.
2. When the changes in load current are tens of milliamperes or more in Fig. 36-1, the changes in load _____ become too large for most applications.

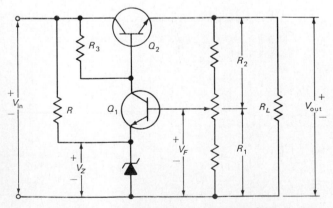

Fig. 36-3. Negative-feedback voltage regulator.

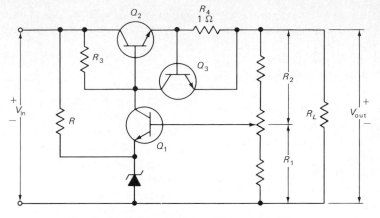

Fig. 36-4. Voltage regulator with current limiting.

3. To increase the current capacity of a regulator, we can add an _____ follower. This increases the current capacity by a factor of _____ .

4. If $R_2 = 30$ kΩ and $R_1 = 10$ kΩ, then the closed-loop gain of Fig. 36-3 equals _____ .

5. If $A_{CL} = 3$, $V_Z = 6$ V, and $V_{BE} = 0.7$ V, then the regulated output voltage of Fig. 36-3 is _____ V.

6. The current-limiting resistor R_4 of Fig. 36-4 is changed to 4.7 Ω. Current limiting will occur for a load current between _____ and _____ .

MATERIALS REQUIRED

- Power supply: Adjustable from 15- to 25-V
- Equipment: Voltmeter (digital if available)
- Resistors: 33-, 680-Ω; 2.2-k, 3.3-k, 10-kΩ ½-W
- Potentiometer: 5-kΩ
- Decade resistance box (if unavailable, use 100-kΩ potentiometer)
- One zener diode: 1N753 (or any with V_Z near 6.2 V)
- Three transistors: 2N3904 (or almost any small-signal NPN silicon transistor)

PROCEDURE

Calculations

1. In Fig. 36-5, what is the value of A_{CL} when the wiper is all the way up? When it is all the way down? Record your answers in Table 36-1.

2. If V_Z is 6.2 V, calculate the minimum and maximum load voltages. Record the values in Table 36-1. (Use a V_{BE} of 0.7 V.)

3. Suppose Q_3 turns on when its V_{BE} is 0.66 V. What is the value of load current where current limiting begins? Record this answer as I_{max} in Table 36-1.

TABLE 36-1. Calculations

$A_{CL(min)} =$	
$A_{CL(max)} =$	
$V_{out(min)} =$	
$V_{out(max)} =$	
$I_{max} =$	

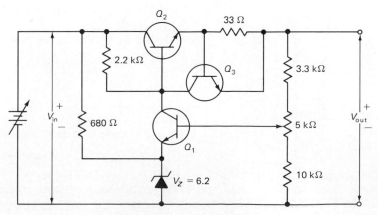

Fig. 36-5. Experimental circuit.

Adjustment Range

4. Connect the circuit of Fig. 36-5.
5. Adjust the power supply to a source V_{in} of 20 V.
6. Adjust the 5-kΩ potentiometer to get minimum V_{out}. Record the value of this minimum load voltage in Table 36-2. Similarly, adjust to get maximum load voltage and record in Table 36-2.

Effect of Load Change

7. Adjust the potentiometer to get a V_{out} of 10 V. Connect the decade resistance box across the load terminals. Change the resistance from 100 to 1 kΩ. As you do this, the voltage will decrease slightly. Record the change as $\Delta V_{out(load)}$ in Table 36-2.

Current Limiting

8. Decrease the decade resistance until the load voltage starts dropping off. Current limiting is now taking place. The lower you make the resistance, the lower the voltage drops.
9. Set the decade resistance to 1 kΩ. Throw a short circuit

TABLE 36-2. Measurements

$V_{out(min)}$ =	
$V_{out(max)}$ =	
$\Delta V_{out(load)}$ =	
$\Delta V_{out(line)}$ =	

across the load terminals and notice how the voltage drops to 0. Remove the short circuit and notice how the voltage returns to 10 V.

Effect of Line Change

10. Vary the supply voltage from 15 to 25 V. The load voltage will change slightly when you do this. Record the change as $V_{out(line)}$ in Table 36-2.
11. Set V_{in} to 20 V, V_{out} to 10 V, and decade resistance to 1 kΩ. Measure the dc voltages on each transistor terminal. Write these values on the schematic diagram (Fig. 36-5) next to each transistor terminal.

QUESTIONS

1. What was the minimum A_{CL} recorded in Table 36-1?
2. What was the maximum A_{CL} recorded in Table 36-1?
3. What were the minimum and maximum load voltages recorded in Table 36-1?
4. Compare the minimum and maximum load voltages of Table 36-2 with the theoretical values of Table 36-1.
5. When the decade resistance is changed from 100 to 1 kΩ, the load voltage changes only slightly. Explain why this is important in a voltage regulator.
6. When the line voltage was changed from 15 to 25 V, The

load voltage changed by a much smaller amount. Explain the importance of this.
7. In step 11, you recorded all transistor voltages. Approximately how much collector current was there in Q_2? What was the power dissipation in this transistor?

Answers to Self-Test

1. current
2. voltage
3. emitter; β
4. 4
5. 20.1 V
6. 0.128; 0.149 A

THREE-TERMINAL IC REGULATORS

OBJECTIVES

1. To experiment with a fixed IC regulator
2. To connect an adjustable IC regulator
3. To test a current regulator

INTRODUCTORY INFORMATION

Early IC Regulators

In the late 1960s, IC manufacturers began producing a voltage regulator on a chip. These first-generation devices include a zener diode, a high-gain amplifier, current limiting, and other useful features. You supply the unregulated input voltage (typically from a bridge rectifier); the IC regulator then produces an almost constant output voltage.

Figure 37-1a shows the pin diagram of the LM300 (metal-can package). By connecting an external resistor to pin 1, we can set the level where current limiting occurs. Pin 2, the booster output, is where an external pass transistor is connected to increase the load-current capability. The unregulated input goes to pin 3 while pin 4 is grounded. Pin 5 is for an external capacitor to bypass the zener diode; this reduces the noise it generates. Feedback voltage V_F goes to pin 6. We need to connect a compensating capacitor to pin 7; this prevents oscillations. (Oscillations are undesirable ac signals that are internally generated. Later experiments tell you more about oscillations.) The regulated output voltage comes out of pin 8.

Figure 37-1b is a basic regulator that can handle load currents in tens of milliamperes. The 10-Ω current-limiting resistor sets the current limit at approximately 25 mA. The 0.1-μF capacitor bypasses the zener diode and reduces its noise. And the 47-pF compensating capacitor prevents the unwanted oscillations.

To increase the load current to hundreds of milliamperes,

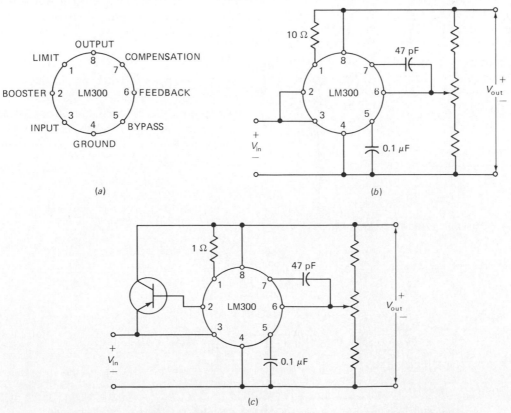

Fig. 37-1. *(a)* LM300 pinout; *(b)* simple voltage regulator; *(c)* boosting current capacity.

we need to add an external pass transistor as shown in Fig. 37-1c. Here, the 1-Ω resistor sets the current limit to around 250 mA. A circuit like this can hold the regulated output voltage to within 1 percent despite changes in the line and load conditions.

The disadvantage of these first-generation IC regulators is the need for external components, plus eight terminals or pins that have to be connected in various ways to get what we want. Ideally, we should not have to connect many external components. Furthermore, an IC regulator ought to be a simple three-terminal device: one pin for the unregulated input, another for the regulated output, and the third for common (ground).

Three-Terminal IC Regulators

The second-generation IC regulators are three-terminal devices that can supply load currents from 100 mA to more than 3 A. These new IC regulators, available in plastic or metal packages, are easy to use and are virtually blowout-proof.

The LM340 series is typical of second-generation IC regulators. Figure 37-2 shows a block diagram. A built-in reference voltage drives the noninverting input of an amplifier. The feedback voltage comes from an internal voltage divider, preset to give output voltages from 5 to 24 V. Specifically, the following voltages are available in the LM340 series: 5, 6, 8, 10, 12, 15, 18, and 24 V. The chip includes a pass transistor that can handle more than 1.5 A of load current. Also included are current limiting and *thermal shutdown*.

Thermal shutdown occurs when the internal temperature reaches 175°C. At this point, the regulator turns off and prevents any further increase in chip temperature. Thermal shutdown is a precaution against excessive power dissipation. Because of the thermal shutdown and the current limiting, the LM340 series is almost indestructible.

A Simple Regulator

Figure 37-3a shows the LM340 in its simplest configuration, a fixed-voltage regulator. In this case, we have a bridge rectifier driving an LM340-5, a device with a 5-V output (tolerance is typically ±2 percent). Pin 1 is the input, pin 2 is the output, and pin 3 is ground. A circuit like this not only regulates output voltage, it attenuates ripple. The LM340-5 has a typical ripple rejection of 80 dB, equivalent to 10,000. In other words, any ripple at the input is 10,000 times smaller at the output.

When the IC is more than a few inches from the supply filter capacitance, the inductance of the connecting lead may produce unwanted oscillations. For this reason, a bypass capacitor C_1 is often used on pin 1 (Fig. 37-3b). Typical values for this bypass capacitor are from 0.1 to 1 μF. To improve the transient response of the LM340, an output bypass capacitor C_2 may be connected; it is typically around 1 μF.

An LM340-5 will regulate over an input range of 7 to 20 V. On the other hand, an LM340-24 holds its output at 24 V for an input range of 27 to 38 V. In general, any device in the LM340 series needs an input voltage at least 2 to 3 V greater than the regulated output; otherwise, it stops regulating.

Adjustable IC Regulator

Figure 37-4 shows external components added to an LM340 to get an adjustable output voltage. The common terminal of the LM340 is not grounded, but rather is connected to the top of R_2. This means the regulated output V_{REG} is across R_1. A quiescent current I_Q flows into pin 3. Therefore, the total current through R_2 is

$$I_2 = I_Q + \frac{V_{REG}}{R_1}$$

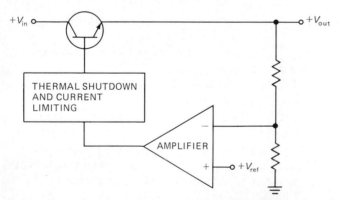

Fig. 37-2. Equivalent circuit for the LM340 series.

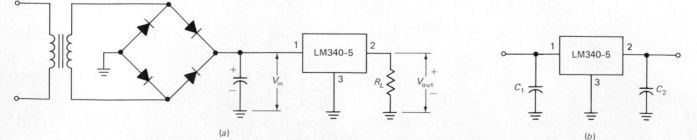

(a)   (b)

Fig. 37-3. *(a)* Bridge circuit drives IC regulator. *(b)* Bypass capacitors may be needed.

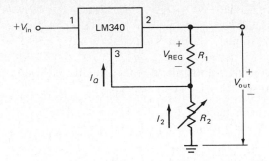

Fig. 37-4. Adjustable voltage regulator.

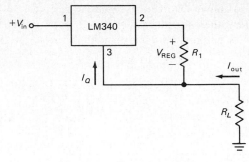

Fig. 37-5. Current regulator.

The output voltage is

$$V_{out} = V_{REG} + \left(I_Q + \frac{V_{REG}}{R_1}\right) R_2 \qquad (37\text{-}1)$$

For the LM340 series, I_Q has a maximum value of 8 mA and varies only 1 mA over all line and load changes.

Often, I_Q is negligible compared to V_{REG}/R_1, and Eq. (37-1) reduces to

$$V_{out} = \left(\frac{R_2}{R_1} + 1\right) V_{REG} \qquad (37\text{-}2)$$

This means the circuit of Fig. 37-4 acts like an adjustable voltage regulator.

Current Regulator

Figure 37-5 is another application for the LM340. Here a load resistor takes the place of R_2. As before, the quiescent current I_Q flows through R_L. Also, the current through R_1 flows through R_L. Therefore, the load current is

$$I_{out} = I_Q + \frac{V_{REG}}{R_1} \qquad (37\text{-}3)$$

When I_Q is negligible, the foregoing reduces to

$$I_{out} = \frac{V_{REG}}{R_1} \qquad (37\text{-}4)$$

As an example, suppose $V_{REG} = 5$ V and $R_1 = 10\ \Omega$. Then Eq. (37-3) gives an I_{out} of 508 mA, while Eq. (37-4) gives 500 mA. This output current is essentially constant and independent of R_L. This means we can change R_L and still have a fixed output current.

SUMMARY

1. The second-generation IC regulators are three-terminal devices.
2. These devices can hold the output voltage constant.

3. The LM340 series is typical of second-generation IC regulators.
4. The regulated voltages of the LM340 series are from 5 to 24 V.
5. LM340 devices include current limiting and *thermal shutdown*.
6. When an IC regulator is more than a few inches from the supply, it may be necessary to connect a bypass capacitor across the regulator input.
7. The input voltage to an LM340 device should be at least 2 or 3 V greater than the regulated output.

SELF-TEST

Check your understanding by answering these questions.

1. The second-generation of IC regulators has only _____ pins or terminals.
2. The _____ series is typical of the second-generation IC regulators.
3. An LM340 device not only regulates the dc output voltage, it also attenuates any _____ at the input.
4. If an IC regulator is more than a few inches from the unregulated supply, a _____ capacitor may be needed across the regulator input to prevent oscillations.
5. The _____ voltage to an IC regulator ought to be at least 2 to 3 V greater than the regulated output.
6. If $R_2 = 100\ \Omega$ and $R_1 = 50\ \Omega$ in Fig. 37-4, then an LM340-6 would produce a V_{out} of _____ V.
7. If $R_1 = 50\ \Omega$ in Fig. 37-5, then an LM340-8 would produce an I_{out} of _____ mA.

MATERIALS REQUIRED

- Power supply: Adjustable from 1- to 25-V
- Oscilloscope
- Resistors: 47-, 100-, 150-, 180-Ω ½-W
- Capacitor: 1-μF
- Equipment: VOM (digital multimeter if available)
- IC regulator: LM340-8

Fixed-Voltage Regulator

1. Connect the circuit of Fig. 37-6.
2. Use the dc-coupled input of the oscilloscope to look at the output of the regulator (pin 2).
3. Slowly increase the adjustable supply from 1 to 15 V and observe the display. All you should see is a solid dc voltage. (If oscillations are present in the output, connect a 1-μF capacitor across the input of the regulator.)
4. Adjust the input supply to the values listed in Table 37-1. Measure and record the regulated output voltages.

Adjusting the Output Voltage

5. Connect the circuit of Fig. 37-7 with $R_2 = 47\ \Omega$.
6. Measure and record V_{out} (Table 37-2).

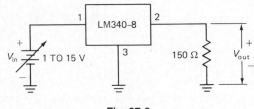

Fig. 37-6.

TABLE 37-1.   Fixed-Voltage Regulator

V_{in}, V	V_{out}, V
1	
5	
10	
11	
12	
13	
14	
15	

7. Change R_2 to the values listed in Table 37-2 and record V_{out} for each value.

Current Regulation

8. Measure and record the current through R_2 for each value listed in Table 37-3.

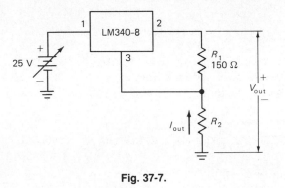

Fig. 37-7.

TABLE 37-2.   Adjusting Voltage

R_2, Ω	V_{out}, V
47	
100	
180	

TABLE 37-3.   Current Regulation

R_2, Ω	I_{out}
47	
100	
180	

QUESTIONS

1. What does the data of Table 37-1 tell you?
2. Refer to Table 37-1. What is the minimum input voltage where the regulator was still working properly?
3. Use Eq. (37-1) and $I_Q = 8$ mA to calculate the V_{out} in Fig. 37-7 for an R_2 of 100 Ω.
4. How does the calculation of question 3 compare to the data of Table 37-2?
5. What does the data of Table 37-3 tell you?

6. If an LM340-12 is used in Fig. 37-7, what is the value of V_{out} for an R_2 of 47 Ω? (Use an I_Q of 8 mA.)
7. If an LM340-15 is used in Fig. 37-7, what is the value of I_{out} for an R_2 of 47 Ω? (Use an I_Q of 8 mA.)

Answers to Self-Test

1. 3
2. LM340
3. ripple
4. bypass
5. input
6. With Eq. (37-1), 18.8 V; with Eq. (37-2), 18 V
7. 160 mA

THE HARTLEY OSCILLATOR

OBJECTIVES

1. To connect a Hartley oscillator and observe and compare the collector and base waveforms
2. To check the frequency of the oscillator

INTRODUCTORY INFORMATION

Oscillatory "Tank" Circuit

An oscillator is an electronic device for generating an ac signal voltage. The frequency of the generated signal depends on the circuit constants. Oscillators are used in radio and TV receivers, in radar, in all transmitting equipment, and in military and industrial electronics.

Oscillators may generate sinusoidal or nonsinusoidal waveforms, from very low frequencies up to very high frequencies. The local oscillator in most present-day broadcast-band AM superheterodynes will "cover" a range of frequencies from 1000 through 2100 kHz (approx).

An oscillation is a back-and-forth motion. In mechanics a pendulum or swing illustrates the principle of oscillation. Once a pendulum is started, it would continue swinging indefinitely if it were not for the energy lost in overcoming friction. It is necessary to add energy periodically to offset this loss and keep the pendulum moving.

In a parallel *LC* circuit, electrons oscillate when the circuit is excited. In the circuit of Fig. 38-1, when S_1 is closed, capacitor C will charge to the battery voltage V. If S_1 is then opened and S_2 closed, C will discharge through L, creating an expanding magnetic field about L. After C has discharged, the magnetic field collapses and induces a voltage in L which tends to maintain electron flow through L in the same direction as when C was discharging. This electron flow charges C in the opposite polarity. After the magnetic field has collapsed, C again tries to neutralize its charge. Electron flow through L is now in the opposite direction. An expanding magnetic field again appears around L, but this time it is in the opposite direction. This process continues back and forth, causing electrons to oscillate in the tuned circuit, also called *tank circuit*. However, owing to the resistance in the circuit and the resulting heat losses (I^2R), the amplitude of oscillation is damped, as in Fig. 38-1, although the period of every cycle is the same. The frequency of oscillation is

$$f = \frac{1}{2\pi\sqrt{LC}} \qquad (38\text{-}1)$$

Overcoming Losses in an Oscillatory Tank Circuit

When the energy fed into the circuit has been used up, it is necessary to supply more energy by recharging capacitor C from the power supply and again permitting it to discharge through L. By switching S_1 and S_2 at the proper time, we can maintain oscillation. Moreover, a sine wave of constant amplitude may be generated. In this process dc energy is used by the circuit to offset losses.

Another method of maintaining oscillations in the *LC* tank circuit is to connect the tank circuit in the output of an amplifier, as in Fig. 38-2. The transistor amplifier is cut off by V_{BB}, which reverse-biases the base-emitter circuit. A sine wave is injected into the base circuit with such amplitude that collector current flows at the peak of the negative alternation. This shock-excites the *LC* circuit in the collector of Q, and the tank circuit oscillates. If the input sine wave has the same frequency as the frequency of oscillation of the tank circuit, the oscillation in the *LC* circuit is maintained.

Tickler-Coil Oscillator

By using positive feedback, the circuit can provide its own input signal. In Fig. 38-3, L_1 is inductively coupled to L. When power is first applied, collector current begins to flow. As this current flows through L, it induces a negative voltage

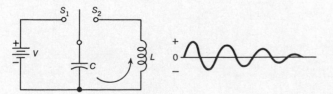

Fig. 38-1. Exciting a parallel *LC* circuit into oscillation.

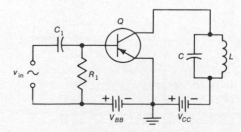

Fig. 38-2. Amplifiers with resonant *LC* circuits.

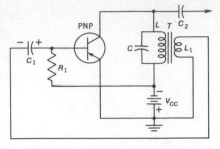

Fig. 38-3. Tickler-coil oscillators.

in L_1. The negative voltage returns to the base of the transistor, forcing more base and collector current to flow. The increased collector current then shock-excites the LC tank into oscillation.

The feedback voltage to the base is a sine wave of the same frequency as the signal in the LC circuit. The feedback voltage to the base therefore eliminates the need for an input signal, and the LC tank will oscillate as previously described.

When the power is first applied, there is no bias since the base and emitter are at the same potential. On the first half-cycle, the induced voltage returning to the base is negative. This causes base current to flow, charging capacitor C_1 with the polarity shown. On the other half-cycle, the returning voltage is positive; this cuts off the base current. The capacitor now discharges through R_1. Because of the long time constant, the capacitor loses only a small part of its charge. On the next half-cycle, the induced voltage is again negative, and base current flows briefly to replace the small charge lost through R_1.

The net effect is to bias the base-emitter circuit to cutoff, with the base current flowing only at the peaks of the negative half-cycle. During the brief time base current flows, collector current also flows. The current pulses in the collector circuit provide the energy to overcome the losses in the tank circuit, which therefore keeps oscillating. L and C determine the frequency of oscillation.

A variation of the tickler-coil oscillator is shown in Fig. 38-4. A difference between this and the one shown in Fig. 38-3 is the elimination of C_1, which changes the class of operation. Another difference is the bottom of L_1 returns to the negative terminal of the battery rather than to ground.

Series-Fed Hartley Oscillator

Figure 38-5 shows a series-fed Hartley oscillator. In this circuit, tickler coil L_1 is part of L, which becomes an auto-transformer. The NPN transistor used in Fig. 38-5 is biased as a conventional amplifier, with forward bias on the base-emitter circuit and reverse bias on the emitter-collector circuit. Collector current flows through L_1 and produces a regenerative current in L which is fed to the base. By design, the tap on the autotransformer L is at the proper point to sustain oscillation in the tank circuit. L-L_1 and C determine the resonant frequency.

R_1 sets the base-emitter bias. C_1 charges because of the current in the base-emitter circuit. When an NPN transistor is used in Fig. 38-5, the polarity of charge on C_1 is as shown. The base is maintained at a negative potential with respect to the emitter, biasing the transistor to cutoff, except during the positive peaks of the oscillations. If a PNP transistor were used in a similar circuit, the charge on C_1 would be opposite in polarity, and the transistor base would be maintained positive with respect to the emitter.

This type of oscillator is called *series-fed* because the RF and dc paths are the same, just as they would be in a series circuit.

Parallel or Shunt-Fed Hartley Oscillator

Figure 38-6*a* is an example of a *shunt-fed* Hartley oscillator, so called because the RF path is in parallel with the dc path. This type of circuit is frequently used as the local oscillator in superheterodyne receivers. The capacitor C_2 and inductor L_1 form the path for RF current in the collector-to-ground circuit. RF current through L_1 induces a voltage in L of the proper phase and amplitude to sustain oscillation. The position of the emitter tap on the coil (junction of L_1 and L) determines how much signal is fed back to the base circuit.

The capacitor C and the autotransformer (L and L_1) make up the resonant circuit which determines the frequency of oscillation. C can be made variable for "tuning" the oscillator to various frequencies. C_1 and R_1 form the RC circuit which develops the bias voltage at the base. The RF choke in the collector keeps the RF signal out of the V_{CC} supply.

The circuit may be dc-stabilized by use of a resistor R_2 connected in the emitter circuit (Fig. 38-6*b*). R_2 is bypassed by C_3 to prevent ac (signal) degeneration.

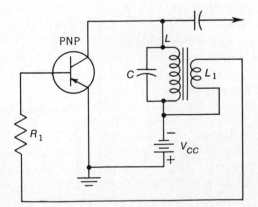

Fig. 38-4. Variation of tickler-coil oscillator.

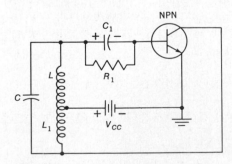

Fig. 38-5. Series-fed Hartley oscillators.

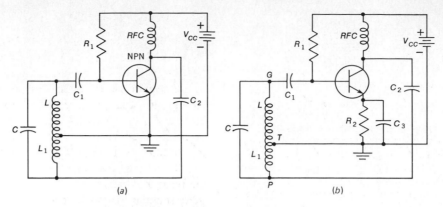

Fig. 38-6. Parallel-fed Hartley oscillators.

The Hartley oscillator coil has three connections. These are usually coded on the coil. If they are not, it is generally possible to identify them by a resistance check. The resistance between the tap T and P is small compared with the resistance between T and G (Fig. 38-6b). If the coil connections are not made properly, the oscillator will not operate.

Checking Oscillator Frequency

Oscilloscope with Calibrated Time Base as Frequency Standard

The approximate frequency of an oscillator may be calculated from the LC constants using the equation

$$F = \frac{1}{2\pi\sqrt{LC}}$$

The frequency of an oscillator may also be measured in other ways. Several methods will be discussed here and applied in the experimental procedure which follows.

One method uses an oscilloscope with a calibrated time base to measure the period of the waveform. From the period, the frequency is calculated. The periodic waveform whose frequency is to be measured is observed on the oscilloscope. The Time/cm controls are set at "calibrated," and the width of the waveform is measured horizontally along the time base. Suppose for example that the width of the sine wave in Fig. 38-7 is 4 cm and that the time base is calibrated at 1 ms $(1 \times 10^{-3}\,\text{s})/\text{cm}$. The period of the waveform may be calculated by multiplying the width in centimeters by the Time/cm setting of the scope. In this case the period t in seconds is $4 \times 1 \times 10^{-3}$. The frequency F may now be calculated using the equation

$$F = \frac{1}{t} \tag{38-2}$$

where F is given in Hertz and t in seconds. For the example above

$$F = \frac{1}{4 \times 1 \times 10^{-3}} = 250\ \text{Hz}$$

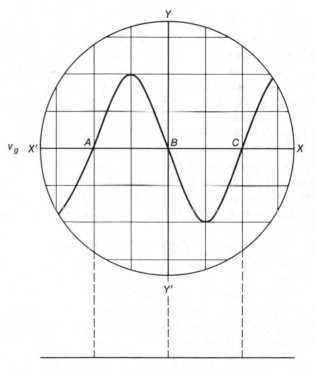

Fig. 38-7. Width of sine wave is 4 cm.

Heterodyne Frequency-Meter Method

Speaker as Null Indicator. Frequency can also be checked experimentally by the use of a heterodyne frequency meter, called a *beat-frequency* meter.

The operation of the frequency meter involves the heterodyning, or beating together, of two signal frequencies. When these signals are mixed in a nonlinear device, such as rectifier or diode, beat notes are created. The newly created frequencies consist of the difference and sum of the two frequencies and harmonic combinations. When the difference in frequency between the two signals is in the audio range, that is, 30 to 15,000 Hz, the beat note can be heard by using a pair of earphones or a loudspeaker. As the two signals approach the same frequency, the frequency difference approaches 0. The two signals are assumed to be equal in frequency at zero beat.

The Hartley Oscillator **221**

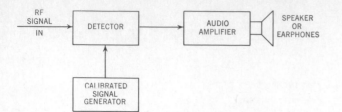

Fig. 38-8. Heterodyne frequency meter.

(a)   *(b)*   *(c)*

Fig. 38-10. Heterodyne beat patterns on oscilloscope.

Zero beat is not heard but is located between the two points on the frequency-meter dial which produces the lowest-audible frequency.

Figure 38-8 is the block diagram of a beat-frequency meter. Included in this device are a calibrated signal generator, a detector, an AF amplifier, and speaker. The unknown RF signal is fed into the meter and the frequency-calibrated dial is adjusted for zero beat. The unknown RF signal frequency is read from the meter dial.

Oscilloscope as Null Indicator. Heterodyne frequency measurements require a null, or zero, indicator. In the method just described a speaker was used as the null indicator. It is possible to employ an oscilloscope as the indicator instead of a speaker. In this case the output of the audio amplifier, in Fig. 38-8, would be fed to an oscilloscope rather than to a speaker. This presupposes the use of an oscilloscope with a heterodyne frequency meter. As a matter of fact, it is possible to dispose of the heterodyne frequency meter when an oscilloscope is used, because the oscilloscope has built in two of the elements in a frequency meter, namely, the amplifier and null indicator. What is required therefore are an external detector and calibrated RF generator.

An external *demodulator probe* acts as the detector. Figure 38-9 shows the circuit diagram of a demodulator probe. A 1N34 diode is connected as an RF rectifier. The *R* and *C* values have been chosen to given RF filtering. The demodulator probe replaces the vertical input probe of the oscilloscope. The probe tip is connected to the unknown frequency signal source. Also coupled to the probe tip is the output of an accurately calibrated signal generator. The generator is varied until zero beat is indicated on the oscilloscope.

The heterodyne-beat patterns on the oscilloscope are shown in Fig. 38-10. Zero beat is shown at (*b*). (This is an idealized drawing. Some signal will be observed at zero beat.) At (*a*) and (*c*) are shown audio sine waveforms, which

are observed on either side of zero beat. As the generator frequency is moved further from zero beat, the frequency of the beat note increases.

The frequency of an oscillator may also be checked using a frequency meter, or an oscilloscope with a calibrated time base.

SUMMARY

1. Electronic oscillators generate sinusoidal and non-sinusoidal ac voltages.
2. Oscillators are used in every branch of electronics, such as radio, television, radar, and transmitters.
3. An *LC* circuit, such as that in Fig. 38-1, may be shock-excited into oscillation, but unless the losses of the circuit are replaced, the oscillations are damped and die out.
4. Special arrangements of feedback circuits, using transistors in conjunction with coils and capacitors, may supply the energy required to overcome resistance losses in the tank circuit and sustain oscillation.
5. One type of oscillatory circuit employs "tickler" coils to feedback energy in proper phase, from the output to the input of the oscillator. Examples are shown in Figs. 38-3 and 38-4.
6. A Hartley oscillator uses a three-terminal auto-transformer-type tickler coil to sustain oscillation; for example, the coil L-L_1 in Fig. 38-6.
7. Figures 38-6*a* and *b* show transistor Hartley oscillators. Figure 38-6*b* is dc-stabilized by the inclusion of emitter resistor R_2 and bypass capacitor C_3.
8. Figure 38-6 shows parallel or *shunt-fed* Hartley oscillators, because the generated ac signal current path is in parallel with the dc current path.

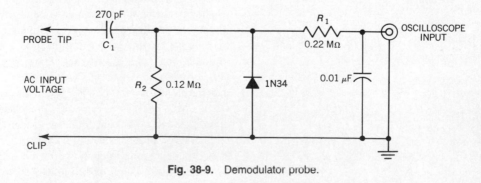

Fig. 38-9. Demodulator probe.

9. A *series-fed* Hartley oscillator is shown in Fig. 38-5. Here the dc and ac paths are the same.
10. The oscillators in Figs. 38-1 through 38-6 generate sine waves.
11. The frequency of an *LC* oscillator depends on the inductance and capacitance of the resonant or tank circuit and may be found using the equation

$$F = \frac{1}{2\pi\sqrt{LC}}$$

where *F* is in hertz, *L* in henrys, and *C* in farads.
12. The frequency *F* of an oscillator may be measured by an oscilloscope with a calibrated time base.
13. *F* may also be measured by use of heterodyne frequency meters.
14. A frequency meter or an oscilloscope with a calibrated time base may also be used to measure the frequency of the oscillator.

SELF-TEST

Check your understanding by answering these questions.

1. In Fig. 38-1, when S_2 is closed and *C* discharges through *L* a damped oscillation is generated. The waveform is damped because of the _____ in the circuit.
2. In Fig. 38-2, *Q* conducts during the most _____ (*negative, positive*) part of the input cycle.
3. In Fig. 38-3, when power is first applied and the transistor starts to conduct, the top of winding L_1 must be driven _____ (*negative, positive*) to start the oscillatory cycle.
4. The Hartley oscillator in Fig. 38-6 uses a _____ terminal coil to sustain oscillation.
5. In Fig. 38-6, the bias developed by C_1R_1 after oscillation begins keeps the transistor OFF except during the top of the _____ (*positive, negative*) alternation of the sine wave.
6. In Fig. 38-6 the voltage measured from base to emitter, when the oscillator is ON, will be _____ (*positive, negative*).
7. In measuring the voltage from base to emitter in Fig. 38-6, a _____ (*high, low*)-impedance voltmeter should be used.
8. In Fig. 38-6, an oscilloscope connected from collector to emitter will show a _____ .
9. The purpose of the RFC in Fig. 38-7 is to keep the _____ out of the _____ .
10. An oscilloscope with a calibrated time base is used to measure the frequency of an oscillator. If the Time/cm control is set at 100 microseconds per centimeter (μs/cm), and if the width of one cycle of the sine wave is 2 cm, the frequency of the oscillator is _____ Hz.

MATERIALS REQUIRED

- Power supply: Variable regulated low-voltage dc source; high-voltage source for optional vacuum-tube-oscillator experiment, RF signal generator
- Equipment: Oscilloscope (time-calibrated base) and demodulator probe; EVM; 0–10-mA milliammeter
- Resistors: 390-, 15,000-Ω; 270-kΩ ½-W plus others as required for extra-credit experiments
- Capacitors: 15-, 47-, 250-pF; 0.001-, 0.01-μF
- Semiconductors: 2N6004 transistor (other choice: 2N3904)
- Miscellaneous: 30-mH RF choke; Hartley oscillator coil (Miller #2021 or equivalent—broadcast band); SPST switch

PROCEDURE

1. Connect the circuit of Fig. 38-11. *L* is a broadcast-band tapped oscillator coil. *M* is a dc milliammeter or a VOM set on the 10-mA current range. Check circuit connections before applying power.
2. **Power on.** With an oscilloscope whose Time/cm switch is set at 1 μs/cm (or a nontriggered scope set on the highest sweep-frequency range), observe and measure the waveform at TP 1 and 2. Record the results in Table 38-1.

   Note that on narrowband oscilloscopes, peak-to-peak measurements will be relative because of high-frequency signal attenuation by the amplifiers in the oscilloscope.
3. With an EVM measure V_{BE}, base-to-emitter dc voltage, and V_{CE}, collector-to-emitter dc voltage, at TP 1 and 2, respectively. Record in Table 38-1.

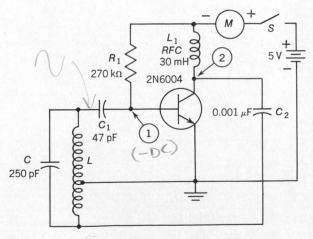

Fig. 38-11. Experimental Hartley oscillator.

TABLE 38-1. Experimental Oscillator Checks

Step	TP 1			TP 2			
	Waveform	V p-p	V_{BE}	Waveform	V p-p	V_{CE}	I_T
2, 3, 4		9 v	−2.3		9 v	2.8	8.7m
6		8 v	−1.9		4.7 v	4	4.6m
7		7.5 v	(+? .15)		3.1 v	4.3	3.3m

7) 526 K   7) 555 K

4. Measure the circuit current and record in Table 38-1.
5. **Power off.** Add the bias stabilization network, R_2, and C_3 to the experimental circuit, as in Fig. 38-12.
6. **Power on.** Repeat steps 2, 3, and 4.
7. Open C_3 and repeat steps 2, 3, and 4.

Measuring Oscillator Frequency (Oscilloscope with Calibrated Time Base)

8. Using a scope with a calibrated time base, measure the width of one cycle of the oscillator waveform. Convert this width into time t by multiplying cycle width (in centimeters) by the setting of the Time/cm switch. Determine the oscillator frequency F by using the formula

$$F = \frac{1}{t} \qquad \frac{1}{1.8 \cdot 1\mu s} = 555 K$$

where F is in Hertz and t is in seconds. Record in Table 38-2. Show all your computations.

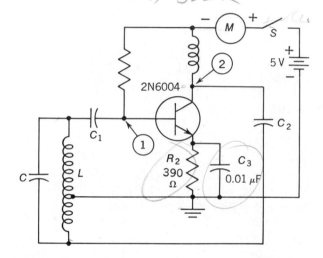

Fig. 38-12. Adding bias stabilization to experimental Hartley oscillator.

Measuring Oscillator Frequency

9. Replace the oscilloscope probe with an RF demodulator probe. Bypass the oscilloscope input with a 0.01-μF capacitor. Connect the probe tip to TP 1 in Fig. 38-12. Observe the waveform at TP 1. Record this waveform in Table 38-2. If a demodulator probe is not available, the circuit of Fig. 38-10 will serve as a demodulator.
10. Connect a 15-pF capacitor in series with the "hot" lead of a calibrated RF signal generator and inject the output

of the generator into TP 1. Set the generator at 600 kHz. Vary the generator frequency control on either side of 600 kHz until zero beat is observed on the oscilloscope. Generator output control should be set at the lowest level which will permit a null indication to be seen on the oscilloscope. Oscilloscope sensitivity is set at maximum. In Table 38-2 record the waveforms at zero beat (b) and on either side of zero beat (a) and (c). Record also the generator frequency at zero beat. This is the oscillator frequency.

TABLE 38-2. Measuring Oscillator Frequency by Zero Beat

Step	RF Waveform	Zero-beat Waveforms			Oscillator Frequency, 555 kHz
		(a)	(b)	(c)	
10		X	X	X	
9, 10					

QUESTIONS

1. What are the conditions required for oscillation?
2. How are the requirements for oscillation met in the circuit of Fig. 38-11?
3. Explain the effects of removing C_3 in the circuit of Fig. 38-12.
4. Compare and explain current readings in Table 38-1 for steps 4 and 7.
5. What is the purpose of capacitor C_2 in Fig. 38-12?
6. Will the oscillator operate with C_2 open?
7. (a) When is a "null" observed in using the zero-beat method of determining oscillator frequency? (b) What does the null mean?
8. Why did you encounter spurious zero-beat indications when measuring oscillator frequency?
9. (a) How did the measured oscillator frequency compare in the two methods you used experimentally (oscillo- scope with calibrated time base and zero-beat method)? (b) Explain any difference in the two frequencies.
10. How can you determine whether an oscillator circuit is oscillating?

Extra Credit

11. What are the requirements for the frequency characteris- tics of the vertical and horizontal amplifiers in an oscilloscope used to determine the oscillator frequency in this experiment?

Answers to Self-Test

1. losses
2. negative
3. negative
4. 3
5. positive
6. negative
7. high
8. sine wave
9. RF; dc supply
10. 5000

PHASE-SHIFT OSCILLATOR

OBJECTIVES

1. To determine the range of frequency variation of an *RC* phase-shift oscillator
2. To compare the phase of output and feedback voltages in the oscillator

INTRODUCTORY INFORMATION

Phase-Shift Feedback

In the Hartley oscillator (Experiment 38), the *L* and *C* of the tank circuit determine the frequency of oscillation. The phase-shift oscillator in this experiment uses resistors and capacitors (*R* and *C*) as the frequency-determining constants.

Recall that the requirements for oscillation include (1) an amplifier with (2) feedback from the output to the input circuit in proper phase to overcome the circuit losses and sustain oscillation. In the Hartley oscillator, an increase in collector current resulted in feeding back a positive voltage to the base to sustain that current increase, whereas a decrease in collector current resulted in feedback of a negative voltage to the base to sustain the collector-current decrease. The voltage fed back to the base must be in phase with the voltage on the base to sustain oscillation.

The manner in which feedback is accomplished is unimportant, as long as it is in proper phase and of sufficient amplitude to overcome the energy losses in the circuit. The feedback path simply distinguishes one type of oscillator from another.

In Fig. 39-1 the feedback network is shown in block form. Q_1 is an amplifier with a resistive collector load. There is a 180° phase shift between base and collector under normal operating conditions. The feedback network must introduce another 180° phase shift from the collector back to the base, in order to accomplish oscillation. A triple-section *RC* network, as in Fig. 39-2, can do this. Point *A* is connected to

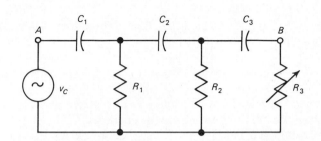

Fig. 39-2. *RC* phase-shift network.

the collector of Q_1 in Fig. 39-1, and *B* is connected to the base of Q_1.

Consider a single section, R_1C_1 (Fig. 39-3*a*) of this feedback network, and assume that the signal v_C which is coupled to C_1R_1 is a sine wave. C_1R_1 is a capacitive circuit, and the current leads the voltage by an angle which is defined as the "phase angle" of the circuit. The phase angle θ depends on the frequency of v_C and on the values of *R* and *C* and is given by the equation

$$\tan \theta = \frac{X_C}{R} = \frac{1}{2\pi fCR} \qquad (39\text{-}1)$$

$$\theta = \arctan \frac{X_C}{R} \qquad (39\text{-}2)$$

Now C_1 and R_1 may be chosen so that for a desired frequency *f*, θ = 60°. Figure 39-3*b* shows this 60° phase shift. The voltage v_{R1} across R_1 leads v_C, the input voltage, by 60°.

C_2 and R_2 can now be chosen to introduce an additional 60° phase shift between v_{R1} and v_{R2}, so that v_{R2} leads v_C by 120°. Similarly, C_3 and R_3 are selected to introduce another 60° phase shift, and as a result, v_{R3} leads v_C by 180°. Note

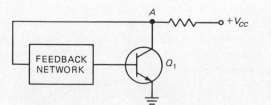

Fig. 39-1. Oscillator feedback network.

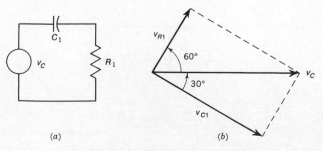

Fig. 39-3. Phase shift in *RC* network.

that there will be a 180° phase shift for just *one* frequency, as determined by the values of C and R in the feedback network.

If any of the selected values of C and R, say, R_3 in Fig. 39-2, is varied, the frequency for which there is a 180° phase shift will change. The phase angles introduced by each section of the feedback network will change because of the new frequency. Oscillation takes place at the frequency for which the total phase shift equals 180°.

Transistor Phase-Shift Oscillator

A transistor phase-shift oscillator must introduce in-phase feedback from the output to the input to sustain oscillation. If a common-emitter amplifier is used, with a resistive collector load, there is a 180° phase shift between the base and collector. Hence the phase-shift feedback network between collector and base must introduce an additional 180° phase shift, at some frequency, if oscillation is to take place. Here also a three-section RC network may be employed.

A transistor connected as a phase-shift oscillator is shown in Fig. 39-4. In this common-emitter amplifier, feedback is from the collector to the base, that is, from the output to the input. The three-section phase-shift network consists of C_1, R_1, C_2, R_2, C_3, and R_3 in series with R_{in}, the input resistance. So that each section may introduce a 60° phase shift (approx) at the resonant frequency, the values $C_1 = C_2 = C_3$ and $R_1 = R_2 = R_3 + R_{in}$. By analysis it can be shown that the frequency of oscillation for these conditions may be expressed by the equation

$$f = \frac{1}{2\pi C \sqrt{6R_1{}^2 + 4R_1 R_L}} \tag{39-3}$$

A necessary condition for sustained oscillation in the RC phase-shift oscillator is given by the equation

$$h_{fe} = 23 + \frac{29R_1}{R_L} + \frac{4R_L}{R_1} \tag{39-4}$$

where h_{fe} is the small-signal forward-current transfer ratio of the transistor.

Example: A numerical example will illustrate the use of Eqs. (39-3) and (39-4). What is the predicted frequency of oscillation of Fig. 39-4, if $R_1 = 2200 \ \Omega$, $R_L = 12,000 \ \Omega$, and $C_1 = 0.1$? Will a transistor with the given values of R_1,

R_L, and C_1 whose $h_{fe} = 40$ provide sufficient feedback for oscillation?

Solution:

(a)
$$h_{fe} = 23 + \frac{29R_1}{R_L} + \frac{4R_L}{R_1}$$

$$= 23 + \frac{29(2200)}{12,000} + \frac{4(12,000)}{2200}$$

$$= 23 + 5.32 + 21.8$$

$$= 50.12$$

A transistor whose $h_{fe} = 40$ will not permit oscillation in this circuit. The h_{fe} of the required transistor must be greater than 50.12.

(b)

$$f = \frac{1}{2\pi C_1 \sqrt{6R_1{}^2 + 4R_1 R_L}}$$

$$= \frac{1}{6.28(0.1 \times 10^{-6})\sqrt{6(2200)^2 + 4(2200)(12,000)}}$$

$$= \frac{10^4}{6.28\sqrt{6(2.2)^2 + 4(2.2)(12)}} = \frac{10^4}{6.28\sqrt{134.6}}$$

$$= 137 \text{ Hz}$$

If a proper transistor is chosen, the frequency of oscillation for the values given in the example will therefore be approximately 137 Hz.

Figure 39-5 is a practical variation of the circuit of a phase-shift oscillator. Note the changed position of the frequency control R_4, and the change in the bias circuit. Bias resistor R_3 is connected to the collector for bias stabilization. The purpose of C_4 is to bypass the base to eliminate parasitic oscillations. Since this is not a perfectly balanced phase-shift circuit, Eq. (39-3) is not directly applicable. However, if rheostat R_4 were set for *zero* resistance and if R_3 were made *variable*, R_3 could be adjusted for phase-shift balance and Eq. (39-3) would then apply.

SUMMARY

1. Oscillators require an amplifying device, such as a vacuum tube or transistor, with feedback in phase from the output back to the input to sustain oscillation.
2. The feedback network in the phase-shift oscillator of Fig. 39-4 consists of a three-section RC circuit (see Fig. 39-2).
3. In Fig. 39-4, the amplifier introduces a 180° phase shift between the signal on the base and the signal at the collector. If for some frequency the RC feedback network introduces an additional 180° phase shift, then the phase shift of the signal fed back equals 180° + 180°, or 360°. This is the same as 0° or an in-phase condition, and the circuit meets one of the requirements for oscillation.
4. The variable resistor R_3 is used to unbalance the feedback network, in order to change the frequency of oscillation.
5. Not only must the phase-shift signal have the proper phase, but it must also be of sufficient amplitude to

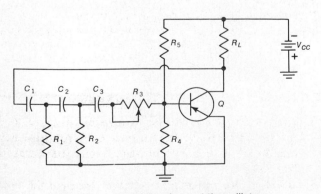

Fig. 39-4. Transistor phase-shift oscillator.

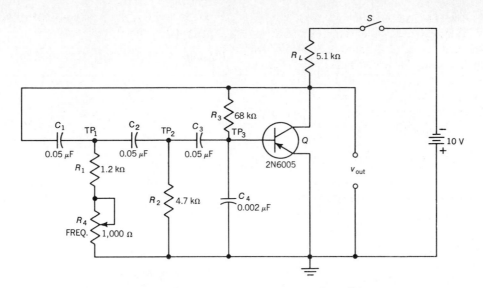

Fig. 39-5. Experimental transistor phase-shift oscillator.

overcome the circuit losses. To meet this requirement the small-signal forward-current transfer ratio (h_{fe}) or gain of the transistor must be such that

$$h_{fe} = 23 + \frac{29R_1}{R_L} + \frac{4R_L}{R_1}$$

6. The frequency of a *symmetrical* three-section phase-shift oscillator, determined by the values of R and C, is given by the equation

$$f = \frac{1}{2\pi C \sqrt{6R_1{}^2 + 4R_1R_L}}$$

where C is the value of each capacitor in farads, R_1 is the value of each resistor in ohms, and f is given in hertz.

SELF-TEST

Check your understanding by answering these questions.

1. In Fig. 39-1, if there were no feedback network and the transistor were simply connected as a signal amplifier, the signal voltage at the collector would be _____ degrees out of phase with the signal voltage at the base.
2. When the circuit operates as a phase-shift oscillator, the feedback network (Fig. 39-4) introduces an additional

_____° phase shift, so that the signal feedback from plate to grid is _____ _____ with the signal on the grid.
3. In the first section of the *RC* phase-shift network in Fig. 39-2, the signal voltage v_1 across R_1 _____ (*leads, lags*) the applied signal voltage v_C.
4. In Fig. 39-2, the size of the phase-shift angle θ of the signal across R_1 depends on the _____ of the signal and on the values of _____ and

_____ .
5. In the phase-shift oscillator of Fig. 39-4 frequency may be changed by varying _____ .
6. Any PNP transistor can be used in place of Q_1 in the transistor phase-shift oscillator of Fig. 39-4.

_____ (*true, false*)

MATERIALS REQUIRED

- Power supply: Variable regulated low-voltage dc source; regulated high-voltage dc source for extra-credit steps
- Equipment: Oscilloscope; EVM (or VOM); frequency counter
- Resistors: 1200-, 4700-, 5100-, 68,000-Ω ½-W
- Capacitors: Three 0.047-, 0.002-μF
- Semiconductors: 2N6005 or the equivalent
- Miscellaneous: SPST switch; 1000-Ω 2-W potentiometer

PROCEDURE

1. Connect the circuit of Fig. 39-5. Set R_4, the frequency control, for maximum resistance.
2. **Power on.** With an oscilloscope observe and measure the waveform and its peak-to-peak voltage at the collector and base of Q. Record the data in Table 39-1. Measure and record the dc voltage at the collector, emitter, and base.

3. Using a frequency counter or an oscilloscope with a calibrated time base, determine the oscillator frequency with R_4 set at maximum resistance. Record the data in Table 39-1.
4. Adjust R_4 for minimum resistance. Measure the new oscillator frequency. Record this frequency in Table 39-1.

TABLE 39-1. *RC Oscillator Frequency Range*

Test Point	Waveform	V p-p	DC, V	Frequency Minimum	Frequency Maximum
Collector					
Emitter	X	X			
Base					

5. Calculate the frequency f of the oscillator when the phase-shift circuit is balanced. [Use Eq. (39-3).] Show your computations. Record this data in Table 39-2.
6. Observe and measure the peak-to-peak amplitude of the output waveform v_{out}, and, if possible, the feedback-voltage waveform at TP 1, 2, and 3. Record this data in Table 39-2.
7. Using an oscilloscope with a calibrated time base, preferably a dual-trace oscilloscope, measure the phase shift

introduced by C_1R_1. Record the data in the Table 39-2. HINT: Use the sine wave at the collector of Q as the reference voltage and the sine wave at TP 1 as the phase-shift signal voltage.
8. If possible, also measure and record the phase relationship between the output signal v_{out} at the collector and the input phase-shift signal v_{in} at the base (or at the junction of C_3 and R_3).

TABLE 39-2. *RC Oscillator Phase-Shift Relationships*

Test Point	Waveform	V p-p	Phase Shift, Degrees Introduced by C_1R_1	Phase Shift, Degrees Between v_{out} and v_{in}
Collector				
TP 1				
TP 2				
TP 3				

QUESTIONS

1. (*a*) What are the requirements for oscillation? (*b*) Do these requirements apply equally to transistor and vacuum-tube circuits?
2. How is feedback accomplished in the *RC* oscillator of Fig. 39-4?
3. Explain in detail a method for making frequency measurements. Draw an interconnection diagram of equipment and identify the two signals involved.
4. How did you measure the phase shift introduced by C_1R_1? Draw an interconnection diagram of equipment and identify the two signals involved.
5. Compare the measured phase-shift signal at TP 1 with

what it should be theoretically if the phase-shift circuit were balanced.
6. Compare the measured phase-shift signal between v_{out} and v_{in} with what it should be theoretically and account for any discrepancy.
7. Why is there a difference in amplitude between the feedback voltage at TP 1, 2, and 3?
8. Refer to Fig. 39-5. How is bias stabilization achieved?

Answers to Self-Test

1. 180
2. 180; in phase
3. leads
4. frequency; R_1; C_1
5. R_3
6. false

OBJECTIVES

1. To build and test a Wien-bridge oscillator
2. To build and test a twin-T oscillator

INTRODUCTORY INFORMATION

Lead-Lag Network

Figure 40-1*a* shows a *lead-lag network*. In a circuit like this, the phase angle leads for low frequencies and lags for high frequencies. In other words, the phase shift looks like Fig. 40-1*b*. Especially important, there is a frequency f_0 where the phase shift equals 0.

Besides information about phase shift, a mathematical analysis using complex numbers indicates the voltage gain of the circuit reaches a maximum of one-third, as shown in Fig. 40-1*c*. The frequency where the gain maximizes and the phase shift equals 0 is given by

$$f_0 = \frac{1}{2\pi RC} \qquad (40\text{-}1)$$

A lead-lag network like Fig. 40-1*a* is a resonant circuit because the voltage gain reaches a maximum at f_0 and the phase shift goes to 0 at f_0. For this reason, f_0 is called the *resonant frequency* of the lead-lag network.

Wien-Bridge Oscillator

The Wien-bridge oscillator is the most widely used oscillator for frequencies in the range of 5 Hz to about 1 MHz. Figure 40-2 shows a Wien-bridge oscillator using an op amp to provide the necessary gain. The Wien bridge consists of a lead-lag network on the left and a voltage divider on the right.

The circuit of Fig. 40-2 uses positive and negative feedback. The positive feedback is through the lead-lag network to the noninverting input. The negative feedback is through the voltage divider to the inverting input. In the voltage divider, R_1 is usually a miniature tungsten lamp.

When power is first applied, the tungsten lamp has low resistance and the voltage gain of the divider is less than one-third. Because the lead-lag network has a voltage gain of one-third and a phase shift of 0 at f_0, the error voltage is large and oscillations begin. As the oscillations build up at frequency f_0, the tungsten lamp heats, and this increases its resistance.

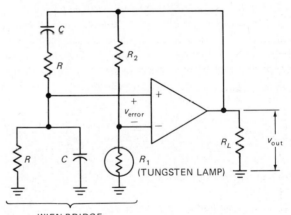

Fig. 40-2. Wien-bridge oscillator using op amp.

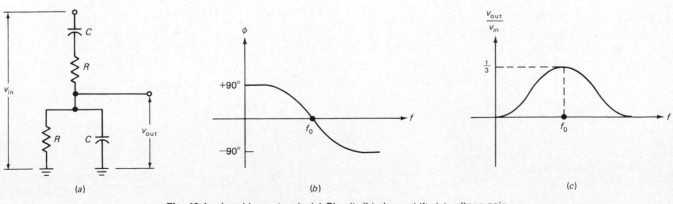

Fig. 40-1. Lead-lag network. *(a)* Circuit; *(b)* phase shift; *(c)* voltage gain.

When the lamp resistance approaches $R_2/2$, the voltage divider has a gain near one-third. At this point, the error voltage approaches 0 and the oscillations level off. In other words, the peak-to-peak output becomes constant when the lamp resistance equals $R_2/2$ (approx).

Twin-T Filter

Figure 40-3a illustrates a *twin-T filter*. A complex-number analysis of this circuit shows it acts like a lag-lead network with a phase angle, as shown in Fig. 40-3b. Again, there is a frequency f_0 where the phase shift equals 0. The voltage gain equals unity at low and high frequencies; at frequency f_0 the voltage gain drops to 0 (Fig. 40-3c).

The twin-T filter is sometimes called a *notch filter* because it can notch out or attenuate those frequencies near f_0. Frequency f_0, known as the *notch frequency*, is given by

$$f_0 = \frac{1}{2\pi RC} \qquad (40\text{-}2)$$

Twin-T Oscillator

Figure 40-4 shows a twin-T oscillator. The positive feedback is through the voltage divider to the noninverting input. The negative feedback is through the twin-T filter. When power is first turned on, the lamp resistance R_2 is low and the positive feedback is maximum. As the oscillations build up, the lamp resistance increases and the positive feedback decreases. As the feedback decreases, the oscillations level off and become constant. In this way, the lamp stabilizes the level of output voltage.

In the twin-T filter, resistance $R/2$ is adjusted. This is necessary because the circuit oscillates at a frequency

slightly different from the ideal notch frequency of Eq. (40-2). At the oscillation frequency, the gain of the twin-T filter must be slightly less than the gain of the voltage divider, and the phase shift of the filter is approximately 0. In this way, the error voltage has the correct amplitude and phase to sustain oscillations.

To ensure the oscillation frequency is close to the notch frequency, the voltage divider should have R_2 much larger than R_1. As a guide, R_2/R_1 is in the range of 10 to 1000. This forces the oscillator to operate at a frequency near the notch frequency.

SUMMARY

1. In a *lead-lag network* the phase shift is positive at low frequencies and negative at high frequencies.
2. The phase shift equals 0 in a lead-lag network when the frequency is f_0.
3. The voltage gain of a lead-lag network reaches a maximum value of one-third when the frequency is f_0.
4. A Wien-bridge oscillator has a lead-lag network in the positive-feedback path and a voltage divider in the negative-feedback path.
5. A miniature tungsten lamp is used in the voltage divider to stabilize the output voltage.
6. The output of a Wien-bridge oscillator levels off when the lamp resistance equals $R_2/2$. At this point, the voltage divider has a gain of approximately one-third.
7. A twin-T filter acts like a lag-lead network, producing negative phase shift at low frequencies and positive phase shift at high frequencies.
8. At frequency f_0 the phase shift of a twin-T filter is 0.

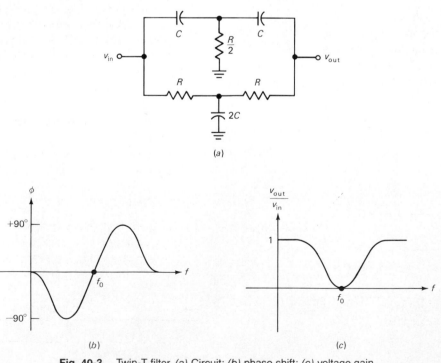

Fig. 40-3. Twin-T filter. *(a)* Circuit; *(b)* phase shift; *(c)* voltage gain.

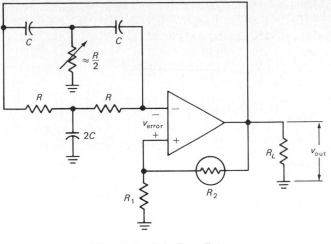

Fig. 40-4. Twin-T oscillator.

9. The voltage gain of a twin-T filter ideally drops to 0 at frequency f_0.
10. A twin-T filter is sometimes called a *notch filter* because it notches or attenuates frequencies near f_0.

SELF-TEST

Check your understanding by answering these questions.

1. At frequency f_0 the phase shift of a lead-lag network equals _____ .

2. Frequency _____ equals the reciprocal of $2\pi RC$.
3. In the Wien-bridge oscillator of Fig. 40-2, the lead-lag network provides _____ feedback and the voltage divider provides _____ feedback.
4. When power is first applied, the tungsten lamp of a Wien-bridge oscillator has _____ resistance and the gain of the divider is less than _____ .
5. If $R = 5\ k\Omega$ and $C = 0.047\ \mu F$ in Fig. 40-2, the frequency of oscillation is approximately _____ kHz.
6. If $R_2 = 1\ k\Omega$ in Fig. 40-2, the tungsten lamp must have a resistance less than _____ Ω for the oscillations to start. The oscillations will stabilize when the lamp resistance equals _____ Ω.
7. At frequency f_0 the phase shift of a twin-T filter is _____ and the voltage gain is _____ .
8. The oscillation frequency of a twin-T oscillator is approximately equal to 1 divided by _____ .

MATERIALS REQUIRED

- Two power supplies: 15-V
- Equipment: Oscilloscope, electronic counter
- Resistors: one 100-Ω, one 270-Ω, two l-kΩ, one 10-kΩ ½-W
- Potentiometer: 1-kΩ
- Capacitors: Four 0.1-μF
- Op amp: 741C

Wien-Bridge Oscillator

1. Calculate the frequency of oscillation in Fig. 40-5. Record this frequency in Table 40-1.
2. Connect the circuit of Fig. 40-5. (NOTE: For convenience, an ordinary resistor is used for R_1 instead of a miniature tungsten lamp.)

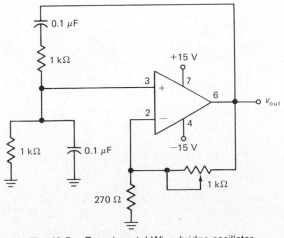

Fig. 40-5. Experimental Wien-bridge oscillator.

3. Look at the output with an oscilloscope; set the Time/cm switch at 0.2 milliseconds per centimeter (ms/cm). Adjust the potentiometer to get as large a sine wave as possible without excessive clipping or distortion. (The signal level should be around 15 V p-p.)
4. Use an electronic counter to measure the output frequency. Record the frequency in Table 40-1. Next, use the counter to measure the period. Record the period. (If a counter is not available, use the oscilloscope to measure the period. Then calculate the frequency. Record both values in Table 40-1.)

Twin-T Oscillator

5. Calculate the approximate frequency of oscillation in Fig. 40-6. Record the value in Table 40-2.
6. Connect the circuit of Fig. 40-6. (Use two 0.1-μF capacitors in parallel for the 0.2-μF capacitor.)

TABLE 40-1. Wien-Bridge Oscillator

$f_{0(calc)}$	$f_{(meas)}$	$T_{(meas)}$

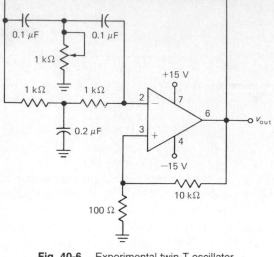

Fig. 40-6. Experimental twin-T oscillator.

TABLE 40-2. Twin-T Oscillator

$f_{0(calc)}$	$f_{(meas)}$	$T_{(meas)}$

7. Look at the output with an oscilloscope (Time/cm switch at 0.2 ms/cm). Adjust the potentiometer to get a large undistorted sine wave (approximately 20 to 25 V p-p).
8. Use the electronic counter to measure the frequency and period of the output. Record the values in Table 40-2.

QUESTIONS

1. Compare the measured frequency of oscillation in Fig. 40-5 to the theoretical value. Suggest reasons why the two may differ.
2. In Fig. 40-5, what is the phase angle of the signal at pin 3 compared to the signal at pin 6?
3. Examine the measured frequency and period in Table 40-1. How are these related mathematically?
4. We want the Wien-bridge oscillator of Fig. 40-1 to oscillate at 5 kHz. What components may we change to accomplish this?
5. Compare the measured frequency of oscillation in Fig. 40-6 to the calculated frequency. Give some reasons why these two may differ.
6. If the capacitors are held constant in Fig. 40-6, what are the required resistance values to get an oscillation frequency of 10 kHz?

Answers to Self-Test

1. 0
2. f_0
3. positive; negative
4. low; one-third
5. 677 Hz
6. 500; 500
7. 0; 0
8. $2\pi RC$

OBJECTIVES

1. To observe collector and base waveforms in a free-running collector-coupled multivibrator
2. To measure multivibrator frequency
3. To synchronize the multivibrator

INTRODUCTORY INFORMATION

Classes of Multivibrators

There is a class of *RC*-coupled oscillators called *multivibrators* which generate nonsinusoidal waveforms, such as square, rectangular, triangular, or sawtooth waves. These waveforms find many applications. The most familiar is the sawtooth wave which is employed as a time base, or sweep, in oscilloscopes.

The multivibrator is one form of relaxation oscillator. Multivibrators may be *self-excited* (also called *free-running*), requiring no external excitation. Or they may be driven oscillators whose operation and frequency are controlled by external driving or triggering voltages. Free-running multivibrators are called *astable*. The frequency of astable multivibrators may be controlled by external synchronizing pulses.

Driven multivibrators are not strictly oscillators, for they require external driving pulses to start their operation and maintain it. There are two classes of driven multivibrators: the monostable (single shot), which has *one* stable state; and the bistable (flip-flop, or trigger), which has two stable states.

Transistors may be used as the amplifying device for multivibrator circuits. Transistor multivibrators are either collector-coupled or emitter-coupled. In this experiment we shall be concerned with the operation of a free-running, collector-coupled multivibrator.

The properties of transistors which explain multivibrator action are as follows:

1. An increase in base current causes an increase in collector current, and conversely a decrease in base current causes a decrease in collector current.
2. An increase in forward bias in the base-emitter junction increases base current; a decrease in forward bias reduces base current.
3. In a PNP transistor, forward bias is increased in the base-emitter junction as the base is driven more negative relative to the emitter. However, a point may be reached where

a further increase in bias will not result in an increase in collector current. At that bias level, collector saturation has been reached.
4. In an NPN transistor forward bias in the base-emitter junction is increased by driving the base more positive relative to the emitter.
5. An increase in collector current flowing in the collector load resistor causes the voltage at the collector to decrease. Therefore an increase in collector current in an NPN transistor will cause the collector voltage to go more *negative*, while an increase in collector current in a PNP transistor will cause the collector voltage to go more *positive*.
6. A *decrease* in current in the collector load resistor will cause the voltage at the collector to increase. Therefore a decrease in collector current in an NPN transistor will cause the collector voltage to go more *positive*, while in a PNP transistor the collector voltage will go more *negative*.
7. The voltage across a capacitor cannot change instantaneously; that is, it takes time to charge or discharge a capacitor. A measure of this time is called the *time constant*. In an *RC* circuit, one time constant in seconds is equal to the product of the resistance *R* (through which the capacitor is charging or discharging) in ohms times the capacitance *C* in farads. Ohms times microfarads, of course, yields microseconds.

Operation of Collector-Coupled Multivibrator

Figure 41-1 is the circuit of a collector-coupled multivibrator. Q_1 and Q_2 are NPN transistors whose collectors are connected to the positive voltage source V_{CC}, through load

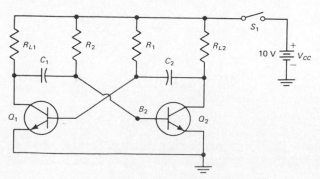

Fig. 41-1. Collector-coupled multivibrator.

resistors R_{L1} and R_{L2}, respectively. The bases of Q_1 and Q_2 are biased, respectively, by resistors R_1 and R_2. Any change in voltage at the collector of Q_1 is coupled by C_1 to the base of Q_2, and similarly any voltage change at the collector of Q_2 is coupled by C_2 to the base of Q_1. It will be shown that C_1 and C_2 provide regenerative feedback to sustain oscillation in this *RC*-coupled oscillator.

To analyze circuit operation let us assume that the two circuits are symmetrical; that is, Q_1 and Q_2 are the same type transistor, $R_{L1} = R_{L2}$, $R_1 = R_2$, and $C_1 = C_2$. When switch S_1 is closed, power is applied to both transistors. Current starts to flow in the collectors of Q_1 and Q_2. Since it is not conceivable that both transistors draw *exactly* the same amount of current, assume that the collector of Q_1 draws more current than that of Q_2. Therefore the voltage at the collector of Q_1 will go more negative than the collector voltage of Q_2. This negative voltage change of Q_1 will be coupled by C_1 to the base of Q_2, reducing the forward bias on Q_2. Accordingly collector current in Q_2 will decrease, and the collector voltage of Q_2 will go more positive. This positive change is coupled by C_2 to the base of Q_1, increasing Q_1 base current and collector current. The voltage at the collector of Q_1 will go more negative, driving the base of Q_2 more negative, reducing Q_2 collector current further, thus driving the collector voltage of Q_2 even more positive. In very short order this regenerative feedback from the collector of one transistor to the base of the next will cause Q_1 collector current to reach saturation while Q_2 collector current is cut off. We say that Q_1 is ON and Q_2 is OFF. The action is practically instantaneous.

How long does Q_1 remain ON and Q_2 OFF? The time constant C_1R_2 (approximately) determines this. Let us see how. Assume (1) that $V_{CC} = +10$ V and (2) that, after power is applied and Q_1 saturates the collector voltage, V_{C1} of Q_1 drops practically to 0; that is, the collector-to-emitter resistance of Q_1 approaches 0. This 10-V negative change at the collector looks like a negative 10-V battery to C_1, and at the instant of change the circuit to C_1 "looks" like that in Fig. 41-2. Since the voltage across C_1 cannot change instantaneously, the entire 20 V must appear across R_2, and the voltage with respect to ground at the junction of C_1 and R_2 (point B_2, the base of Q_2) must measure -10 V, cutting off Q_2. Capacitor C_1 starts discharging through the emitter-

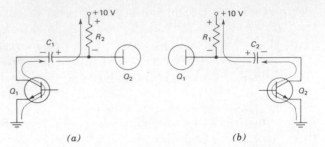

Fig. 41-3. *(a)* Capacitor C_1 discharging through R_2. Q_1 is ON, Q_2 is OFF. *(b)* Capacitor C_2 discharging through R_1. Q_2 is ON, Q_1 is OFF.

collector resistance of Q_1 and through R_2 toward V_{CC} (see Fig. 41-3a). Since the emitter-collector resistance of Q_1, saturated, is very low, the effective-discharge time constant for C_1 is $R_2 \times C_1$. The voltage v_{B2} at the base of Q_2 will change as C_1 discharges, in the manner shown in Fig. 41-4.

When the voltage at the base of Q_2 has increased from -10 to 0 V to a low value of positive voltage, Q_2 is brought out of cutoff to conduction. Q_2 base current and collector current start to flow. The voltage v_{C2} at the collector of Q_2 drops, and this negative voltage change is coupled by C_2 to the base of Q_1, reducing base current and collector current in Q_1. As collector current in Q_1 is reduced, the collector voltage of Q_1, v_{C1}, rises. This positive change at the collector

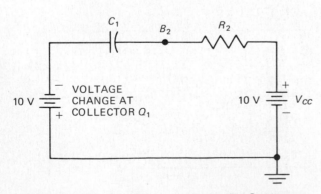

Fig. 41-2. Voltage C_1 "sees" at the instant that Q_1 saturates.

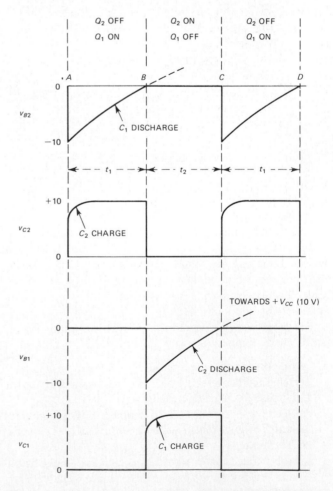

Fig. 41-4. Voltage waveforms in NPN collector-coupled multivibrator.

of Q_1 is coupled by C_1 to the base of Q_2, increasing Q_2 base and collector current. Collector voltage at Q_2 goes more negative, driving the base of Q_1 more negative. Again almost instantaneously, as Q_2 starts conducting, the regenerative feedback action turns Q_2 ON and turns Q_1 OFF. Q_2 remains ON as long as Q_1 is OFF. The length of time Q_1 remains OFF is determined by the C_2R_1 time constant, for the voltage v_{B1} at the base of Q_1 is determined by the discharge of C_2 through the emitter-collector resistance of Q_2 and through R_1, as in Figs. 41-3b and 41-4.

When Q_1 starts to conduct, it switches Q_2 OFF. When Q_2 starts to conduct, it switches Q_1 OFF. The length of time Q_1 conducts is determined by the length of time Q_2 is OFF (see waveform v_{B2}, Fig. 41-4). The length of time Q_2 conducts is determined by the length of time Q_1 is cut off (see v_{B1}, Fig. 41-4).

The idealized waveforms in Fig. 41-4 merit some discussion. We see that as the base voltage v_{B2} of Q_2 is driven 10 V negative during the interval t_1, the collector voltage v_{C2} of Q_2 goes 10 V positive, V_{CC}. We might expect that the voltage at the collector of Q_2 would rise to V_{CC} the moment Q_2 is switched OFF. However, this is not so, since capacitor C_2 cannot charge up instantaneously to V_{CC}. Therefore v_{C2} rises to V_{CC} at the rate determined by the charge of C_2 through R_{L2} in series with the emitter-base resistance of Q_1. See Fig. 41-5 for the charging path of C_2. The exponential rise in voltage of v_{C2} represents the charge of C_2. At the end of the time interval t_1, when v_{B1} has risen from cut off to conduction, the voltage at the collector of Q_2 drops from $+V_{CC}$ to a very low value (close to 0). At that instant Q_1 is switched OFF and the voltage v_{C1}, at the collector of Q_1, starts rising from 0 toward $+V_{CC}$. The exponential rise in v_{C1} is the result of C_1 charging toward $+V_{CC}$ through R_{L1}. At the end of period t_2, when v_{B1} has risen from -10 V to a slight positive voltage, Q_1 is switched ON and Q_2 OFF, and the cycle is repeated.

It will be noted that the ON-OFF time of each transistor is equal, since the circuits are symmetrical. That is, the time duration of the positive and negative alternations of Q_1 and Q_2 are equal, $t_1 = t_2$. The waveforms at the collectors of Q_1

and Q_2 are 180° out of phase. Similarly, the waveforms at the bases of Q_1 and Q_2 are 180° out of phase.

Frequency of Collector-Coupled Multivibrator

The period t of a complete cycle of the "square" wave at the collector of Q_1 or Q_2 is represented by the time interval AC (Fig. 41-4); that is

$$t = t_1 + t_2 \qquad (41\text{-}1)$$

The frequency f of the multivibrator is therefore the inverse of the period t, or

$$f = \frac{1}{t} = \frac{1}{t_1 + t_2} \qquad (41\text{-}2)$$

We can approximate the time intervals t_1 and t_2 and thus arrive at an approximate equation for frequency.

$$f = \frac{1}{0.7C_1R_2 + 0.7C_2R_1} \qquad (41\text{-}3)$$

The reason for Eq. (41-3) becomes apparent when we examine Figs. 41-3 and 41-4. The interval t_1 is determined by the time it takes C_1 to discharge through R_2 from a level of -10 to 0 V (approx). This is a change of 10 V, out of a possible change of 20 V (note that C_1 discharges from -10 toward $+10$ V). This is a 50 percent change in charge. Reference to the universal charge and discharge time-constant chart, Fig. 41-6, shows that a 50 percent charge or discharge requires 0.7 of a time constant. Hence

$$t_1 = 0.7C_1R_2$$

Similarly,

$$t_2 = 0.7C_2R_1$$

Example. A numerical example will illustrate how the frequency of a collector-coupled transistor multivibrator may be calculated using Eq. (41-3).

Suppose that

$$R_1 = R_2 = 500,000 \ \Omega$$
$$C_1 = C_2 = 0.1 \ \mu\text{F}$$

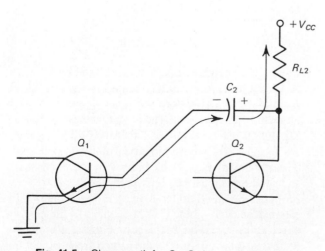

Fig. 41-5. Charge path for C_2. Q_2 is cut off, Q_1 is ON.

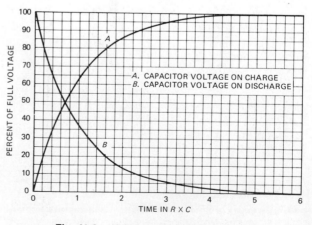

Fig. 41-6. Universal time-constant chart.

A. CAPACITOR VOLTAGE ON CHARGE
B. CAPACITOR VOLTAGE ON DISCHARGE

PERCENT OF FULL VOLTAGE

TIME IN $R \times C$

Then
$$t_1 = 0.7(0.1 \times 10^{-6})(5 \times 10^5)$$
$$= 0.035 \text{ s} = 3.5 \times 10^{-2} \text{ s}$$

Similarly
$$t_2 = 3.5 \times 10^{-2} \text{ s}$$

Therefore

$$t = 3.5 \times 10^{-2} + 3.5 \times 10^{-2} = 7 \times 10^{-2} \text{ s}$$

and
$$f = \frac{1}{t} = \frac{1}{7 \times 10^{-2}} = \frac{10^2}{7} = 14.29 \text{ Hz}$$

Nonsymmetrical Multivibrator

The "square" wave at the collectors of Q_1 and Q_2 may be changed to a "rectangular" wave. Here the positive and negative alternations are not equal in time, that is $t_1 \neq t_2$. This inequality is achieved by making $R_1C_2 \neq R_2C_1$. For example, let $R_1 = 100,000 \ \Omega$ and $R_2 = 1 \ M\Omega$ and $C_1 = C_2 = 0.001 \ \mu\text{F}$. The time constant C_1R_2 is 10 times the value of C_2R_1, and therefore the waveform will be non-symmetrical. However, Eq. (41-3) cannot be used in computing the frequency of an asymmetrical multivibrator. Individual circuit constants and conditions must be considered.

Multivibrator Synchronization

An NPN collector-coupled multivibrator may be synchronized by applying external negative timing pulses to the base of either transistor. The frequency of the synchronizing pulses must be slightly higher than that of the multivibrator, and the amplitude of the pulse must be sufficient to switch the transistors.

Consider Fig. 41-7a. Negative synchronizing pulses are applied through C_3 to the base of Q_1. Assume that pulse 1 appears at the time that Q_1 is conducting. The pulse will be amplified by Q_1 and appear as a positive pulse at the collector of Q_1 (there is a 180° phase shift between base and collector voltages in a transistor amplifier with resistive collector load). The positive pulse is coupled by C_1 to the base of Q_2 and is superimposed on v_{B2}, the C_1 discharge waveform at point X (Fig. 41-7b). If the positive pulse has sufficient amplitude, it brings Q_2 out of cutoff and into conduction at X.

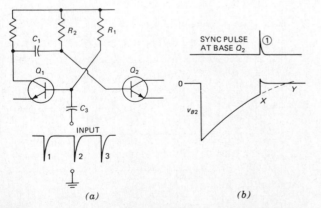

Fig. 41-7. (a) Synchronizing a multivibrator; (b) negative synchronizing pulse is amplified by Q_1 and appears as a positive pulse at base of Q_2, turning Q_2 on.

Q_2 would normally have been switched in at Y in the absence of a synchronizing pulse. If the time interval between pulses 1 and 2 is slightly less than the normal period of the multivibrator, pulse 2, 3, etc., will lock in the multivibrator in a fashion similar to pulse 1.

SUMMARY

1. Multivibrators are *RC*-coupled oscillators which generate nonsinusoidal waveforms.
2. Multivibrators are either *self-excited* (*free-running*, requiring no external triggering) or driven. Driven, or triggered, multivibrators require external pulses to start them and keep them going.
3. Self-excited multivibrators are also called *astable* multivibrators.
4. Transistor astable multivibrators are either collector-coupled or emitter-coupled.
5. Figure 41-1 is the diagram of a transistor collector-coupled multivibrator. Note that Q_1 and Q_2 are cross-coupled. The output of Q_1 feeds Q_2, and the output of Q_2 feeds Q_1.
6. If the time that Q_1 is OFF is called t_1 and the time Q_2 is OFF is called t_2, then the time t of one complete cycle is

$$t = t_1 + t_2$$

7. The frequency f of a multivibrator whose period (time of one complete cycle) is t is

$$f = \frac{1}{t} = \frac{1}{t_1 + t_2}$$

8. A symmetrical collector-coupled multivibrator is one whose equivalent components are equal. Thus if $Q_1 = Q_2$, $R_1 = R_2$, $R_{L1} = R_{L2}$, and $C_1 = C_2$, Fig. 41-1 is symmetrical.
9. In the symmetrical collector-coupled multivibrator of Fig. 41-1, $t = 0.7C_1R_2 = t_2 = 0.7C_2R_1$.
10. The frequency of the symmetrical collector-coupled multivibrator in Fig. 41-1 is

$$f = \frac{1}{0.7C_1R_2 + 0.7C_2R_1} \tag{41-3}$$

where f is in hertz, C in farads, and R in ohms.

11. A nonsymmetrical multivibrator is one in which $R_2C_1 \neq R_1C_2$. Therefore the times t_1 and t_2 of each alternation are not equal. The outputs at the collectors of Q_1 and Q_2 are *rectangular*, rather than square, waves.
12. The frequency of a nonsymmetrical multivibrator cannot be calculated by Eq. (41-3). Individual circuit constants and conditions must be considered.
13. Free-running multivibrators may be synchronized by external negative-going pulses coupled to the base of either transistor. The frequency of the synchronizing pulses must be higher than the natural frequency of the multivibrator.

Check your understanding by answering these questions.

1. Multivibrators are oscillators which generate sine waves. _____ (*true, false*)
2. In Fig. 41-1, the frequency of the multivibrator is determined by the time constants _____ and _____ .
3. For the astable multivibrator in Fig. 41-1 to be symmetrical, the following conditions must be met:
   $Q_1 = $ _____ ; $C_1 = $ _____ ;
   $R_2 = $ _____ ; $R_{L2} = $ _____ .
4. Figure 41-1 is a symmetrical astable multivibrator in which $C_1 = 0.02\ \mu\text{F}$ and $R_2 = 100\ \text{k}\Omega$. The time of one complete cycle is _____ s.
5. The frequency of the multivibrator in question 4 is _____ Hz.
6. As the resistance of R_1 in Fig. 41-1 is increased, the frequency of the multivibrator is _____ .
7. In Fig. 41-8, all the values are as shown except $R_1 = 270\ \text{k}\Omega$. The time during which Q_1 is OFF is _____ (*shorter, longer*) than the time during which Q_2 is OFF.
8. The frequency of a synchronizing pulse delivered to the base of Q_1 (Fig. 41-8) must be _____ (*higher, lower*) than the natural frequency of the multivibrator, for effective synchronization.

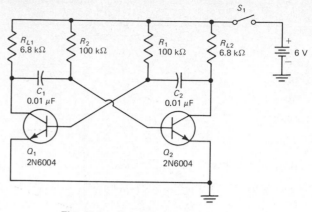

Fig. 41-8. Experimental multivibrator.

MATERIALS REQUIRED

- Power supply: Variable regulated low-voltage dc source
- Equipment: Oscilloscope (with dc vertical amplifiers, if possible); square-wave generator (AF range)
- Resistors: Two 6800-, 10,000-, 22,000-, 33,000-, two 100,000-Ω ½-W
- Capacitors: Two 0.01-μF; 47-, 250-pF
- Semiconductors: Two 2N6004 transistors (other choice: 2N3904); 1N4154 diode or their equivalents
- Miscellaneous: SPST switch; 500,000-Ω 2-W potentiometer; two 2N6005 transistors as required for extra-credit experiment

PROCEDURE

Symmetrical Multivibrator

1. Connect the circuit of Fig. 41-8. **Power on.**
2. *Externally* synchronize your oscilloscope with the signal from the collector of Q_1. Use the dc vertical amplifiers of your oscilloscope, if available, and set the zero level of the trace.
3. Observe the waveforms at each of the test points shown in Table 41-1. Draw these waveforms in proper time phase in the table. Measure also the peak-to-peak amplitude of each waveform. Record the data in Table 41-1. If you are using a dc (vertical amplifier) oscilloscope, show also the polarity of voltage with reference to 0. If you are using a dual-trace oscilloscope, observe and compare the waveforms at the collector and base of Q_1; at the collector of Q_1 and collector of Q_2; and at the base of Q_1 and collector of Q_2.
4. Compute and record the frequency of your multivibrator in column labeled "Frequency, Computed." Show your computations.
5. If the trace of your oscilloscope is time-calibrated, measure and record the time t_1 and t_2 of a complete collector cycle. Using the value of $t = t_1 + t_2$, determine the

frequency; record this frequency in column labeled "Frequency, Measured."
6. If the trace of your oscilloscope is not time-calibrated, measure and record the frequency of the multivibrator by using (for comparison) as a frequency standard a signal from a calibrated, variable-frequency, audio sine- or square-wave generator.

Asymmetrical Multivibrator

7. **Power off.** Replace R_2, the Q_2 base bias resistor, with the series combination of R_3, a 10,000-Ω resistor, and R_4, a 500,000-Ω potentiometer connected as a rheostat (see Fig. 41-9). Set R_4 for *minimum* resistance.
8. **Power on.** Observe the waveforms at the test points shown in Table 41-2. Draw these waveforms in proper time phase in the table. Also measure and record the peak-to-peak amplitude of the waveforms. Measure and record the frequency of the multivibrator.
9. Set R_4 for maximum resistance. Repeat step 8. Use Table 41-3.
10. Reset R_4 for a symmetrical "square" wave. Measure and record its frequency: $f = $ ___724.6___ Hz. Leave R_4 at this setting.

TABLE 41-1. Symmetrical Multivibrator Waveforms and Frequency

Test Point	Waveform				V p-p
Collector of Q_1					6.1
Base of Q_1					6.0
Collector of Q_2					6.1
Base of Q_2					6.0

Time Duration

t_1	t_2	t_1	t_2	X
.77m	.64m	.77m	.64m	

Frequency

Computed	Measured
714.28 Hz	709 Hz

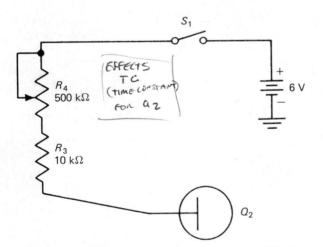

EFFECTS
TC
(TIME CONSTANT)
FOR Q_2

R_4
500 kΩ

R_3
10 kΩ

S_1

+ 6 V −

Q_2

Fig. 41-9. Q_2 base bias circuit for asymmetrical multivibrator.

Synchronizing the Multivibrator

11. Connect the output of a square-wave generator to the input of the circuit in Fig. 41-10. The output of the synchronizing circuit is connected to the base of Q_1 as shown. Set the square-wave-generator frequency 200 Hz higher than the measured frequency f of the multivibrator (see step 10). Set the output of the square-wave generator at 6 V p-p.

12. Externally synchronize the oscilloscope with the output of the square-wave generator. Observe the waveform at the base of Q_1; at the collector of Q_1; at the base of Q_2. Draw these waveforms in Table 41-4.

13. Measure and record the frequency of the multivibrator: _____ Hz.
757.5   6.6 (.2m)

14. Slowly reduce the frequency of the square-wave generator to f. Observe and record effect on shape and frequency of waveform while monitoring the output at the collector of Q_1.
Effect on waveform: Q_1'S WAVEFORM DECREASES
THEN INCREASES TO ORIGINAL VOLTAGE

Effect of frequency: Q_2'S WAVEFORM INCREASES,
THEN DECREASES TO ORIGINAL VOLTAGE

Extra Credit

15. Experimentally determine and explain the operation and purpose of the components in the synchronizing circuit of Fig. 41-10. State your procedure.

16. (a) Experimentally determine what effect, if any, the level of the synchronizing pulse has on synchronization. Explain. (b) What is the minimum-level synchronizing pulse which will permit stable synchronization of the *symmetrical* multivibrator?

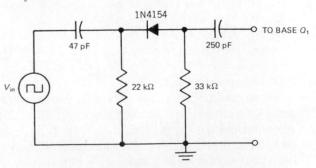

Fig. 41-10. Multivibrator synchronizing circuit.

TABLE 41-2. R_4 Minimum Resistance

Test Point	Waveform	V p-p
Collector of Q_1	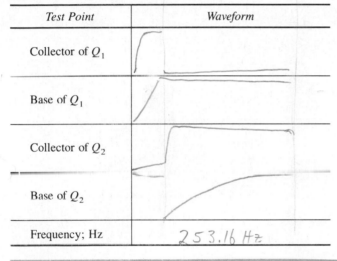	6
Base of Q_1		4
Collector of Q_2		4
Base of Q_2		5.5
Frequency, Hz	2702.7 Hz	

TABLE 41-3. R_4 Maximum Resistance

Test Point	Waveform
Collector of Q_1	
Base of Q_1	
Collector of Q_2	
Base of Q_2	
Frequency; Hz	253.16 Hz

17. Draw the circuit diagram of a symmetrical, collector-coupled multivibrator using 2N6005 (PNP) transistors. Show polarity of battery voltage.

18. Connect the circuit. Observe the waveforms at the collector and base of each transistor. In a specially constructed table draw the waveforms in proper time phase.

TABLE 41-4. Synchronizing the Multivibrator

Test Point	Waveform (synchronized at $f + 200$)
Base of Q_1	
Collector of Q_1	
Base of Q_2	

QUESTIONS

1. Comment on the waveforms in Table 41-1. Do they agree with the theoretical waveforms in Fig. 41-4?
2. In the circuit of Fig. 41-8 what determines how long (*a*) Q_1 is cut off? How long is it OFF? (*b*) Q_1 is conducting? How long is it conducting?
3. How does the computed frequency compare with the measured frequency in Table 41-1? Explain any discrepancies.
4. What is the effect on frequency of the multivibrator as R_4 (Fig. 41-9) is (*a*) increased in value, (*b*) decreased? Why?
5. Is your answer to question 4 confirmed by the data in your experiment? Refer specifically to the data on which your answer is based.
6. Would there be any appreciable change in frequency if the voltage source of the multivibrator in Fig. 41-8 were reduced from 6 to 5 V? Why?
7. What is meant by synchronization of a multivibrator?

8. How is the frequency of a collector-coupled multivibrator varied in this experiment?
9. What other method may be used for synchronizing the multivibrator in Fig. 41-9?
10. (*a*) Can PNP transistors be used instead of NPNs in designing a multivibrator? (*b*) What changes, if any, are required in the circuit?

Extra Credit

11. Will the level of the synchronizing pulse affect synchronization? How?
12. Is it possible to synchronize a multivibrator with a negative pulse whose frequency is *twice* the frequency of the multivibrator? Explain.

Answers to Self-Test

1. false
2. C_1R_2; C_2R_1
3. Q_2; C_2; R_1; R_{L1}
4. 2.8×10^{-3}
5. 357
6. decreased
7. longer
8. higher

LOW

HBH   675.67 Hz

DIFF   833

LOW   714.28 Hz

HIGH   961.53 Hz

           247.25 Hz

DIFF.

OBJECTIVES

1. To trigger a flip-flop from one conduction state to another
2. To observe the effects of a pulse input on the output of a flip-flop

INTRODUCTORY INFORMATION

Description and Use

A flip-flop is a *bistable* multivibrator. A flip-flop is a driven device. Because it is bistable, it can remain in either of two stable states until an *external* signal forces it to take the other state; thus the flip-flop can be used as a digital storage device to retain information. Flip-flops perform many computer functions, such as acting as intermediate storage for information during the computer process and as status indicators to indicate logical conditions existing during an operation. Flip-flops can also be used in a group to function as a binary counter.

Flip-Flop Operation

Figure 42-1 is the circuit diagram of a flip-flop employing NPN transistors, though PNP transistors may also be used. Examination shows that this circuit is almost identical with the collector-coupled multivibrator in Experiment 41. The basic difference is that the flip-flop is direct-coupled, with R_3 and R_5 replacing the coupling capacitors C_1 and C_2 found in Fig. 41-1.

To understand the operation of this circuit, assume that when power is first applied, Q_1 conducts somewhat more heavily than Q_2. This drives the collector of Q_1 more negative than that of Q_2. This negative voltage is coupled by R_5 to the base of Q_2, reducing the forward base-emitter bias of Q_2, and as a result, reducing Q_2 collector current. Accordingly, the voltage at the collector of Q_2 is driven more positive. This positive change is coupled by R_3 to the base of Q_1, increasing collector current in Q_1, driving the base of Q_2 more negative. The regenerative feedback action continues, as above, and in a very brief time interval Q_1 is conducting at saturation, with its collector voltage close to 0. As a result the effective voltage divider providing base-emitter bias for Q_1 (see Fig. 42-2) drives the base of Q_2 negative with respect to its emitter, and cuts off Q_2. The circuit action stops; that is, this is one stable state, with Q_1 ON and Q_2 OFF.

It is possible to reverse the state of conduction, that is, turn Q_2 ON and Q_1 OFF, by applying an external positive triggering pulse to the base of Q_2, the off transistor, of sufficient amplitude to cause Q_2 collector to start drawing current. When that happens, the collector voltage at Q_2 is driven in a negative direction. This negative voltage change is coupled by R_3 to the base of Q_1, reducing collector current in Q_1 and driving the Q_1 collector voltage more positive. This positive voltage change is coupled by R_5 to the base of Q_2, increasing collector current in Q_2. Again, by a similar action to that previously described, in a very brief time interval Q_2 is turned on and Q_1 is turned off.

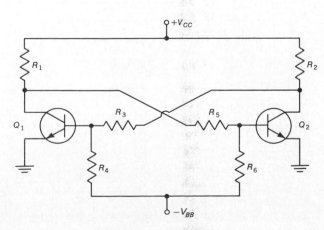

Fig. 42-1. Transistor flip-flop.

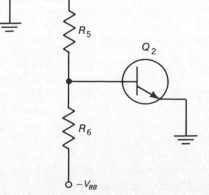

Fig. 42-2. Voltage divider provides reverse bias-emitter bias for Q_2 when Q_1 is conducting.

It is also possible to reverse the states of the two transistors by applying a negative pulse to the base of the normally on transistor. Thus when Q_2 is ON, a negative pulse at the base of Q_2 is amplified and appears as a positive pulse at its collector. If this positive pulse is of sufficient amplitude, it will, when coupled by R_3 to the base of Q_1, switch the two transistors, turning Q_1 ON and Q_2 OFF.

Transistor Switching Modes

Transistor switching circuits employed in computers generally fall into one of two types of operation. One is called *voltage mode,* and the other is called *current switching*.

Voltage-mode circuits operate with relatively large swings in signal level. Transistors in voltage-mode circuits are usually driven into saturation in one operating state and are cut off in the other operating state. Voltage-mode circuits are limited in their operating speed because turnoff delay is increased when a transistor is driven into saturation. For this reason, voltage-mode circuits are employed where speed is not essential. The circuits in this experiment and in Experiments 41 and 44 illustrate voltage-mode operation.

When high switching speed is essential, circuits using the current-switching technique are employed. This system performs digital logic functions by switching well-defined currents, such as 6 mA, from one part of a circuit to another.

SUMMARY

1. A flip-flop is a driven bistable multivibrator which *remains* in one of two stable states until acted on by an external pulse. It is not a free-running or self-excited multivibrator.
2. The circuit of the collector-coupled flip-flop in Fig. 42-1 closely resembles the circuit of the collector-coupled free-running multivibrator in Fig. 41-1, except that the flip-flop is direct-coupled and the free-running multivibrator is *RC*-coupled.
3. A negative pulse applied to the base of the on transistor changes the state of conduction of the two transistors; that is, it will flip the flip-flop.
4. There are large voltage changes at the collectors of the two transistors when the flip-flop is flipped. For example, when Q_1 is ON its collector is practically at ground (0) potential. When Q_1 is OFF its collector rises to $+V_{CC}$. These voltage changes also appear, in reverse, at the collector of Q_2. It is these large voltage changes coupled from collector of one to the base of the other transistor that cause the flip-flop to flip. Therefore this type of circuit is called a *voltage-mode trigger.*
5. Flip-flop circuits find many applications in digital computers. They may act as intermediate storage circuits, or they may be used as logic indicators during an operation.

SELF-TEST

Check your understanding by answering these questions.

1. In the transistor flip-flop (Fig. 42-1), when power is first applied Q_1 conducts more heavily than Q_2. As a result, _____ becomes the ON transistor, _____ the OFF transistor.
2. Because Fig. 42-1 is a self-excited multivibrator, the states of conduction of Q_1 and Q_2 are reversed at periodic intervals without any external trigger pulses. _____ (*true, false*)
3. In the circuit of Fig. 42-1, if Q_1 is ON, a _____ (*positive, negative*) pulse applied to the base of Q_1 will cause Q_1 to go off and Q_2 to go on.
4. The circuit of Fig. 42-1, $V_{CC} = +10$ V, $V_{BB} = -10$ V. When Q_1 is ON, the dc voltage at the collector of Q_1 is approximately _____ (*10, 0, -10*) V.
5. For the conditions in question 4, the collector voltage of Q_2 is approximately _____ V.

MATERIALS REQUIRED

- Power supply: Regulated variable *dual voltage* dc source (NOTE: If a dual voltage supply is not available, a single supply may be used, with a voltage divider across it, connected as in Fig. 42-3. The experimental flip-flop is then connected to the proper taps on the voltage divider.)
- Equipment: EVM; 0–10-mA milliammeter
- Resistors: Two 3300-, two 8200-, two 47,000-Ω ½-W; two 100-Ω 1-W (NOTE: If a dual supply is not available, an additional 220-Ω 1-W resistor will be required.)
- Semiconductors: Two 2N6005 transistors and two 1N5625 diodes or equivalents
- Miscellaneous: SPST switch

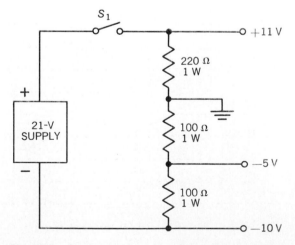

Fig. 42-3. Single power-supply voltage divider for flip-flop.

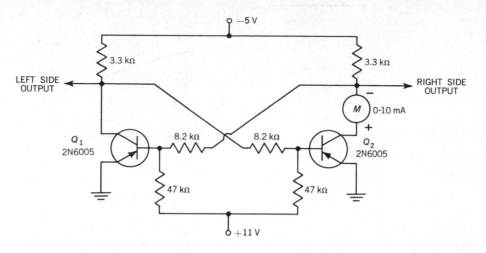

Fig. 42-4. Experimental voltage-mode flip-flop.

1. Connect the circuit of Fig. 42-4. **Power on.**

NOTE: If a dual supply is not available, connect the voltage divider of Fig. 42-3. Set the output level of the dc supply at 21 V. The negative terminal of the dc supply must not be *line-grounded*, to eliminate the possibility of conflicting grounds when using other instruments in the circuit.

2. With an EVM measure the dc voltages at the collector and base of Q_1 and Q_2. Record them in Table 42-1. Indicate which transistor, if any, is ON, which OFF. This is trial 1.
3. **Power off.** After about 1 min turn power **on** again. With an EVM measure and record for trial 2 the dc voltages at the base and collector of Q_1 and Q_2. Indicate also which transistor is ON, which OFF.
4. Repeat step 3 for trials 3 and 4.
5. To flip your flip-flop from one state to the other, alternately ground one transistor base and then the other.

NOTE: Grounding the base of the ON transistor causes the flip-flop to flip. Observe the voltage at one of the outputs

while flipping the trigger. Does the trigger exhibit bistable characteristics? _____

6. When you have determined that the flip-flop is operating properly, fill in Table 42-2 for the conditions shown. Measure and record the dc output voltages.
7. The "signal" output level should vary between 0 (up) and -5 V (down). From your preceding measurements you will note that the down level is no longer -5 V due to the cross-coupling, which forms a voltage divider between -5 and $+11$ V. The output signal *down level* can be "restored" by making the modifications in Fig. 42-5.

TABLE 42-2.   Flip-Flop Output Level

Condition	Right Output	Left Output
Q_1 ON		
Q_2 ON		

TABLE 42-1.   Transistor Flip-Flop Turn-On State

Trial	Transistor	DC, V		Transistor Condition	
		Base	Collector	On	Off
1	Q_1				
	Q_2				
2	Q_1				
	Q_2				
3	Q_1				
	Q_2				
4	Q_1				
	Q_2				

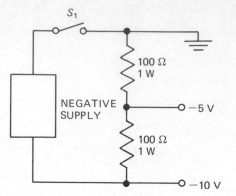

Fig. 42-5. Voltage divider for negative supply to develop −5 V.

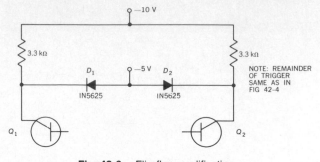

Fig. 42-6. Flip-flop modification.

Returning the collector resistor to a −10-V source causes the output down level to fall below −5 V. The purpose of the diode connected between each output and the −5-V source is to "clamp" the down level of the output to −5 V.

8. If you are using a dual voltage supply, connect a voltage divider across the −10-V source, as in Fig. 42-5. The −5-V tap will be used in Fig. 42-6.

   If you are using a single voltage source, the divider of Fig. 42-3 already has a −5-V tap.

   When you have modified your flip-flop to conform with Fig. 42-6, check again the output levels and record them in Table 42-3.

TABLE 42-3. Clamped Flip-Flop Output

Condition	Right Output	Left Output
Q_1 ON		
Q_2 ON		

QUESTIONS

1. From your first four trials, steps 2 through 4, does it appear that the same transistor will always be an ON transistor when power is first applied in the circuit of Fig. 42-4? Why?

2. What effect does grounding the base of the OFF transistor have on flip-flop operation? Why?

3. What effect does grounding the base of the ON transistor have on flip-flop operation? Why?

4. Assume that the ON transistor is saturated and that the collector voltage is 0 measured with reference to the emitter. Compute the base voltage which would appear at the OFF transistor. Indicate polarity of voltage. Show your computations.

5. How does the computed value, in answer to question 4, compare with the measured value? Explain any discrepancy.

6. How do diodes D_1 and D_2 "clamp" the down level at −5 V?

Answers to Self-Test

1. Q_1; Q_2 4. 0
2. false   5. +10
3. negative

THE ONE-SHOT (MONOSTABLE) MULTIVIBRATOR AND SCHMITT TRIGGER

OBJECTIVES

1. To dertermine experimentally the operating characteristics of a monostable multivibrator
2. To determine experimentally the operating characteristics of a Schmitt trigger

INTRODUCTORY INFORMATION

One-Shot Multivibrator

The one-shot multivibrator circuit is driven, not free-running. Unlike the bistable trigger in Experiment 42, the monostable multivibrator has one stable state to which it returns after an external driving pulse has caused it to execute its cycle. The one-shot multivibrator is often used to widen a pulse or delay its trailing edge.

Figure 43-1a shows two transistors, resistors, and a capacitor connected to operate as a one-shot. C_1 is connected between the collector of Q_1 and the base input of Q_2, R_1 is added, and the collector of Q_2 is connected to the base input of Q_1.

Study of Fig. 43-1a suggests that since the Q_2 base-emitter circuit is forward-biased (by R_1), Q_2 will be saturated and the collector of Q_2 will therefore be at, or close to, ground potential. Since the base of Q_1 is returned to the collector of

Q_2, the base of Q_1 is therefore at ground potential. Note that the emitter of Q_1 is at ground. Hence, no forward bias exists in the base emitter of Q_1, and Q_1 is cut off. Therefore the voltage at the collector of Q_1 is $+V_{CC}$, the positive collector source. This represents the stable state of the circuit: Q_1 cut off and Q_2 conducting at saturation level.

A positive pulse of sufficient amplitude to start current in Q_1 is now applied to the base of Q_1. Current flows in Q_1, and the collector voltage drops. This *negative* change is coupled by C_1 to the base of Q_2, reducing current in Q_2. The collector voltage of Q_2 therefore rises, driving the base of Q_1 positive. More current flows in Q_1, driving the collector of Q_1 more negative, therefore driving the base of Q_2 more negative. In short order, this regenerative action cuts off Q_2, leaving Q_1 conducting very heavily. The resulting change in voltage at the collector of Q_1, from $+V_{CC}$ to 0 (approx), causes capacitor C_1 to discharge through R_1. At the start of the discharge cycle, the full voltage change at the collector of Q_1 appears as $-V_{CC}$ at the base of Q_2. As C_1 continues to discharge toward $+V_{CC}$ (the positive voltage source to which R_1 is returned), the voltage at the base of Q_2 rises exponentially (see waveform v_{B2}). The discharge path for C_1 is shown in Fig. 43-2a.

When the base of Q_2 has risen above 0 V and is slightly positive, Q_2 starts conducting again. The voltage at the collector of Q_2 drops, driving the base of Q_1 in a negative direction. Collector current through Q_1 decreases, causing

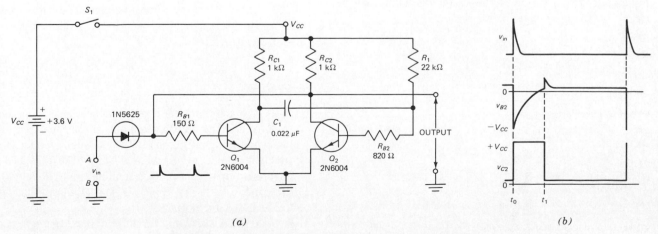

(a)　　　　　　　　　　　　　　　　　　*(b)*

Fig. 43-1. *(a)* Monostable (one-shot) multivibrator; *(b)* waveforms.

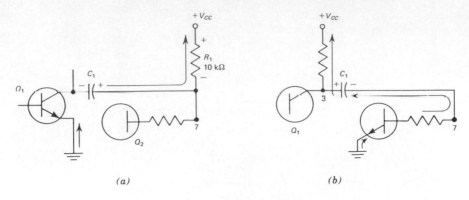

Fig. 43-2. *(a)* Discharge; *(b)* charge path for C_1.

the voltage at the collector of Q_1 to go toward $+V_{CC}$. Capacitor C_1 therefore starts charging. Its charge path (see Fig. 43-2*b*) is through the emitter-base resistance of Q_2, through R_{B2} and through R_{C1}, toward $+V_{CC}$. The action rapidly turns Q_2 on and turns Q_1 off. Q_2 remains ON until a positive pulse causes the one-shot multivibrator to execute its cycle again.

Of interest are the waveforms v_{B2} at the base of Q_2 and v_{C2} at the collector. Note that the positive pulse at the collector of Q_2 in the interval $t_0 - t_1$ represents the time during which Q_2 is cut off.

The positive pulse necessary to start the trigger action can be obtained by differentiating a square wave by the circuit of Fig. 43-3.

Schmitt Trigger

Figure 43-4*a* shows an emitter-coupled binary, also called a *Schmitt trigger* after its inventor. Note how the preceding circuit was modified to make a trigger whose characteristics are completely different from the one-shot.

The Schmitt trigger is a binary trigger because two stable states occur. Q_1 may be ON and Q_2 OFF, or Q_2 may be ON and Q_1 OFF.

Note that Fig. 43-4*a* differs from a collector-coupled binary trigger in that the usual trigger network coupling the collector of Q_2 to the base of Q_1 is replaced by common-emitter coupling across R_4. The switching action may be started by raising or lowering the bias on Q_1.

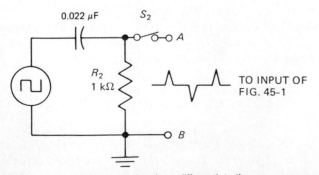

Fig. 43-3. The capacitor and resistor differentiate the square wave. Only the positive pulse triggers the one-shot multivibrator.

In the absence of an input to Q_1, the divider network R_3, R_2, together with the collector resistor in Q_1, maintains the base of Q_2 slightly positive relative to the emitter, and Q_2 operates in the saturation region. The positive voltage developed across the common-emitter resistor R_4 resulting from current flow in Q_2 maintains NPN transistor Q_1 cut off because the base of Q_1 is at ground potential, hence negative, relative to its emitter (therefore reverse-biased). The stable state in the absence of a signal, then, is Q_2 ON and Q_1 OFF.

When a sine wave is applied to the input circuit, Q_1 is turned on during the positive alternation when the base of Q_1 is driven more positive than the positive voltage on the emitter, point P_1 in Fig. 43-4*b*. Q_1 is driven to conduction in the saturation region and is held there as long as the input voltage at the base is more positive than the emitter voltage. Q_2 is turned off. At P_2, the voltage at the base of Q_1 drops below the emitter voltage and switching occurs, with Q_2 being turned on and Q_1 turned off.

The reason that Q_2 turns off when Q_1 goes on is that with Q_1 ON and saturated the collector of Q_1 drops to a low potential. The network R_2, R_3 couples this low Q_1 collector voltage to the base of Q_2 and eliminates the forward bias on Q_2. Therefore Q_2 goes off.

Figure 43-4*b* shows the input and output waveforms. Between the points P_1 and P_2 on the positive alternation of the input sine wave, Q_2 is cut off, and its collector is therefore at $+V_{CC}$. Between P_2 and P_3, Q_2 is ON, and its collector voltage drops close to the emitter voltage. The output wave looks square (or rectangular). Hence the Schmitt trigger is called a *squaring* circuit.

SUMMARY

1. A one-shot multivibrator is monostable; that is, it has *one stable state*. The one-shot remains in this state until an *external* driving pulse causes it to execute its cycle, after which it returns to its original state. The one-shot is therefore a driven, *not* a free-running, multivibrator.
2. In the monostable MV in Fig. 43-1, Q_2 is the normally on transistor, Q_1 the normally off.
3. An external *positive* driving pulse applied to the base of Q_1 turns Q_1 ON and Q_2 OFF. The length of time that Q_2 is

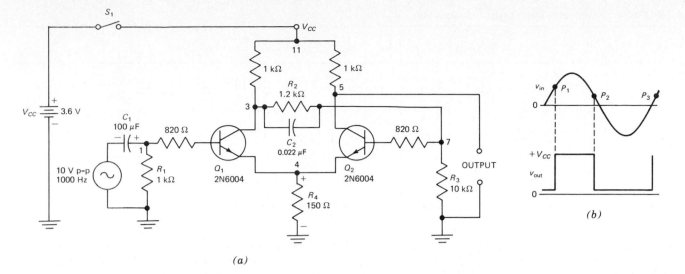

Fig. 43-4. *(a)* Schmitt trigger produces a square-wave output from a sine-wave input; *(b)* waveforms.

OFF is determined by the time constant C_1R_1. The longer the time constant, the wider the pulse developed at the collector of Q_2 (see Fig. 43-1*b*). The discharge of C_1 through R_1 maintains the base of Q_2 negative during the pulse interval.

4. The one-shot multivibrator is often used to delay the trailing edge of a pulse or to widen a pulse. In digital circuits it may be used as a *no-bounce* switch to enter a binary 1 into an adder.

5. The Schmitt trigger (Fig. 43-4*a*) is an emitter-coupled binary circuit which converts a sine-wave input into a rectangular wave. It is therefore called a *squaring* circuit.

SELF-TEST

Check your understanding by answering these questions.

1. In Fig. 43-1*a*, what keeps Q_2 ON in the absence of an input driving pulse is the forward bias provided Q_2 by _____ and _____ .

2. In Fig. 43-1*a*, when Q_2 is ON, Q_1 is _____ .

3. A _____ (*positive, negative*) pulse on the base of Q_1 (Fig. 43-1*a*) will switch the conduction of Q_1 and Q_2.

4. In Fig. 43-1*a*, when Q_1 is OFF, its collector voltage is _____ . When Q_1 is turned on, it saturates and its collector voltage drops to _____ V (approx).

5. In Fig. 43-1*a*, the negative change in voltage which occurs at the collector of Q_1 when Q_1 is turned on is

coupled by _____ and _____ to the base of Q_2.

6. In Fig. 43-1*a*, the width of the pulse developed at the collector of Q_2 is determined by the time constant _____ .

7. In the Schmitt trigger (Fig. 43-4*a*), in the absence of an input signal, the base of Q_2 is _____ (*positive, negative*) relative to its emitter.

8. In Fig. 43-4*a*, Q_1 is normally _____ (OFF, ON) in the absence of an input signal.

9. In Fig. 43-4*a*, when the input signal is of sufficient amplitude to switch the conduction of Q_1 and Q_2, the collector of Q_2 rises to _____ .

10. If the amplitude of the input sine wave is increased, but its frequency is the same, the width of the output pulse (Fig. 43-4*a*) is _____ (*increased, decreased, unaffected*).

MATERIALS REQUIRED

- Power supply: Variable regulated low-voltage dc
- Equipment: Oscilloscope; EVM; AF sine/square-wave generator (with ext. sync output jack, if possible)
- Resistors: 150-, two 820-, three 1000-, 10,000-, 22,000-Ω ½-W
- Capacitors: Two 0.022-, 100-µF 50-V
- Semiconductors: Two 2N6004 transistors (other choice: 2N3904); 1N5625 diode; or equivalent
- Miscellaneous: Two SPST switches

PROCEDURE

One-Shot Multivibrator

1. Connect the circuits in Figs. 43-1*a* and 43-3, connecting *A* to *A* and *B* to *B*. S_1 and S_2 are open. Set V_{CC} at $+3.6$ V. Set square-wave generator at 500 Hz, output at 5 V p-p.

2. Externally trigger/sync your oscilloscope with signal at the sync output jack on the generator, or with its square-wave output.

3. With your oscilloscope, observe the square wave at the

TABLE 43-1. Monostable Multivibrator Characteristics

Step	Test Point	Waveform	V p-p	Pulse Width
3	Generator output		5	X
4	Across R_2			X
5	Collector of Q_2			X
6	Base of Q_2			X
7	Collector of Q_2			
9	Collector of Q_2			
10	Collector of Q_2	X		

output of the generator and set the oscilloscope controls for two waveforms, properly centered, as in Table 43-1.

4. Switch S_2 is still open. Observe the waveform across R_2, the 1000-Ω input resistor. In Table 43-1 draw this wave in time phase with the input square wave. Measure and record the peak-to-peak amplitude of the waveform.

5. *Close S_1*, applying power to the circuit. S_2 is still open. Observe the waveform, if any, at the collector of Q_2 and record in Table 43-1.

6. *Close S_2*. Both switches are now closed. Observe, measure, and record the waveform at the base of Q_2, in proper phase with the input.

7. Observe, measure, and record the output waveform at the collector of Q_2, in proper phase with the input. Measure and record the width of the positive pulse.

NOTE: If there is a spike riding at the leading edge of the output pulse, gradually reduce the output of the generator until the spike disappears.

8. **Power off.** Substitute a 10,000-Ω resistor for R_1.
9. **Power on.** Observe and measure the amplitude and pulse width of the output waveform at the collector of Q_2. Draw the waveform in proper phase with the input. Record its amplitude and pulse width.

NOTE: If you do not have an oscilloscope with a calibrated time base, measure the pulse width in centimeters or inches.

10. Reduce the frequency of the square-wave generator to 350 Hz. Measure and record the width of the positive output pulse. Has the pulse width changed from that in step 9? What has changed?

TABLE 43-2. Characteristics of a Schmitt Trigger

Step	Test Point	Waveform	V p-p	Pulse Width
13	Generator Output		10	X
14	Collector of Q_2			
15	Collector of Q_2			

Schmitt Trigger

11. Connect the circuit of Fig. 43-4a. V_{CC} is set at $+3.6$ V. Close S_1, applying power to the circuit. Set the *sine-wave* generator frequency at 1000 Hz, output at 10 V p-p.
12. Externally trigger/sync the oscilloscope as in step 2.
13. With your oscilloscope observe the sine wave at the output of the generator and adjust the oscilloscope controls for two waveforms, properly centered as in Table 43-2.
14. Observe the output waveform (collector of Q_2). Measure and record in Table 43-2, in proper phase with the input, its peak-to-peak amplitude and the width of the positive pulse.
15. Reduce the sine-wave generator output to 4 V p-p and repeat step 14.

QUESTIONS

1. Why is the waveform across R_2, step 4, Table 43-1, no longer a square wave?
2. In which step of the procedure did you determine that the circuit of Fig. 43-1 is not a free-running multivibrator? Refer specifically to the data in Table 43-1 which confirm your answer.
3. In the one-shot multivibrator, which time constant yielded a wider pulse: 0.022 μF $\times$ 10,000 Ω or 0.022 μF $\times$ 22,000 Ω? Why?
4. What is the relationship between the waveform on the base of Q_2 and the width of the pulse at the collector of Q_2 in Fig. 43-1a? Why?
5. (a) In Fig. 43-1a, what turns Q_2 off? (b) Why doesn't it remain off all the time?
6. In the Schmitt trigger, what effect does the amplitude of the input sine wave have on the width of the positive alternation at the collector of Q_2 in Fig. 43-4a? Why?
7. What effect does the frequency of the input sine wave have on the width of the positive alternation at the collector of Q_2 (Fig. 43-4a)? Why?

Answers to Self-Test

1. R_{B1}; R_1
2. OFF
3. positive
4. $+V_{CC}$; 0
5. C_1; R_1
6. $C_1 \times R_1$
7. positive
8. OFF
9. $+V_{CC}$
10. increased

SAWTOOTH (RAMP-FUNCTION) GENERATOR

OBJECTIVE

1. To observe the waveforms in an experimental sawtooth generator
2. To determine the effects of frequency and the time constant on the amplitude and linearity of the sawtooth output

INTRODUCTORY INFORMATION

The Sawtooth Wave

The sawtooth wave is one of the most important periodic waveforms in electronics. It is used extensively as a time base in oscilloscopes, in radar indicators, and in countless other applications.

Figure 44-1 is the graph of a periodically changing voltage varying linearly with respect to time. The voltage increases from 0 to the level V in t_1 s and decreases from the level V to 0 in t_2 s. The cycle is then repeated. Because the shape of the waveform resembles the teeth of a saw, it is called a *sawtooth* wave.

The concept of linear variation is an important one in electronics. Therefore let us examine it more closely. As an example, assume that a car is traveling at a constant rate of speed, say 30 miles per hour (mi/h). In 1 h the distance d which the car has covered is 30 mi; in 2 h d is 60 mi; in 3 h d is 90 mi; etc. The equation which relates distance d, rate r, and time t is

$$d = r \times t \qquad (44\text{-}1)$$

For the example cited the equation may be written as

$$d = 30t \qquad (44\text{-}2)$$

where d is miles covered and t is the time traveled in hours.

Since the rate r is assumed to be constant, Eqs. (44-1) and (44-2) are said to be *first-degree equations in d and t*. That is, the exponents of the variables d and t are 1, not 2 or 3, etc. Figure 44-2 is a graph of Eq. (44-2). Some points satisfying Eq. (44-2), namely, (0,0), (1,30), and (2,60), are joined, and it is evident that they lie along a straight line. The graph of Eq. (44-2) is therefore a straight line, and we may say that there is a *linear* relationship between d and t. In the case of Eq. (44-2), a linear relationship between d and t simply means that equal increments of time t result in equal increments of distance d.

It can be shown that the graph of any first-degree equation is a straight line, and, conversely, that the equation of every straight line is of the first degree.

Refer now to Fig. 44-1. It is evident that in the time interval t_1, also called the *rise* time, voltage increases linearly with respect to time so that equal voltage increments take place in equal time increments.

What happens to the voltage wave of Fig. 44-1 in the time interval t_2? The voltage falls from its maximum level V to 0, and it is evident that the *fall* time (t_2) is much shorter than the *rise* time (t_1).

Generating a Sawtooth Wave

A sawtooth wave may be generated by charging and discharging a capacitor utilizing proper RC time constants. A sawtooth-wave generator is also called a *ramp-function* generator. In the circuit of Fig. 44-3a, Q is an NPN transistor amplifier which is conducting heavily in the absence of a signal. A rectangular pulse v_{in} (Fig. 44-3b) is applied to the input. The time constant $C_1 R_1$ is long compared with the

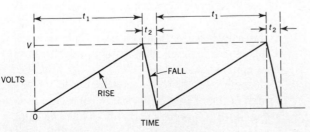

Fig. 44-1. Sawtooth wave in which voltage varies linearly with time.

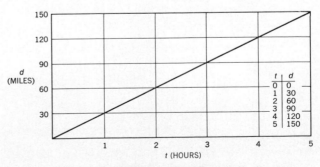

Fig. 44-2. A straight-line graph of Eq. (44-2): $d = 30t$.

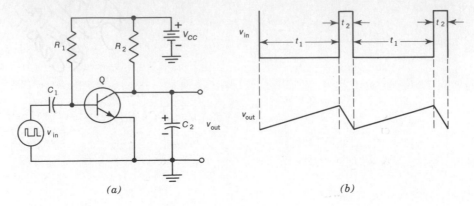

(a)   (b)

Fig. 44-3. *(a)* An amplifier is used to generate a sawtooth wave. *(b)* Input and output waveforms.

period $(t_1 + t_2)$ of the pulse. Hence the rectangular waveform is coupled by C_1R_1 to the base. During the negative alternation, time t_1, the base is driven negative relative to the emitter, cutting off base and collector current. The voltage at the collector of Q, which was close to 0 prior to the pulse, now tries to rise to the collector source voltage $+V_{CC}$. However, it cannot do so instantaneously because the voltage across capacitor C_2 cannot charge instantaneously. Hence, the voltage at the collector rises toward V_{CC} at the rate determined by the charge of C_2 through R_2. If the time constant C_2R_2 is *long* compared with the interval t_1, C_2 will charge only part of the way toward V_{CC}.

During the time interval t_2, the positive portion of v_{in} is applied to the base, turning Q on. Current flows in the collector circuit, discharging capacitor C_2 and dropping the voltage at the collector to its level prior to the injection of v_{in}. The discharge time constant for C_2 is relatively short because it consists of $C_2 \times R_{EC}$, where R_{EC} is the low emitter-to-collector resistance while the transistor is conducting heavily.

A long charge time constant C_2R_2 and a short discharge time constant C_2R_{EC} produce a sawtooth wave, v_{out}, as in Fig. 44-3b. The *rise* portion of v_{out}, the output waveform at the collector, is strictly speaking not *linear* because this voltage rise depends on the charge characteristic of a capacitor, which is exponential. However, reference to the charge curve in the universal time constant chart (Fig. 41-6) shows that if the early portion of the capacitor charge curve is used, it is relatively linear. For example, if the time constant

$C_2R_2 = 5t_1$, during the interval t_1, the capacitor has only one-fifth of a time constant within which to charge. Therefore, though it will only charge up approximately 20 percent, the rate of charge will be relatively linear. The longer the time constant C_2R_2 is, the more linear is the rise time of v_{out}, but also the *lower* is the output voltage v_{out}.

Another means of generating a sawtooth wave is shown in Fig. 44-4. Here, in the absence of a signal, transistor Q is cut off because there is no forward base-to-emitter bias, since the base and emitter are both at ground potential. When v_{in} is applied to the base, Q is turned on during the short *positive* (t_2) pulse time, discharging C_2 through R_{EC}. During the long negative period (t_1) of the pulse, Q is again turned off, and C_2 charges through R_2 toward the collector source voltage.

In the circuits of both Fig. 44-3 and 44-4, Q is gated on or off by the input rectangular pulse. The *rise* portion of the sawtooth wave developed at the collector occurs during the time that the transistor is cut off.

PNP transistors may also be used to develop sawtooth waves.

The rectangular pulse required for gating the sawtooth generator may be supplied by a multivibrator.

The peak-to-peak amplitude of the output waveform can be no greater than the source voltage V_{CC} in Figs. 44-3 and 44-4. For low-voltage transistors this means that v_{out} is relatively low in amplitude. If higher peak-to-peak outputs are required, high-voltage transistors must be used.

SUMMARY

1. A sawtooth waveform is a periodic waveform whose voltage varies linearly with respect to time (Fig. 44-1).
2. A complete cycle consists of a *rising* portion and a *falling* portion.
3. A sawtooth wave is used as the time base in oscilloscopes, in radar cathode-ray-tube indicators, and in other applications.
4. A sawtooth wave may be generated by utilizing the linear portion of the charge or discharge curve of a capacitor.
5. In Fig. 44-3, C_2 is the charge-discharge capacitor.
6. Another sawtooth generator is shown in Fig. 44-4.

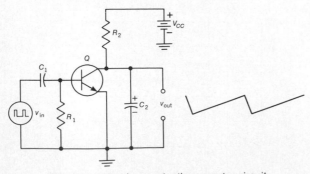

Fig. 44-4. Another sawtooth generator circuit.

7. The time constant C_2R_2 in Figs. 44-3 and 44-4 determines whether the sawtooth waveform is linear. If the time constant C_2R_2 is *long* compared with the period of the incoming wave v_{in}, then the waveform is linear. Otherwise, the sawtooth wave is nonlinear.

8. In Figs. 44-3 and 44-4, if all other parameters are held to the same value but C_2 is varied, both the linearity and amplitude of the output waveform are affected.

9. The rectangular pulse v_{in} required for turning Q (in Figs. 44-3 and 44-4) on and off may be supplied by a nonsymmetrical multivibrator.

10. The peak-to-peak amplitude of the sawtooth wave in Figs. 44-3 and 44-4 can be no higher than the source voltage V_{CC}.

SELF-TEST

Check your understanding by answering these questions.

1. The sum of the rise time and fall time of a periodic sawtooth wave constitutes the period or time of a cycle of that wave. _____ (*true, false*)

2. The amplifier Q in Fig. 44-3 is normally _____ (ON, OFF) in the absence of any input signal.

3. Q in Fig. 44-4 is normally _____ (ON, OFF) in the absence of any input signal.

4. The positive-going portion (rise) of the sawtooth wave in Fig. 44-4 is generated during the time when Q is turned _____ (*on, off*) by the input signal.

5. In Fig. 44-3, if the time t_1 of the negative alternation of v_{in} is decreased, the peak-to-peak voltage of v_{out} will _____ (*increase, decrease, remain the same*), assuming no other changes in the circuit.

6. For the conditions in question 5, the sawtooth wave will become _____ (*less, more*) linear.

MATERIALS REQUIRED

- Power supply: $+250$-V regulated dc source; variable regulated low-voltage dc source
- Instruments: Oscilloscope; EVM; AF sine-wave generator
- Resistors: 1000-, 1200-, two 6800-, 10,000-, 15,000-, 33,000-, 56,000-, 100,000-Ω; 1-MΩ ½-W
- Capacitors: Two 0.01-, 0.0047-, 0.047-, two-100µF 50-V; 25-µF 50-V
- Semiconductors: Two 2N6004, 2N6005, and 2N3440 transistors
- Miscellaneous: two SPST switches; 50,000- and 500,000-Ω 2-W potentiometers

PROCEDURE

Low-Level Sawtooth Wave

1. Connect the circuit of Fig. 44-5.

NOTE: Q is a PNP transistor connected as an emitter follower to isolate the collector-coupled multivibrator from C_{out}, the sawtooth capacitor. If C_{out} were connected from the collector of Q_1 to ground, the multivibrator would cease to operate. Try it.

Set the *frequency* control for *minimum* Q_1 base resistance (100,000 Ω). *Close* S_1 and S_2, thus applying power to the circuit.

2. With an oscilloscope externally synchronized by the output waveform of Q_1 at TP 1, observe and measure the waveforms at TP 1, 2, and 3. Draw these waveforms in proper time phase in Table 44-1. Measure the peak-to-peak amplitude and the frequency. Record these measurements in Table 44-1.

3. Adjust the frequency control for maximum Q_1 base resistance. Observe and measure the output waveform at TP 3. Draw it in Table 44-1. Record peak-to-peak amplitude in Table 44-1. Measure also and record the frequency. Indicate and record effect on linearity.

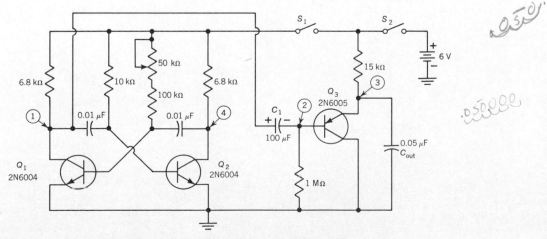

Fig. 44-5. Experimental sawtooth generator, low level.

TABLE 44-1. Low-Level Sawtooth Wave Generator

Step	Test Point	Waveform	Linearity	V p-p	Frequency, Hz
	1	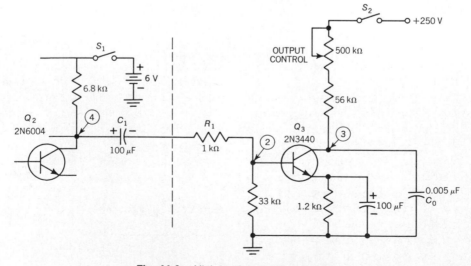	X	6	
2	2		X	6	1.6 kHz
	3		X	2	
3	3		X	2.6	1.13 kHz
4	3		X		X
5	1		X		X
	3		X		

4. Replace C_{out} with a 25-μF capacitor polarized properly. Draw the output waveform, and record its peak-to-peak amplitude and effect on linearity.

5. *Open* S_1. S_2 remains *closed*. Observe and draw the waveforms, if any, at TP 1 and 3.

High-Level Sawtooth Wave

6. *Open* S_2. S_1 and S_2 are now both open. Remove the circuit of Q_3 but leave connected the multivibrator circuit of Q_1 and Q_2. Connect capacitor C_1 as shown in Fig.

44-6, and add the circuit of Q_3, connected as shown in Fig. 44-6. Note that the collector of Q_3 is fed from a separate $+250$-V power supply.

7. *Close* S_2, applying power to Q_3, but leave S_1 *open*. With an EVM measure the dc voltage at the collector, emitter, and base of Q_3. Record the voltages in Table 44-2. With your oscilloscope observe the waveform, if any, at the collector, TP 3. Draw the waveform, if any, in Table 44-2. Measure and record its peak-to-peak amplitude.

8. *Close* S_1. Power is now ON for all transistors. Readjust the frequency control in base of Q_1 for minimum

Fig. 44-6.   High-level sawtooth generator.

45
EXPERIMENT
OP AMP GENERATORS

OBJECTIVES

1. To measure the trip points of an op amp Schmitt trigger
2. To generate a square wave by driving a Schmitt trigger with a sine wave
3. To generate a square wave with an op amp relaxation oscillator
4. To generate a triangular wave by integrating a square wave

INTRODUCTORY INFORMATION

Op Amp Schmitt Trigger

Figure 45-1a shows an op amp Schmitt trigger. Because of the positive feedback to the noninverting input, the output is saturated in either the positive or negative direction. Assume the output is positively saturated. Then a positive voltage is fed back to the noninverting input. This positive voltage is called the *upper trip point* (UTP). As long as the inverting input voltage is less than the UTP, the output voltage remains positively saturated. This means the operating point is somewhere along the upper part of the graph in Fig. 45-1b.

If we slowly increase the input voltage, we eventually reach a point where it is slightly more positive than the UTP. When this happens, the error voltage changes polarity, driving the op amp into negative saturation (Fig. 45-1b). With the output now negative, the voltage divider feeds back a negative voltage to the noninverting input. This negative voltage is referred to as the *lower trip point* (LTP).

The output remains in negative saturation as long as the input voltage is more positive than the LTP. The only way to change the output is to decrease the input voltage until it is slightly more negative than the LTP. Then the error voltage

changes polarity and the output switches back to positive saturation as shown in Fig. 45-1b.

The formulas for the trip points are

$$UTP = \frac{R_1}{R_1 + R_2} V_{SAT} \qquad (45\text{-}1)$$

$$LTP = \frac{-R_1}{R_1 + R_2} V_{SAT} \qquad (45\text{-}2)$$

Generating Square Waves

One way to generate square waves is to drive a Schmitt trigger with a sine wave whose positive peak is greater than the UTP and whose negative peak is less than the LTP.

For instance, Fig. 45-2a shows a Schmitt trigger with equal resistors in the voltage divider. This means the UTP is

$$UTP = \frac{R}{2R} V_{SAT} \quad = \frac{1}{2} V_{SAT}$$

Similarly, the LTP equals $-V_{SAT}/2$. If V_{SAT} is 10 V, then UTP is $+5$ V and LTP is -5 V. If the sine wave shown in Fig. 45-2a has a positive peak greater than $+5$ V and a negative peak less than -5 V, then the Schmitt trigger produces the square-wave output shown in Fig. 45-2b.

The shape of the input signal is immaterial. Instead of a sine wave, any periodic signal with sufficient amplitude to drive the Schmitt trigger will result in an output square wave like Fig. 45-2b.

Relaxation Oscillator

Figure 45-3a is an example of a *relaxation oscillator*, a circuit whose output frequency depends on the charging and discharging of a capacitor (or inductor). Here is how the circuit works. To begin with, notice the circuit is similar to a Schmitt

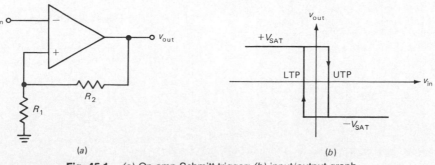

(a)

(b)

Fig. 45-1. *(a)* Op amp Schmitt trigger; *(b)* input/output graph.

259

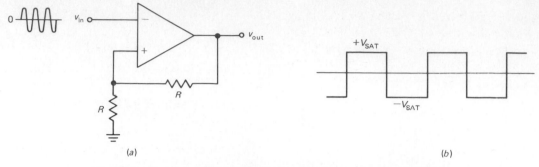

Fig. 45-2. Sine wave driving Schmitt trigger produces square-wave output.

trigger. Since the voltage divider has equal resistors, the UTP is $+V_{SAT}/2$ and the LTP is $-V_{SAT}/2$. Instead of being externally driven like the usual Schmitt trigger, the circuit of Fig. 45-3a supplies its own input voltage by way of the RC circuit.

Assume the output voltage is positively saturated when power is first applied. Because of the voltage divider, half of this positive voltage appears at the noninverting input. On the other hand, the inverting input initially starts at 0 V because the capacitor is uncharged. The capacitor will charge exponentially through resistor R. As long as the capacitor voltage is less than the UTP, the output remains in positive saturation.

As capacitor charges, its voltage eventually goes slightly more positive than the UTP. When this happens, the error voltage reverses polarity and the output voltage switches to negative saturation as shown in Fig. 45-3b.

The capacitor now discharges exponentially (Fig. 45-3c). The output voltage will remain negative as long as the capacitor voltage is more positive than the LTP. Eventually, the capacitor voltage becomes slightly more negative than $-V_{SAT}/2$. Then the error voltage reverses polarity and the output switches back to the positive state.

As you can see in Fig. 45-3b and c, the output is a square wave and the input is an exponential wave. A mathematical analysis shows the frequency of the output square wave is

$$f = \frac{0.455}{RC} \quad (45\text{-}3)$$

For instance, if $R = 6.8$ kΩ and $C = 0.02$ μF, then the frequency is

$$f = \frac{0.455}{6.8(10^3) \times 0.02(10^{-6})} = 3.35 \text{ kHz}$$

Op Amp Integrator

Figure 45-4a shows an op amp *integrator*. To understand how it works, let us first discuss the concept of a *virtual ground*. In Fig. 45-4a, the op amp has such a high voltage gain that the inverting input voltage approaches 0. Also, because of the

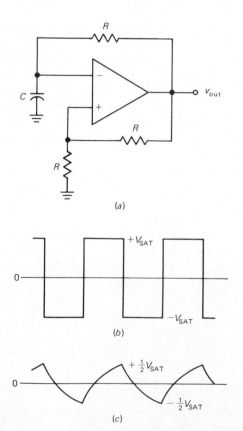

Fig. 45-3. Op amp relaxation oscillator. (a) Circuit; (b) output; (c) capacitor voltage.

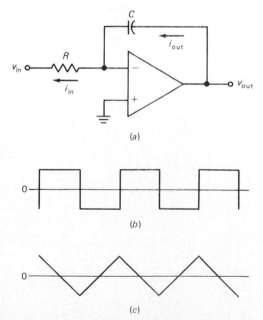

Fig. 45-4. Op amp integrator. (a) Circuit; (b) input; (c) output.

high input impedance, the inverting input has approximately zero current. Since it has approximately zero voltage and current, the inverting input is called a *virtual ground*; this means it is a ground to voltage, but not to current. (An ordinary ground has zero voltage and can sink any amount of current.)

Because the inverting input acts like a virtual ground, the input current of Fig. 45-4a is

$$i_{in} = \frac{v_{in}}{R}$$

Since no current enters the inverting input, all of the input current must go to the capacitor. That is

$$i_{out} = i_{in} = \frac{v_{in}}{R}$$

This says the charging current of the capacitor equals the input voltage divided by the resistance.

When a capacitor has a constant charging current, its voltage is a ramp (linear change). Why? Because capacitance is defined as

$$C = \frac{Q}{V}$$

which can be rearranged as

$$V = \frac{Q}{C}$$

When the current is constant, Q increases linearly. Since V is directly proportional to Q, it too increases linearly. In other words, a constant charging current produces a ramp of voltage across the capacitor.

When a square wave like Fig. 45-4b drives an op amp integrator, the output is a triangular wave like Fig. 45-4c. A mathematical analysis shows the peak-to-peak output voltage is

$$v_{out(p\text{-}p)} = \frac{v_{in(p\text{-}p)}}{4fRC} \qquad (45\text{-}4)$$

As an example, suppose $f = 1$ kHz, $R = 10$ kΩ, and $C = 0.2$ μF. If the input square wave has a peak-to-peak value of 12 V, then the output triangular wave has a peak-to-peak value of

$$v_{out(p\text{-}p)} = \frac{12\text{ V}}{4 \times 10^3 \times 10^4 \times 0.2(10^{-6})} = 1.5\text{ V}$$

Practical Op Amp Integrator

Because a capacitor appears open at zero frequency, the input offset in Fig. 45-4a may saturate the op amp. To avoid this, a resistor is usually shunted across the capacitor as shown in Fig. 45-5. Typically, this resistance R_2 is 5 to 10 times the input resistance R_1. The shunt resistance has virtually no effect on the output, provided the input frequency is much greater than

$$f = \frac{1}{2\pi R_2 C} \qquad (45\text{-}5)$$

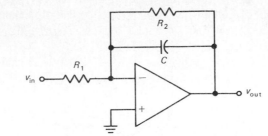

Fig. 45-5. Shunt resistor across capacitor to minimize effect of input offset.

Sine-Square-Triangular Generator

Figure 45-6 shows a block diagram for a signal generator that produces a sine wave, a square wave, and a triangular wave. A Wien-bridge oscillator produces the sine wave. This sine wave is one of the outputs. The sine wave also drives a Schmitt trigger to produce a square wave. This square wave is a second output. Finally, the square wave drives an integrator to produce a triangular wave.

SUMMARY

1. The positive voltage fed back to the noninverting input of an op amp Schmitt trigger is called the *upper trip point* (UTP). The negative voltage that is fed back is the *lower trip point* (LTP).
2. A *relaxation oscillator* is a circuit whose frequency depends on the charging and discharging of a capacitor (or inductor).
3. The op amp relaxation oscillator produces a square wave output.
4. The inverting input of an op amp is a *virtual ground*. This means it has approximately zero voltage and current.
5. Because the inverting input acts like a virtual ground, the charging current to the capacitor equals the input current in an op amp integrator.
6. A constant charging current produces a ramp of voltage across a capacitor.
7. A shunt resistor is connected across the capacitor of an op amp integrator to prevent input offset from saturating the output.

SELF-TEST

Check your understanding by answering these questions.

1. The input voltage to a Schmitt trigger must be more positive than the _____ to drive the output into negative saturation.

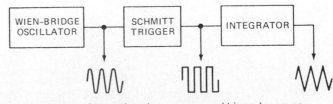

Fig. 45-6. Generating sine, square, and triangular waves.

2. If the resistors are equal in the voltage divider of a Schmitt trigger, the UTP equals _____ and the LTP equals _____ .

3. The relaxation oscillator discussed earlier is a circuit whose frequency is controlled by the charging and discharging of a _____ .

4. A relaxation oscillator is like a _____ trigger except that it supplies its own input voltage through an RC circuit.

5. If $R = 33$ kΩ and $C = 0.047$ μF, the output frequency in Fig. 45-3 is _____ kHz.

6. The amount of current into a virtual ground is _____ .

7. In an op amp integrator the output current to the capacitor equals the _____ current.

8. A constant charging current produces a _____ of voltage across a capacitor. .

9. If $R = 33$ kΩ and $C = 0.047$ μF in Fig. 45-4a, then a square wave with a peak-to-peak value of 15 V and a frequency of 2 kHz produces a triangular peak-to-peak output of _____ V.

MATERIALS REQUIRED

- Two power supplies: 15 V
- Equipment: Oscilloscope, electronic counter, voltmeter, ac generator
- Resistors: 1-, two 3.3-, three 10-, 12-, one 100-kΩ ½-W
- Potentiometer: 1-kΩ
- Capacitors: Two 0.022-, 0.047-, 0.1-μF
- Op amp: 741C

PROCEDURE

Op Amp Relaxation Oscillator

1. Calculate the frequency of the output signal in Fig. 45-7 for each value of C listed in Table 45-1. Record the values under f_{calc} in Table 45-1.
2. Connect the circuit of Fig. 45-7 using a C of 0.022 μF.
3. Look at the output with an oscilloscope. The signal should be approximately a square wave. Measure the peak-to-peak value. Record this under v_{out} in Table 45-1.
4. Use an electronic counter to measure the frequency of the output. (If a counter is not available, use the oscilloscope to measure the period, then calculate the frequency.) Record the frequency under f_{meas} in Table 45-1.
5. Use the oscilloscope to look at the signal on pin 2. You should see an exponential wave. Measure the peak-to-peak value. Record under v_{in} (Table 45-1).
6. Repeat steps 3 through 5 for the other values of C shown in Table 45-1.

TABLE 45-1. Relaxation Oscillator

C, μF	f_{calc}	v_{out}, V	f_{meas}	v_{in}, V
0.022				
0.047				
0.1				

Op Amp Integrator

7. Connect the circuit of Fig. 45-8 using a C of 0.022 μF.
8. Look at the output with an oscilloscope. You should see a triangular waveform. Measure and record its peak-to-peak voltage (Table 45-2).
9. Repeat step 8 for the other values of C listed in Table 45-2.

TABLE 45-2. Integrator

C, μF	v_{out}, V
0.022	
0.047	
0.1	

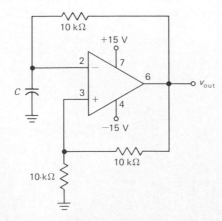

Fig. 45-7. Experimental relaxation oscillator.

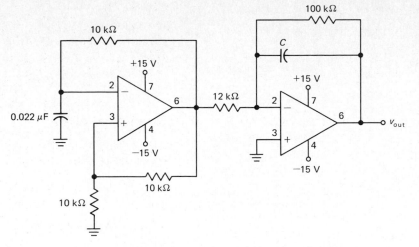

Fig. 45-8. Experimental circuit for producing triangular wave.

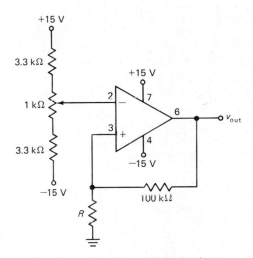

Fig. 45-9. Experimental Schmitt trigger.

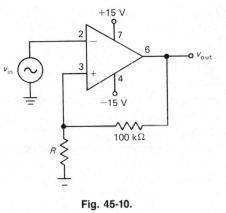

Fig. 45-10.

Schmitt Trigger

10. Connect the circuit of Fig. 45-9 with an R of 10 kΩ.
11. Look at the output with an oscilloscope. Adjust the input voltage until the output goes positive.
12. Slowly adjust the input voltage until the output just goes negative. Measure the input voltage. Record this value under UTP in Table 45-3.
13. Again, slowly adjust the input voltage to switch the output to positive. Measure the input voltage and record under LTP (Table 45-3).
14. Connect the circuit of Fig. 45-10 with an R of 10 kΩ. Set the signal generator to 1 kHz. Increase the signal level until a square wave appears at the output of the Schmitt trigger.
15. Slowly reduce the signal level until the square wave stops.
16. Measure the peak-to-peak value of the input sine wave. Record this value under v_{in} (Table 45-3).
17. Repeat steps 14 through 16 for an R of 1 kΩ.

Extra Credit

18. Connect the circuit of Fig. 45-11.
19. Adjust the signal generator to get an input sine wave of 5 V peak-to-peak at 1 kHz.
20. Vary the 1-kΩ potentiometer and observe the output of the Schmitt trigger with an oscilloscope (Time base = 0.2 ms/cm). Make notes of what you observe.

TABLE 45-3. Schmitt Trigger

R, kΩ	UTP	LTP	v_{in}, V
10			
1			

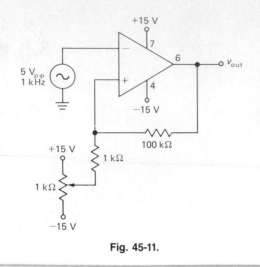

Fig. 45-11.

QUESTIONS

1. Give one reason why the measured frequency may differ from the calculated frequency in Table 45-1.
2. Why is the input voltage approximately half the output voltage in Table 45-1?
3. In Fig. 45-7, the 10-Ω resistor that is grounded is changed to 5 kΩ. What will happen to the input and output voltage?
4. In Fig. 45-8, what is the approximate frequency driving the integrator?
5. Explain why the integrator output (Table 45-2) decreases when capacitor C increases.
6. Explain why the trip points are closer together when R is 1 kΩ in Table 45-3.
7. How is v_{in} related to the trip points in Table 45-3?

Extra Credit

8. What happened to the output waveform as you varied the potentiometer? Explain why this happened.

Answers to Self-Test

1. UTP	6. 0
2. $+V_{SAT}/2$; $-V_{SAT}/2$	7. input
3. capacitor	8. ramp
4. Schmitt	9. 1.21 V
5. 293 Hz	

OBJECTIVES

1. To measure the frequency and duty cycle of an astable 555 timer
2. To measure the pulse width out of a monostable 555 timer
3. To examine the signal out of voltage-controlled oscillator
4. To build a sawtooth generator using a 555 timer

INTRODUCTORY INFORMATION

Flip-Flop Review

Before discussing the 555 timer, we need to review the flip-flop introduced in Experiment 42. Figure 46-1 shows a simplified version of the earlier flip-flop. Each collector drives the opposite base through a 100-kΩ resistor. On power-up, the transistor with the higher β saturates and the other cuts off.

For instance, suppose the right-hand transistor is saturated. Then its collector voltage is approximately 0 V. This means no base drive for the left-hand transistor, so it goes into cutoff and its collector voltage approaches $+15$ V. This high voltage produces more than enough base current in the right-hand transistor to sustain its saturation. In other words, the overall circuit is latched or stuck in this state: left-hand transistor cutoff and right transistor saturated. In this case, output Q is approximately 0 V.

The foregoing analysis works just as well as the other way. The left-hand transistor can be saturated, and the right-hand transistor cut off. For this state, output Q is approximately 15 V.

Q can be either low or high, approximately 0 or 15 V. A circuit like this is called a *memory element* because it can store binary information. Computers use thousands of memory elements.

RS Flip-Flop

Figure 46-2 shows one way to trigger a flip-flop. A high *set* (S) input forces the left-hand transistor to saturate if it is not already saturated. As soon as the left-hand transistor saturates, the overall circuit latches and

$$Q = 15 \text{ V}$$

A high S input therefore sets the output to 15 V. Here it remains, even though you later remove the S input.

A high *reset* (R) input drives the right-hand transistor into saturation if it is not already saturated. Once this happens, the overall circuit remains latched even though you remove the R input. In this state,

$$Q = 0 \text{ V}$$

Figure 46-2 is an example of how to build an *RS flip-flop*, a transistorized memory element with set and reset inputs. Q is the output. Incidentally, a complementary output $\overline{Q}$ is available from the collector of the left-hand transistor. This may or may not be used, depending on the application.

Figure 46-3 shows the schematic symbol for an RS flip-flop of any design. Whenever you see this symbol, remember the action: The circuit latches in either of two states. A high S input sets Q to high; a high R input resets Q to low. Output Q remains in a given state until triggered into the opposite state.

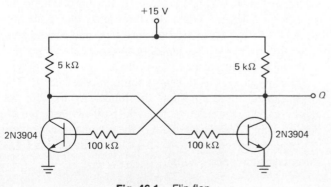

Fig. 46-1. Flip-flop.

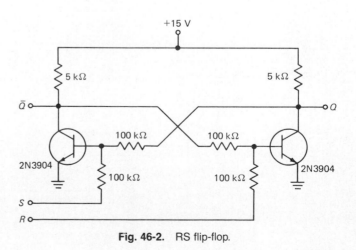

Fig. 46-2. RS flip-flop.

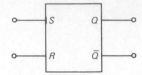

Fig. 46-3. Symbol for RS flip-flop.

Basic Timing Concept

Figure 46-4a illustrates some basic ideas needed in our later discussion of the 555 timer. Assume output Q is high. This saturates the transistor and clamps the capacitor voltage at ground. In other words, the capacitor is shorted and cannot charge.

The noninverting input voltage of the op amp is called the *threshold* voltage, and the inverting input voltage is referred to as the *control* voltage. With the RS flip-flop set, the saturated transistor holds the threshold voltage at 0. The control voltage, on the other hand, is fixed at $+10$ V because of the voltage divider.

Suppose we apply a high voltage to the R input. This resets the RS flip-flop. Output Q goes to 0 and this cuts off the transistor. Capacitor C is now free to charge. As the capacitor charges, the threshold voltage increases.

Eventually, the threshold voltage becomes slightly greater than the control voltage ($+10$ V). The output of the op amp then goes high, forcing the RS flip-flop to set. The high Q output saturates the transistor and this quickly discharges the capacitor.

Notice the two waveforms in Fig. 46-4b. An exponential rise is across the capacitor, and a positive-going pulse appears at the $\bar{Q}$ output.

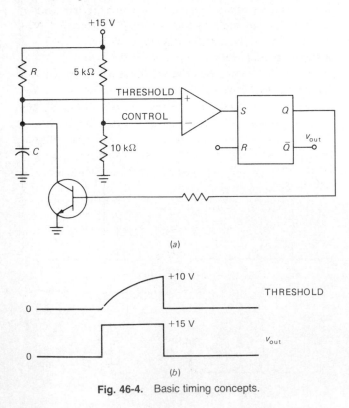

Fig. 46-4. Basic timing concepts.

555 Block Diagram

The NE555 timer introduced by Signetics is an 8-pin IC that can be connected to external components for either astable or monostable operation. Figure 46-5 shows a simplified block diagram. Notice the upper op amp has a threshold input (pin 6) and a control input (pin 5). In most applications, the control input is not used, so that the control voltage equals $+2V_{CC}/3$. As before, whenever the threshold voltage exceeds the control voltage, the high output from the op amp will set the flip-flop.

The collector of the *discharge* transistor goes to pin 7. When this pin is connected to an external timing capacitor, a high Q output from the flip-flop will saturate the transistor and discharge the capacitor. When Q is low, the transistor opens and the capacitor can charge as previously described.

The complementary signal out of the flip-flop goes to pin 3, the output. When the external reset (pin 4) is grounded, it *inhibits* the device (prevents it from working). This on-off feature is useful sometimes. In most applications, however, the external reset is not used and pin 4 is tied directly to the supply voltage.

Notice the lower op amp. Its inverting input is called the *trigger* (pin 2). Because of the voltage divider, the noninverting input has a fixed voltage of $+V_{CC}/3$. When the trigger input voltage is slightly less than $+V_{CC}/3$, the op amp output goes high and resets the flip-flop.

Finally, pin 1 is the chip ground, while pin 8 is the supply pin. The 555 timer will work with any supply voltage between 4.5 and 16 V.

Monostable Operation

Figure 46-6a shows the 555 timer connected for monostable (one-shot) operation. It works as follows. When the trigger input is slightly less than $+V_{CC}/3$, the lower op amp has a high output and resets the flip-flop. This cuts off the transistor, allowing the capacitor to charge.

When the threshold voltage is slightly greater than $+2V_{CC}/3$, the upper op amp has a high output, which sets the flip-flop. As soon as Q goes high, it turns on the transistor; this quickly discharges the capacitor.

Figure 46-6b shows typical waveforms. The trigger input is a narrow pulse with a quiescent value of $+V_{CC}$. The pulse must drop below $+V_{CC}/3$ to reset the flip-flop and allow the capacitor to charge. When the threshold voltage slightly exceeds $+2V_{CC}/3$, the flip-flop sets; this saturates the transistor and discharges the capacitor. As a result, we get one rectangular output pulse.

The capacitor C has to charge through resistance R. The larger the RC time constant, the longer it takes for the capacitor voltage to reach $+2V_{CC}/3$. In other words, the RC time constant controls the width of the output pulse. Solving the exponential equation for capacitor voltage gives this formula for the pulse width

$$W = 1.1RC \qquad (46\text{-}1)$$

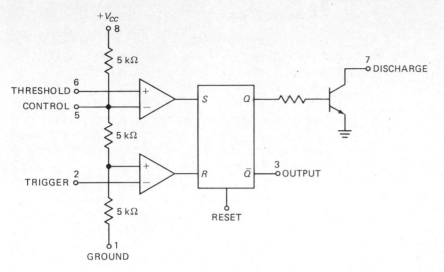

Fig. 46-5. Block diagram of 555 timer.

For instance, if $R = 22$ kΩ and $C = 0.068$ µF, then the output of the monostable 555 timer is

$$W = 1.1 \times 22(10^3) \times 0.068(10^{-6}) = 1.65 \text{ ms}$$

Normally, a schematic diagram does not show the op amps, flip-flop, and other components inside the 555 timer. Rather, you will see a schematic diagram like Fig. 46-7 for the monostable 555 circuit. Only the pins and external com-

ponents are shown. Incidentally, notice that pin 5 (control) is bypassed to ground through a small capacitor, typically 0.01 µF. This provides noise filtering for the control voltage.

Recall that grounding pin 4 inhibits the 555 timer. To avoid accidental reset, pin 4 is usually tied to the supply voltage as shown in Fig. 46-7.

Astable Operation

Figure 46-8a shows the 555 timer connected for astable operation. When Q is low, the transistor is cut off and the capacitor is charging through a total resistance of $R_A + R_B$. Because of this, the charging time constant is $(R_A + R_B)C$. As the capacitor charges, the threshold voltage increases.

Eventually, the threshold voltage exceeds $+2V_{CC}/3$; then the upper op amp has a high output and this sets the flip-flop. With Q high, the transistor saturates and grounds pin 7. Now the capacitor discharges through R_B. Therefore, the discharging time constant is $R_B C$. When the capacitor voltage drops slightly below $+V_{CC}/3$, the lower op amp has a high output and this resets the flip-flop.

Figure 46-8b illustrates the waveforms. As you see, the timing capacitor has an exponentially rising and falling voltage. The output is a rectangular wave. Since the charging time constant is longer than the discharging time constant, the output is not symmetrical; the high state lasts longer than the low state.

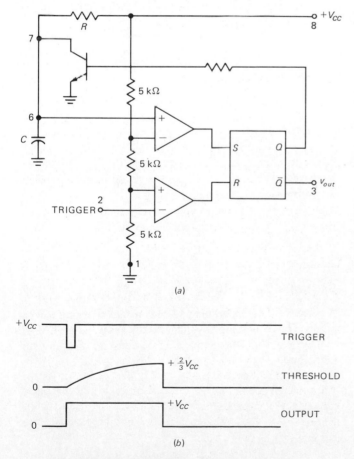

(a)

(b)

Fig. 46-6. (a) Monostable operation; (b) waveforms.

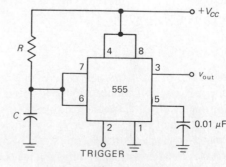

Fig. 46-7. Monostable 555 timer.

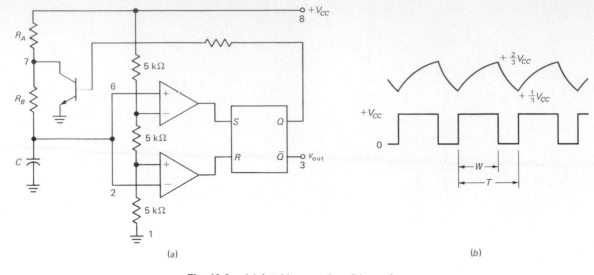

(a)   (b)

Fig. 46-8. *(a)* Astable operation; *(b)* waveforms.

To specify how unsymmetrical the output is, we will use the duty cycle defined as

$$D = \frac{W}{T} \times 100\% \qquad (46\text{-}2)$$

As an example, if $W = 2$ ms and $T = 2.5$ ms, then the duty cycle is

$$D = \frac{2 \text{ ms}}{2.5 \text{ ms}} \times 100\% = 80\%$$

Depending on resistances R_A and R_B, the duty cycle is between 50 and 100 percent.

A mathematical solution to the charging and discharging equations gives the following formulas. The output frequency is

$$f = \frac{1.44}{(R_A + 2R_B)C} \qquad (46\text{-}3)$$

and the duty cycle is

$$D = \frac{R_A + R_B}{R_A + 2R_B} \times 100\% \qquad (46\text{-}4)$$

If R_A is much smaller than R_B, the duty cycle approaches 50 percent.

Figure 46-9 shows the astable 555 timer as it usually appears. Again notice how pin 4 (reset) is tied to the supply voltage and how pin 5 (control) is bypassed to ground through a 0.01-μF capacitor.

Voltage-Controlled Oscillator

Figure 46-10*a* shows a *voltage-controlled* oscillator (VCO). Recall that pin 5 (control) connects to the inverting input of the upper op amp. Normally, the control voltage is $+2V_{CC}/3$ because of the internal voltage divider. In Fig. 46-10*a*, however, the voltage from an external potentiometer overrides the internal voltage. In other words, by adjusting the potentiometer, we can change the control voltage.

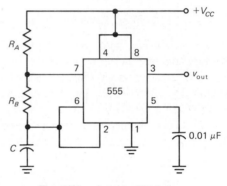

Fig. 46-9. Astable 555 timer.

Figure 46-10*b* illustrates the voltage across the timing capacitor. Notice it varies between $+V_{\text{control}}/2$ and $+V_{\text{control}}$. If we increase V_{control}, it takes the capacitor longer to charge and discharge; therefore, the frequency decreases. As a result, we can change the frequency of the circuit by varying the control voltage.

Incidentally, the control voltage may come from a potentiometer or it may be the output of another transistor circuit, op amp, etc.

Sawtooth Generator

As discussed in Experiment 45, a constant charging current produces a linear ramp of voltage across a capacitor. The PNP transistor of Fig. 46-11*a* produces a constant charging current equal to

$$I_C = \frac{V_{CC} - V_E}{R} \qquad (46\text{-}5)$$

where

$$V_E = V_{BE} + \frac{R_2}{R_1 + R_2} V_{CC} \qquad (46\text{-}6)$$

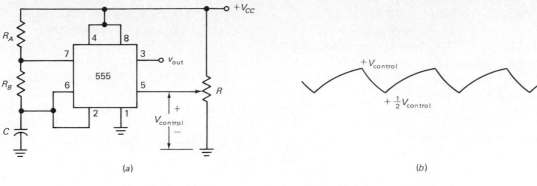

Fig. 46-10. *(a)* Voltage-controlled oscillator; *(b)* timing waveform.

For instance, if $V_{CC} = 15$ V, $R = 20$ kΩ, $R_1 = 5$ kΩ, $R_2 = 10$ kΩ, and $V_{BE} = 0.7$ V, then

$$V_E = 0.7 \text{ V} + 10 \text{ V} = 10.7 \text{ V}$$

and

$$I_C = \frac{15 \text{ V} - 10.7 \text{ V}}{20 \text{ k}\Omega} = 0.215 \text{ mA}$$

When a trigger starts the monostable 555 timer of Fig. 46-11*a*, the PNP current source forces a constant charging current into the capacitor. Therefore, the voltage across the capacitor is a linear ramp as shown in Fig. 46-11*b*. The *slope* S of the linear ramp is defined as the rise over the run, or

$$S = \frac{V}{T} \tag{46-7}$$

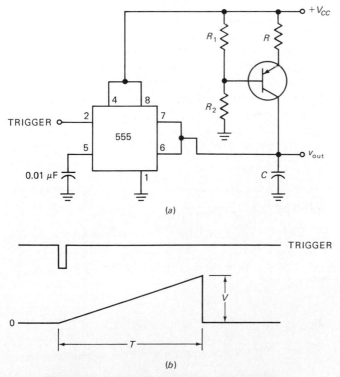

(a)

(b)

Fig. 46-11. *(a)* Sawtooth generator; *(b)* waveforms.

For instance, if $V = 10$ V and $T = 2$ ms, then the slope is

$$S = \frac{10 \text{ V}}{2 \text{ ms}} = 5 \text{ V/ms}$$

This says the ramp rises 5 V/ms.

Since the basic capacitor equation is

$$V = \frac{Q}{C}$$

we can divide both sides by T to get

$$\frac{V}{T} = \frac{Q/T}{C}$$

When the charging current is constant, this reduces to

$$S = \frac{I}{C} \tag{46-8}$$

In other words, you can predict the slope of a linear ramp using the ratio of charging current to capacitance. If the charging current is 0.215 mA (found earlier) and the capacitance is 0.022 μF, the ramp will have a slope of

$$S = \frac{0.215 \text{ mA}}{0.022 \text{ } \mu\text{F}} = 9.77 \text{ V/ms}$$

SUMMARY

1. A high *set* (S) input sets the output of an *RS flip-flop* to the high state. A high *reset* (R) input resets the output to the low state.
2. In a 555 timer the noninverting input of the upper op amp is called the *threshold* voltage; the inverting input is the *control* voltage.
3. When the threshold voltage exceeds the control voltage, the RS flip-flop is set. This saturates the discharge transistor.
4. The inverting input of the lower op amp in a 555 is called the *trigger*.
5. When trigger voltage is less than $+V_{CC}/3$, the RS flip-flop is reset. This cuts off the *discharge* transistor.
6. The 555 timer can be connected for astable or monostable operation.

7. Normally, the control voltage of a 555 timer equals $+2V_{CC}/3$ because of the internal voltage divider. In VCO applications, however, an external voltage is applied to the control pin to override the voltage from the internal voltage divider.

8. By using a PNP current source, the 555 timer can produce linear ramps.

SELF-TEST

Check your understanding by answering these questions.

1. In Fig. 46-4a, the control voltage equals _____ V.

2. To saturate the transistor of Fig. 46-4a, the Q output must be _____ .

3. To set the RS flip-flop of Fig. 46-5, the threshold voltage must be slightly greater than the _____ voltage.

4. If $V_{CC} = +15$ V in Fig. 46-5, the trigger voltage must be slightly less than _____ V to reset the RS flip-flop.

5. In Fig. 46-7, $R = 68$ kΩ and $C = 0.047$ μF. The pulse width of the output is _____ ms.

6. $R_A = 27$ kΩ, $R_B = 68$ kΩ, and $C = 0.22$ μF in Fig. 46-9. The frequency of the output is _____ Hz and the duty cycle is _____ percent.

7. To get an astable output whose duty cycle approaches 50 percent in Fig. 46-9, R_A should be much _____ than R_B.

8. $V_{CC} = 15$ V, $R_1 = 3.9$ kΩ, $R_2 = 8.2$ kΩ, $R = 1$ kΩ, and $C = 0.15$ μF in Fig. 46-11a. The constant charging current is _____ mA and the slope is _____ V/ms.

MATERIALS REQUIRED

- One power supply: 15-V
- Equipment: Oscilloscope, ac generator
- Resistors: Two 1-kΩ, one 4.7-kΩ, two 10-kΩ, one 22-kΩ, one 33-kΩ, one 47-kΩ, one 68-kΩ, one 100-kΩ ½-W
- Potentiometer: 1-kΩ
- Capacitors: Two 0.01-μF
- Transistor: 2N3906 (other choice: 2N6005)
- Op amp: 741C
- Timer: NE555

PROCEDURE

Astable 555 Timer

1. Calculate the frequency and duty cycle in Fig. 46-12 for the resistances listed in Table 46-1. Record the results under f_{calc} and D_{calc}.

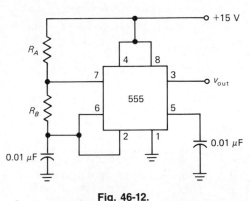

Fig. 46-12.

TABLE 46-1. Astable Operation

R_A, kΩ	R_B, kΩ	f_{calc}	D_{calc}	f_{meas}	D_{calc}
10	100				
100	10				
10	10				

2. Connect the circuit of Fig. 46-12 with $R_A = 10$ kΩ and $R_B = 100$ kΩ.

3. Look at the output with an oscilloscope. Measure W and T. Work out the frequency and duty factor. Record under f_{meas} and D_{meas} in Table 46-1.

4. Look at the voltage across the capacitor (pin 6). You should see an exponentially rising and falling wave between 5 and 10 V.

5. Repeat steps 2 through 4 for the other resistances of Table 46-1.

Voltage-Controlled Oscillator

6. Connect the VCO of Fig. 46-13.

7. Look at the output with an oscilloscope.

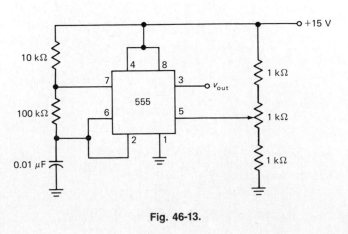

Fig. 46-13.

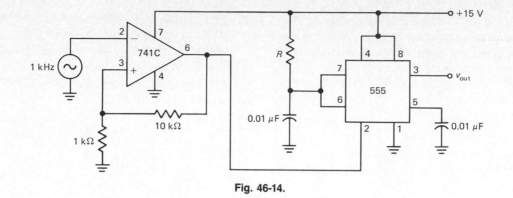

Fig. 46-14.

8. Vary the 1-kΩ potentiometer and notice what happens. Record the minimum and maximum frequencies here:

$$f_{min} = \underline{\hspace{2cm}}$$

$$f_{max} = \underline{\hspace{2cm}}$$

Monostable 555 Timer

9. Figure 46-14 shows a Schmitt trigger driving a monostable 555 timer. Calculate the pulse width for each R listed in Table 46-2. Record the results under W_{calc}.
10. Connect the circuit of Fig. 46-14 with an R of 33 kΩ.
11. Look at the output of the Schmitt trigger (pin 6 of the 741C). Set the frequency of the sine-wave input to 1 kHz. Adjust the sine-wave level until you get a Schmitt-trigger output with a duty cycle of approximately 90 percent.

12. Look at the output of the 555 timer. Measure the pulse width. Record this value under W_{meas} in Table 46-2.
13. Repeat steps 10 through 12 for the remaining R values of Table 46-2.

Sawtooth Generator

14. Calculate the charging current in Fig. 46-15 for each value of R shown in Table 46-3. Record the values.
15. Calculate the slope of capacitor voltage in volts per millisecond. Record under S_{calc} in Table 46-3.
16. Connect the circuit of Fig. 46-15 with an R of 10 kΩ. This is almost the same as Fig. 46-14 except for the PNP current source.
17. Set the ac generator to 1 kHz. Adjust the level to get a duty cycle of approximately 90 percent out of the Schmitt trigger.

TABLE 46-2. Monostable Operation

R, kΩ	W_{calc}	W_{meas}
33		
47		
68		

TABLE 46-3. Sawtooth Generator

R, kΩ	I_{charge}, mA	S_{calc}, V/ms	S_{meas}, V/ms
10			
22			
33			

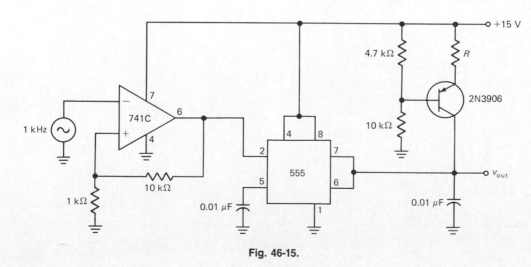

Fig. 46-15.

18. Look at the output voltage; it should be a sawtooth. Measure the ramp voltage and time. Then work out the slope in voltages per millisecond. Record the value under S_{meas} in Table 46-3.
19. Repeat steps 16 through 18 for the remaining values of R in Table 46-3.

Extra Credit

20. We want to control the amplitude of the ramp in Fig. 46-15. Figure out how to do it, then modify and test the circuit.

21. It is possible to simplify the PNP current source of Fig. 46-15 by eliminating the two base resistors and deriving the base voltage from the 555 timer. If you can figure out how to do this, make the required change and measure the slope of the sawtooth for an R of 22 kΩ. Record here:

$$S = \underline{\hspace{2cm}}$$

QUESTIONS

1. How does ratio R_A/R_B affect the duty cycle of an astable 555 timer?
2. What effect does increasing the timing capacitor have on the frequency out of an astable 555 timer?
3. Explain how the V_{CO} of Fig. 46-13 works.
4. How much ac voltage is there at pin 5 in Fig. 46-14? How much dc voltage is there?
5. What happens to the width of the output in Fig. 46-14 if the timing resistor is decreased?
6. What is the output frequency in Fig. 46-15?
7. What effect does R have on the sawtooth of Fig. 46-15?

Extra Credit

8. What did you do to control the amplitude of the sawtooth?
9. How did you change the PNP current source. Why did this work?

Answers to Self-Test

1. 10	5. 3.52
2. high	6. 40.2; 58.3
3. control	7. smaller
4. 5	8. 4.13; 27.6

MIXERS, MODULATORS, AND DEMODULATORS

OBJECTIVES

1. To observe the difference frequency out of a mixer
2. To amplitude-modulate a carrier using a bipolar modulator
3. To demodulate an AM signal

INTRODUCTORY INFORMATION

Nonlinear Distortion

Figure 47-1 shows the *transconductance* curve of a bipolar transistor. This graph of collector current versus base-emitter voltage is nonlinear. Notice how a sinusoidal V_{BE} voltage produces a nonsinusoidal collector current. In other words, the shape of the output current is no longer a true duplication of the input shape. Since the collector current flows through a load resistance, the output voltage will also have nonlinear distortion.

Nonlinear distortion produces *harmonics* (multiples of the input frequency). For instance, if the base-emitter voltage is a pure sine wave with a frequency of 1 kHz, then the collector current in Fig. 47-1 is a distorted sine wave with a *fundamental* (lowest) frequency of 1 kHz and harmonic frequencies of 2 kHz, 3 kHz, 4 kHz, . . . , and so on. In general, if f is the fundamental frequency, the output contains frequencies of f, $2f$, $3f$, . . . , nf.

Figure 47-2 illustrates the idea behind harmonics. A pure sine wave drives a nonlinear circuit. The output of the non-

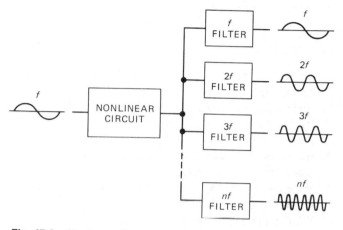

Fig. 47-2. Nonlinear distortion produces harmonics which can be filtered.

linear circuit is a distorted sine wave, which contains harmonics. A bank of filters can separate each harmonic from the others to get frequencies of f, $2f$, $3f$, . . . , nf. As you see, nonlinear distortion creates new sinusoidal signals.

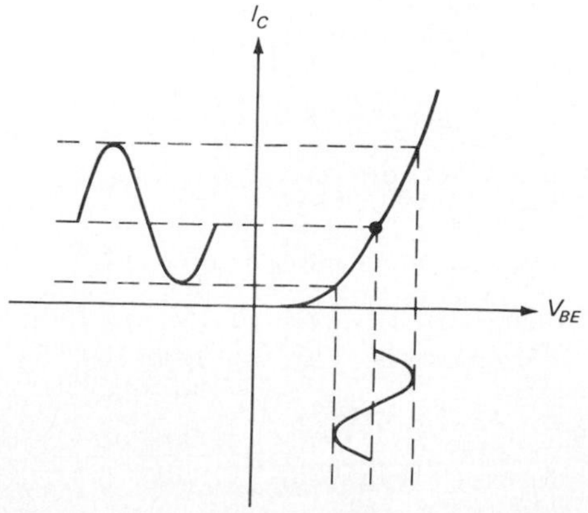

Mixer

A *mixer* is a nonlinear circuit with two input signals and one output signal. The output signal is a distorted combination of the two input signals. In Fig. 47-3, the input signals have frequencies of f_x and f_y. Because of the nonlinear distortion, the output signal contains the original input frequencies, plus their harmonics. For instance, if the two input frequencies are 100 and 101 kHz, the output signal contains these frequencies: 100 kHz, 101 kHz, 200 kHz, 202 kHz, 300 kHz, 303 kHz, . . . , and so on.

A detailed mathematical analysis reveals a startling property of a mixer. In addition to the harmonics that are produced, *new frequencies* appear in the output that are equal to

Fig. 47-1. Transconductance curve of transistor.

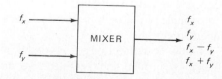

Fig. 47-3. A mixer produces the original frequencies, the sum, and the difference.

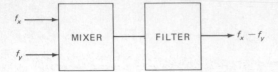

Fig. 47-4. Mixer output is filtered to get difference frequency only.

the *sum and difference* of the two input frequencies. If f_x and f_y are the input frequencies, the new frequencies are

$$\text{Sum} = f_x + f_y \qquad (47\text{-}1)$$

$$\text{Diff} = f_x - f_y \qquad (47\text{-}2)$$

As an example, if the two input frequencies are 100 and 101 kHz, the output of a mixer like Fig. 47-3 contains 100 and 101 kHz, their harmonics, and new frequencies given by

$$\text{Sum} = 101 \text{ kHz} + 100 \text{ kHz} = 201 \text{ kHz}$$
$$\text{Diff} = 101 \text{ kHz} - 100 \text{ kHz} = 1 \text{ kHz}$$

Usually, the output of a mixer is filtered as shown in Fig. 47-4 to allow only the difference frequency to reach the final output. For instance, if $f_x = 1455$ kHz and $f_y = 1000$ kHz, then only the difference frequency

$$\text{Diff} = 1455 \text{ kHz} - 1000 \text{ kHz} = 455 \text{ kHz}$$

reaches the final output of Fig. 47-4.

Figure 47-5 is an example of a transistor mixer. One signal drives the base; the other drives the emitter. One of the input signals is large; this is necessary to ensure nonlinear operation. The other input signal is usually small. One of the reasons this signal is small is because it often is a weak signal coming from an antenna.

The collector tank circuit is usually tuned to the difference frequency. Because of this, the two original frequencies, their harmonics, and the sum frequency are filtered out of the output. Only the difference frequency reaches the final output.

Field-effect transistors (FETs) are also used in mixers. Figure 47-6a shows a junction field-effect transistor (JFET) mixer, and Fig. 47-6b is a dual-gate MOSFET mixer. In each circuit, the output tank circuit is tuned to the difference frequency. This suppresses the two original frequencies, their harmonics, and the sum frequency, so that only the difference frequency appears at the final output.

Amplitude Modulation

Radio, television, and many other electronic systems would be impossible without *modulation*; it refers to a low-frequency signal controlling the amplitude, frequency, or phase of a high-frequency signal. When the low-frequency signal controls the amplitude of the high-frequency signal, we get amplitude modulation (AM).

Figure 47-7a shows a simple modulator. A high-frequency signal v_x is the input to a potentiometer; therefore, the amplitude of the output signal depends on the position of the wiper. If we move the wiper up and down sinusoidally, we get the

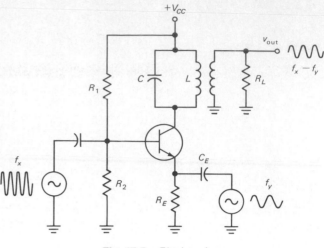

Fig. 47-5. Bipolar mixer.

AM waveform of Fig. 47-7b; the amplitude of the high-frequency signal is varying at a low-frequency rate.

The high-frequency signal is called the *carrier*, and the low-frequency signal is the *modulating signal*. Hundreds of carrier cycles normally occur during one cycle of the modulating signal. For this reason, an AM waveform on an oscilloscope looks like the signal of Fig. 47-7c; the positive peaks of the carrier are so closely spaced they form a solid upper boundary known as the *upper envelope*; similarly, the negative peaks form the *lower envelope*.

Figure 47-8 is an example of a transistor modulator. The carrier signal v_x is the input to a common-emitter amplifier. The circuit amplifies the carrier by a factor of A, so that the output is Av_x. The modulating signal is part of the biasing; therefore, it produces low-frequency variations in emitter current. As shown elsewhere, the voltage gain is proportional to emitter current. For this reason, the amplified carrier signal looks like the AM waveform shown; the peaks of the output signal vary sinusoidally with the modulating signal. Stated another way, the upper and lower envelopes have the shape of the modulating signal.

Percent Modulation

In Fig. 47-9a, the AM waveform has a maximum peak-to-peak value of $2V_{\max}$ and a minimum peak-to-peak value of $2V_{\min}$. The modulation coefficient is given by

$$m = \frac{2V_{\max} - 2V_{\min}}{2V_{\max} + 2V_{\min}} \qquad (47\text{-}3)$$

and the percent modulation by

$$\text{Percent modulation} = m \times 100\% \qquad (47\text{-}4)$$

For instance, suppose we see an AM waveform like Fig. 47-9b on an oscilloscope. This waveform has

$$m = \frac{16 - 4}{16 + 4} = 0.6$$

and

$$\text{Percent modulation} = 60\%$$

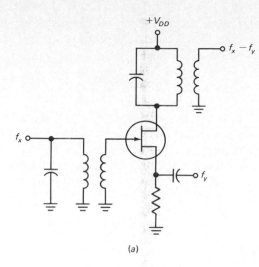

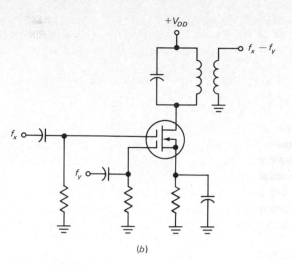

(a)   (b)

Fig. 47-6. (a) JFET mixer; (b) MOSFET mixer.

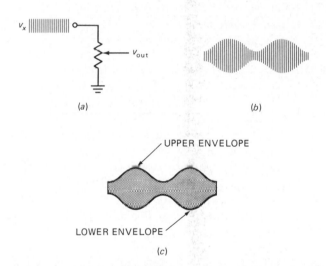

(a)   (b)

UPPER ENVELOPE

LOWER ENVELOPE

(c)

Fig. 47-7. (a) Potentiometer can produce AM; (b) AM signal; (c) AM envelopes.

Side Frequencies

A modulator like Fig. 47-8 is a nonlinear circuit. Because of this, we get sum and difference frequencies, similar to a mixer. As an example, suppose the modulating frequency is 1 kHz and the carrier is 1 MHz. Then,

$$\text{Sum} = 1.001 \text{ MHz}$$

$$\text{Diff} = 999 \text{ kHz}$$

The new frequencies are called *side frequencies*. The sum is the upper side frequency, and the difference is the lower side frequency.

Like a mixer, a modulator produces the two original frequencies, the sum, and the difference frequency. Unlike a mixer, however, the final output of Fig. 47-8 contains the carrier and the two side frequencies.

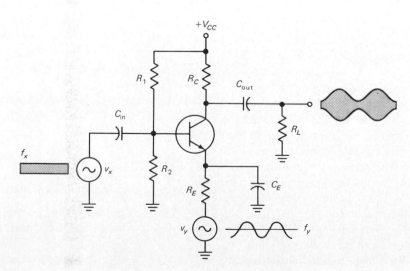

Fig. 47-8. Bipolar amplitude modulator.

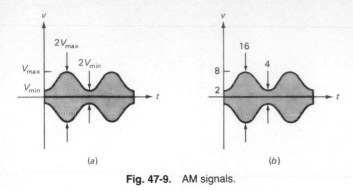

Fig. 47-9. AM signals.

Radio Frequencies

For an antenna to be efficient, its minimum length should be

$$L = \frac{7.5(10^7)}{f} \qquad (47\text{-}5)$$

where length L is in meters and frequency f in hertz. If an antenna is shorter than this, it will not radiate signals efficiently.

Antennas would be immense at audio frequencies (20 to 20,000 Hz). For instance, to radiate 1000 Hz efficiently, we would need an antenna length of

$$L = \frac{7.5 \times 10^7}{1000} = 7.5 \times 10^4 \text{ meters (m)} = 47 \text{ mi}$$

This is much too long. For this reason, it is impractical to transmit audio frequencies directly into space. Instead, communication systems transmit radio frequencies (RF); these are frequencies greater than 20 kHz. With RF signals, reasonable antenna lengths are possible.

Envelope Detector

AM broadcast signals use carrier frequencies between 540 and 1600 kHz. In the studio an audio signal modulates the carrier to produce an AM signal. A transmitting antenna of suitable length then radiates this AM signal into space. Miles away, a receiving antenna picks up the modulated RF signal. After being amplified, this signal is *demodulated* (the audio is recovered).

Figure 47-10a shows one type of demodulator. Basically, it is a peak detector (Experiment 34). Ideally, the peaks of the input signal are detected to recover the upper envelope. For this reason, the circuit is called an *envelope detector*.

During each carrier cycle, the diode turns on briefly and charges the capacitor to the peak voltage of the particular carrier cycle. Between peaks, the capacitor discharges through the resistor. By making the RC time constant much greater than the period of the carrier, we get only a slight discharge between cycles. In this way, most of the carrier signal is removed. The output then looks like the upper envelope with a small ripple as shown in Fig. 47-10b.

But here is a crucial idea. Between points A and C in Fig. 47-10b, each carrier peak is smaller than the preceding one. If the RC time constant is too long, the circuit cannot detect the next carrier peak (see Fig. 47-10c). The hardest part of the

envelope to follow occurs at B in Fig. 47-10b: at this point, the envelope is decreasing at its fastest rate. With advanced mathematics, we can derive this formula for the cutoff frequency of the envelope detector:

$$f_{y(\text{max})} = \frac{1}{2\pi RCm} \qquad (47\text{-}6)$$

where m is the modulation coefficient. If the envelope frequency is greater than $f_{y(\text{max})}$, the detected output drops 20 dB/decade.

By adding a low-pass filter to the output of the Fig. 47-10a, we can remove the small RF ripple that remains on the detected signal (Fig. 47-10b).

SUMMARY

1. The I_C versus V_{BE} graph of a bipolar transistor is nonlinear.
2. A sinusoidal V_{BE} signal produces a nonsinusoidal collector current. The change in signal shape is called *nonlinear distortion*.
3. Nonlinear distortion produces *harmonics f, 2f, 3f, . . . , nf.*
4. A *mixer* is a nonlinear circuit with two input signals. The output of the mixer contains the original frequencies, their harmonics, the sum frequency, and the difference frequency.
5. Usually, the mixer output is filtered so that only the difference frequency reaches the final output.
6. One of the mixer input signals must be large enough to produce nonlinear distortion. The other signal, typically from an antenna, is small.
7. *Modulation* refers to a low-frequency signal controlling the amplitude, frequency, or phase of a high-frequency signal.
8. The high-frequency signal is called the *carrier,* and the low-frequency signal is the *modulating signal.*
9. To produce an AM signal the modulating signal controls the gain of an RF amplifier.
10. An AM signal contains a carrier and two *side frequencies.* The upper side frequency equals the sum of the carrier and modulating frequencies; the lower side frequency equals the difference.
11. After an AM signal is received, it is amplified and demodulated.
12. One way to *demodulate* an AM signal is with an *envelope detector.*

SELF-TEST

Check your understanding by answering these questions.

1. If a sine wave with a frequency of 5 kHz drives a nonlinear circuit, what are the frequencies appearing in the output?
2. A mixer has these input frequencies: $f_x = 32$ MHz and $f_y = 28$ MHz. What do the sum and difference frequencies equal?

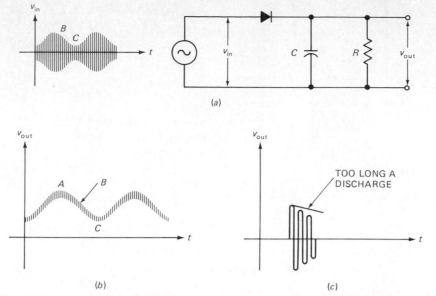

Fig. 47-10. *(a)* Envelope detector; *(b)* envelope with small carrier ripple; *(c)* envelope frequency too high.

3. If the mixer output is filtered in question 2, what is the usual frequency at the output of the filter?

4. In an AM radio the weak signal to a mixer has a frequency between 540 and 1600 kHz. The strong signal comes from a local oscillator (LO), a circuit inside the radio. The LO signal has a higher frequency than the weak signal. If the difference frequency always equals 455 kHz, what is the frequency of the local oscillator when the other frequency is 540 kHz? When it is 1600 kHz?

5. The _____ signal of Fig. 47-8 varies the voltage gain of the circuit. This varies the amplitude of the

_____ .

6. An AM signal has a maximum peak-to-peak voltage of 10 V and a minimum peak-to-peak voltage of 0 V. What is the percent modulation?

7. The modulating signal has a frequency between 20 Hz and 20 kHz. The carrier has a frequency of 1080 kHz. What are the lower side frequencies for 20 Hz? For 20 kHz?

8. What is the minimum length for the transmitting antenna of a radio station whose frequency is 1080 kHz?

9. An envelope detector has $R = 10$ kΩ, $C = 1000$ pF, and $m = 0.5$. What does $f_{y(\max)}$ equal?

MATERIALS REQUIRED

- Two power supplies: 15-V
- Equipment: Oscilloscope, two signal generators (100 to 500 kHz), one audio generator
- Resistors: Two 1-kΩ, one 4.7-kΩ, four 10-kΩ, one 22-kΩ, four 100-kΩ ½-W
- Capacitors: Three 0.001-μF, one 0.01-μF, two 0.1-μF
- Diode: 1N914
- Transistors: Two 2N3904 (other choice: 2N6005)
- Op amp: 741C
- Speaker: 3.2-Ω

PROCEDURE

Mixer

1. If $f_x = 101$ kHz and $f_y = 100$ kHz in Fig. 47-11*a*, what does the difference frequency equal? Record answer here:

$$f_x - f_y = \underline{\hspace{2cm}}$$

2. The mixer output of Fig. 47-11*a* is filtered by two low-pass *RC* circuits. The approximate cutoff frequency of each is given by

$$f_c = \frac{1}{2\pi RC}$$

Calculate f_c and record the answer here:

$$f_c = \underline{\hspace{2cm}}$$

3. Connect the circuit of Fig. 47-11*a*.

4. Turn v_y down to 0. With the oscilloscope, adjust v_x to 0.1 V p-p. Set the frequency to 101 kHz.

5. Next adjust v_y to 1 V p-p and 100 kHz.

6. Look at the final output signal with a vertical sensitivity of 0.1 V/cm (ac input) and a sweep time of 0.2 ms/cm. Vary the frequency of the v_x generator slowly in the vicinity of 101 kHz until you get a 1-kHz output signal.

7. Look at point *B*, the input to the final *RC* filter. Notice the ripple on the 1-kHz signal.

8. Look at point *A*, the input to the first *RC* filter. Use a vertical sensitivity of 2 V/cm. Notice how large the ripple is here.

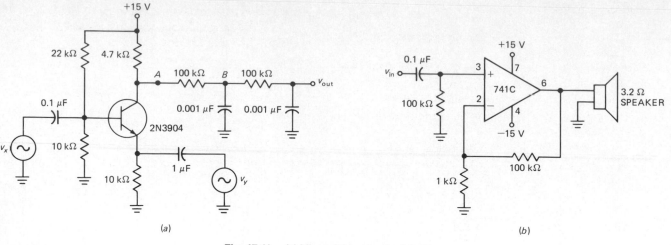

(a)

(b)

Fig. 47-11. *(a)* Mixer; *(b)* beat-note detector.

Beat Note

9. Connect the filtered mixer output (Fig. 47-11*a*) to the input of the op amp circuit (Fig. 47-11*b*).

10. Vary the frequency of either ac generator and notice what happens to the beat note (difference frequency) out of the speaker.

Amplitude Modulator

11. In Fig. 47-12*a*, what are the side frequencies? Record your answers here:

Upper side frequency = _____

Lower side frequency = _____

12. Connect the circuit of Fig. 47-12*a*.

13. Set the audio generator to 200 Hz and the RF generator to 500 kHz.

14. Turn the audio generator down to 0 (do not disconnect). Adjust the RF generator to get a final output v_{out} of 0.3 V p-p.

15. Use a sweep speed of 1 ms/cm. Turn up the audio signal and you will see amplitude modulation.

16. Increase and decrease the audio level and notice how the percent modulation changes.

Envelope Detector

17. In Fig. 47-12*b*, work out the highest frequency the envelope detector can follow without attenuation for a modulation of 100 percent. Record it here:

$$f_{y(max)} = \underline{\hspace{3cm}}$$

18. Connect the output of the amplitude modulator (Fig. 47-12*a*) to the circuit of Fig. 47-12*b*.

19. Use the oscilloscope to look at the input to the envelope detector (across the lower 1-kΩ resistor of Fig. 47-12*b*). Adjust the modulation to 100 percent.

20. Look at the output of the envelope detector. You should have an audio signal.

21. Vary the frequency of the audio generator and notice what happens to the output of the envelope detector.

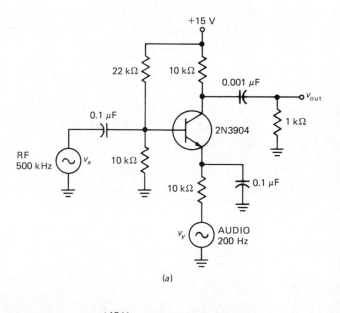

(a)

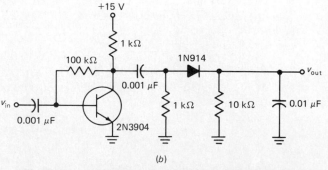

(b)

Fig. 47-12. *(a)* Amplitude modulator; *(b)* envelope detector.

QUESTIONS

1. What did you calculate for the approximate cutoff frequency in step 2?
2. In Fig. 47-11a, why do you not see the original frequencies (100 and 101 kHz) at the output?
3. Explain why the beat note from the speaker of Fig. 47-11b changed frequency when you varied one of the generators.
4. If the beat note out of the speaker is 0 in Fig. 47-11b, what can you say about the two input frequencies in Fig. 47-11a?
5. What were the side frequencies recorded in step 11?
6. What is the frequency of the envelope in Fig. 47-12a?
7. What was the frequency recorded in step 17?

8. Explain why the output of the envelope detector (Fig. 47-12b) changed frequency when the audio frequency of Fig. 47-12a was varied.

Answers to Self-Test

1. 5 kHz, 10 kHz, 15 kHz, . . ., n(5 kHz)
2. 60 MHz; 4 MHz
3. 4 MHz
4. 995 kHz; 2055 kHz
5. modulating (or low frequency); carrier (or high frequency)
6. 100 percent
7. 1.07998 and 1.08002 MHz for 20 Hz; 1.06 and 1.1 MHz for 20 kHz
8. 69.4 meters
9. 31.8 kHz

PHASE-LOCKED LOOP

OBJECTIVES

1. To adjust the center frequency of a phase-locked loop
2. To observe the lock range
3. To measure the maximum and minimum capture frequencies
4. To measure the dc voltage from the FM output

INTRODUCTORY INFORMATION

Phase Detector

Suppose we have a mixer with input frequencies of 50 Hz and 50 kHz. Then the difference frequency is 0, which represents dc. In other words, a dc voltage comes out of a mixer when the input frequencies are equal.

A *phase detector* is a mixer optimized for use with equal input frequencies. It is called a phase detector (or phase comparator) because the amount of dc voltage depends on the phase angle ϕ between the input signals. As the phase angle changes, so too does the dc voltage.

Figure 48-1a illustrates the phase angle between two sinusoidal signals. When these signals drive the phase detector of Fig. 48-1b, a dc voltage comes out. One type of phase detector has a dc output voltage that varies as shown in Fig. 48-1c. When the phase angle ϕ is 0, the dc voltage is maximum. As the phase angle increases from 0 to 180°, the dc voltage decreases to a minimum value. When ϕ is 90°, the dc output is the average of the maximum and minimum outputs.

For example, suppose a phase detector has a maximum output of 10 V and a minimum output of 5 V. When the two inputs are in phase, the dc output is 10 V. When the inputs are 90° out of phase, the dc output is 7.5 V. When the inputs are 180° out of phase, the dc output is 5 V. The key idea is the dc output decreases when the phase angle increases.

VCO

Experiment 46 showed how a 555 timer could be operated as a voltage-controlled oscillator (VCO) by applying a dc voltage to the control input. Recall the idea. When the dc voltage in Fig. 48-2a increases, the frequency of the output signal decreases. In other words, a dc voltage controls the oscillator frequency. Typically, the frequency decreases linearly with an increase in dc voltage (see Fig. 48-2b).

Many other designs are possible for VCOs. For example, one approach uses an *LC* oscillator (Experiment 38) with a varactor (voltage-controlled capacitor). By varying the dc voltage applied to the varactor, we can change the capacitance and control the resonant frequency.

The important thing to remember about any VCO is this: An input dc voltage controls the output frequency. In this experiment, an increase in dc control voltage causes the VCO frequency to decrease.

Phase-Locked Loop

Figure 48-3 is the block diagram of a *phase-locked loop* (PLL). An input signal with a frequency of f_x is one of the inputs to a phase detector. The other input comes from a VCO. The output of the phase detector is filtered by a low-pass filter. This removes the original frequencies, their harmonics, and the sum frequency. Only the difference fre-

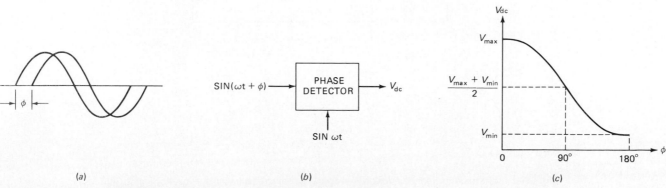

Fig. 48-1. *(a)* Phase angle between signals; *(b)* phase detector; *(c)* output of phase detector.

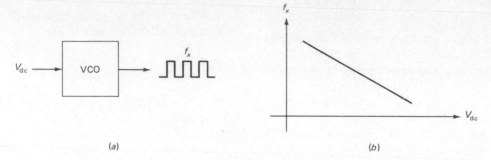

(a)   (b)

Fig. 48-2. *(a)* DC input controls VCO output. *(b)* VCO frequency inversely proportional to dc input.

quency (dc voltage) comes out of the low-pass filter. This dc voltage then controls the frequency of the VCO.

This feedback system locks the VCO frequency on to the input frequency. When the system is working correctly, the VCO frequency equals f_x, the same as the input signal. Therefore, the phase detector has two inputs with equal frequency; the phase angle between these inputs determines the amount of dc output. Figure 48-3b shows the phasors for the input signal and the VCO.

If the input frequency changes, the VCO frequency will track it. For instance, if the input frequency f_x increases slightly, its phasor rotates faster and the phase angle increases as shown in Fig. 48-3c. This means less dc voltage will come out of the phase detector. The lower dc voltage forces the VCO frequency to increase until it equals f_x.

On the other hand, if the input frequency decreases, its phasor slows down and the phase angle decreases as shown in Fig. 48-3d. Now more dc voltage will come out of the phase detector; this causes the VCO frequency to decrease until it equals the input frequency. In other words, the PLL automatically corrects the VCO frequency and phase angle.

Here is a numerical example. Suppose the VCO is locked on to an input frequency of 50 kHz. If the input frequency increases to 51 kHz, the phase detector immediately sends less voltage to the VCO and increases its frequency to 51 kHz. If the input frequency later decreases to 49 kHz, the

phase detector sends more dc voltage to the VCO and decreases its frequency to 49 kHz. In either case, the feedback automatically adjusts the phase angle to produce a dc voltage that locks the VCO frequency on to the input frequency.

The *lock range B_L* is the range of frequencies the VCO can produce, given by

$$B_L = f_{max} - f_{min} \qquad (48\text{-}1)$$

where f_{max} and f_{min} are the maximum and minimum VCO frequencies. For example, if the VCO frequency can vary from 40 to 60 kHz, the lock range equals

$$B_L = 60 \text{ kHz} - 40 \text{ kHz} = 20 \text{ kHz}$$

Once the PLL is locked on, the input frequency f_x can vary from 40 to 60 kHz; the VCO will track this input frequency and the locked output will equal f_x.

Free-Running Mode

Recall the astable 555 timer with no control voltage. It oscillated at a natural frequency determined by the circuit components. The same is true of the VCO in Fig. 48-3a. If the input signal is disconnected, the VCO oscillates in a free-running mode with its frequency determined by its circuit elements.

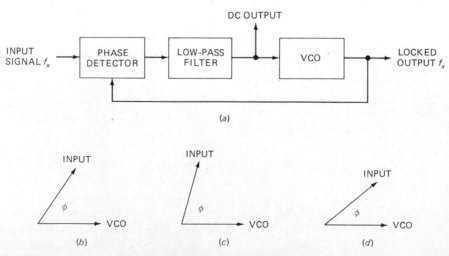

(a)

(b)   (c)   (d)

Fig. 48-3. *(a)* Phase-locked loop; *(b)* phasor diagram; *(c)* increase in input frequency increases phase angle; *(d)* decrease in input frequency decreases phase angle.

Capture and Lock

Assume the PLL is free-running or unlocked. The PLL can lock on to the input frequency if it lies within the *capture range*, a band of frequencies centered on the free-running frequency. The formula for capture range is

$$B_C = f_2 - f_1 \qquad (48\text{-}2)$$

where f_2 and f_1 are the highest and lowest frequencies the PLL can lock into. The capture range is always less than or equal to the lock range and is related to the cutoff frequency of the low-pass filter. The lower the cutoff frequency, the smaller the capture range.

Here is an example. Suppose the PLL can initially lock into a frequency as high as 52 kHz or as low as 48 kHz. Then the capture range is 4 kHz, with a center frequency of 50 kHz. If the lock range is 20 kHz and lock has been acquired, then the input frequency can vary gradually from 40 to 60 kHz without losing lock.

Locked Output

One use for the locked output f_x of a PLL is to synchronize the horizontal and vertical oscillators of TV receivers to the incoming sync pulses. PLLs can also automatically tune each TV channel by locking on to the channel frequency. Still another use for PLLs is locking on to weak signals from satellites and other distant sources; this improves the signal-to-noise ratio.

In general, the locked output is a signal with the same frequency as the input signal. Even though the input signal may drift over a rather large frequency range, the output frequency will remain locked on; this eliminates the need to tune a resonant circuit for maximum output.

FM Output

Figure 48-4a shows an *LC* oscillator with a variable tuning capacitor. If the capacitance is varied, the oscillation frequency changes. Figure 48-4b illustrates the output signal. This is an example of *frequency modulation* (FM). If the capacitance of Fig. 48-4a varies sinusoidally at a rate of 1 kHz, the modulation frequency is 1 kHz.

When an FM signal like Fig. 48-4b is the input to a PLL like Fig. 48-3a, the VCO will track the input frequency as it changes. As a result, a fluctuating voltage comes out of the low-pass filter. This voltage has the same frequency as the modulating signal. In other words, the dc output now represents a demodulated FM output. This is useful in FM receiv-

ers. If the modulating signal is music, the signal out of the FM output will be the same music.

The 565

The NE565 from Signetics is a 14-pin IC that can be connected to external components to form a PLL. Figure 48-5 shows a simplified block diagram. Pins 2 and 3 are a differential input to the phase detector. If a single-ended input is preferred, pin 3 is grounded and the input signal is applied to pin 2. Pins 4 and 5 are usually connected together. In this way, the VCO output becomes an input to the phase detector. In those applications where the locked output is desired, pin 4 is the output pin.

An external timing resistor is connected to pin 8, and an external timing capacitor to pin 9. These two components determine the free-running frequency of the VCO, given by

$$f = \frac{0.3}{R_T C_T} \qquad (48\text{-}3)$$

R_T and C_T are selected to produce a free-running VCO frequency at the center of the input frequency range. If you want to lock on to an input frequency between 40 and 60 kHz, you choose R_T and C_T to produce a free-running VCO frequency of 50 kHz.

Pin 7 is the FM output, used only when an FM signal is driving the phase detector. In FM receivers, a demodulated signal comes out of this pin. This signal then goes to other amplifiers and eventually comes out the loudspeaker.

Notice the filter capacitor C_F between pin 7 and ground. This capacitor and the internal 3.6-kΩ resistor form a low-pass *RC* filter to remove the original frequencies, their harmonics, and the sum frequency. The cutoff frequency of this filter is given by

$$f = \frac{1}{2\pi R_F C_F} \qquad (48\text{-}4)$$

The lower the cutoff frequency of this filter, the smaller the capture range. In some applications, the filter capacitor is omitted and the capture range equals the lock range.

SUMMARY

1. When the input frequencies to a mixer are equal, the output is a dc voltage.
2. Given equal frequency inputs, the output of a *phase detector* is a dc voltage that depends on the phase angle between the input signals.
3. The control voltage to a VCO determines the output frequency.
4. The low-pass filter of a *phase-locked loop* removes the original frequencies, the harmonics, and the sum frequency. This allows the difference frequency, usually a dc voltage, to control the VCO.
5. The phase angle between the input signal and the VCO signal determines the dc voltage out of the phase detector.
6. When a PLL is locked on, the VCO frequency equals the input frequency.

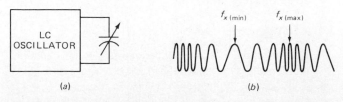

$f_{x\,(min)}$ $f_{x\,(max)}$

(a) *(b)*

Fig. 48-4. Frequency modulation. *(a)* Varying capacitor varies frequency; *(b)* FM signal.

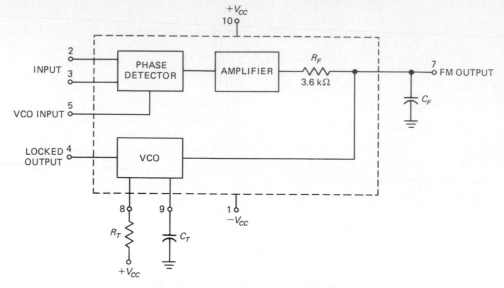

Fig. 48-5. Block diagram of the 565 PLL.

7. The *lock range* is the band of frequencies the VCO can generate.

8. The *capture range,* always less than or equal to lock range, is the band of frequencies that the PLL can lock on when lock does not yet exist.

9. When the input signal is an FM signal, the PLL produces a demodulated audio signal at its FM output.

SELF-TEST

Check your understanding by answering these questions.

1. In the phase detector discussed earlier, a decrease in phase angle produces an _____ in dc output voltage.

2. In the VCO discussed earlier, an increase in dc control voltage produces a _____ in frequency.

3. If the maximum and minimum VCO frequencies are 450 and 350 kHz, the lock range is _____ .

4. A PLL can acquire initial lock for a maximum input frequency of 415 kHz and a minimum input frequency of 385 kHz. The capture range equals _____ .

5. A 565 has a timing resistor of 10 kΩ and a timing capacitor of 0.01 μF. The free-running frequency is _____ .

6. The low-pass filter used with a 565 has an external capacitor of 0.01 μF. The cutoff frequency of the filter is _____ .

MATERIALS REQUIRED

- Two power supplies: 9-V
- Equipment: Oscilloscope, audio generator, electronic counter, voltmeter (digital if available)
- Resistors: Two 1-kΩ, one 4.7-kΩ ½-W
- Potentiometer: 5-kΩ
- Capacitors: Two 0.01-μF, one 0.47-μF, two 0.1-μF, one 1-μF
- PLL: NE565

PROCEDURE

Free-Running Frequency

1. If the ac signal generator of Fig. 48-6 is disconnected, the 565 is in the free-running mode of operation. What is the VCO frequency if the wiper of the potentiometer is all the way up? All the way down? Record the calculated frequencies in Table 48-1.

2. Connect the circuit of Fig. 48-6 with a C_F of 0.2 μF (use two 0.1-μF capacitors in parallel).

3. Temporarily disconnect the ac signal generator to get free-running operation.

4. With the oscilloscope look at the VCO output (pin 4). Use a vertical sensitivity of 5 V/cm and a time base of 0.1 ms/cm. Vary the potentiometer and notice how the frequency changes.

5. Measure the minimum and maximum free-running frequencies with an electronic counter. (If not available, use $f = 1/T$ where T is the period seen on the oscilloscope.) Record the measured frequencies in Table 48-1.

6. Adjust the potentiometer to get a center frequency of approximately 5 kHz.

Lock Range

7. Reconnect the ac signal generator. Look at the signal on pin 2. Adjust the generator to get 0.5 V p-p at 5 kHz.

8. Look at the output (pin 4). Vary the frequency of the generator from 4 to 6 kHz. Notice how the PLL output is locked into the input signal.

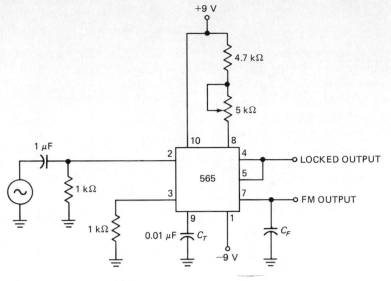

Fig. 48-6. Experimental PLL circuit.

TABLE 48-1.   Free-Running Operation

Wiper	f_{calc}, kHz	f_{meas}, kHz
Up		
Down		

9. Measure maximum and minimum lock frequencies. Record these values in Table 48-2.

Capture Range

10. Calculate the cutoff frequencies of the low-pass filter for the values of C_F given in Table 48-3. Record the results.

11. Measure the maximum and minimum capture frequencies. Record in Table 48-3. Calculate and record the capture range, B_C.

TABLE 48-2.   Lock Range

f_{max}, kHz	f_{min}, kHz

12. Repeat step 11 for the other values of C_F shown in Table 48-3.

FM Output

13. Measure the dc voltage (digital voltmeter if available) at the FM output (pin 7) for an input frequency of 4 kHz. Record the dc voltage in Table 48-4.

14. Repeat step 13 for the remaining frequencies in Table 48-4.

Extra Credit

15. Figure out how you can get a free-running frequency of 50 kHz. Modify the circuit as needed. Then test it by locking the PLL on an input signal.

TABLE 48-4.   FM Output

Frequency, kHz	V_{dc}, V
4	
5	
6	

TABLE 48-3.   Capture Range

C_F, μF	f_{cutoff}, kHz	f_{max}, kHz	f_{min}, kHz	B_C, kHz
0.2				
0.1				
0.047				

QUESTIONS

1. What did you calculate for the minimum and maximum free-running frequencies in step 1?
2. Explain briefly how a PLL works.
3. Use the data of Table 48-2 to work out the lock range.
4. What is the capture range for $C_F = 0.2 \ \mu F$?
5. What effect does the filter capacitor have on the capture range?
6. The data of Table 48-4 proves a PLL can produce a demodulated FM output. Explain why this is so.

Extra Credit

7. What change did you make to get a free-running frequency of 50 kHz?

Answers to Self-Test

1. increase	4. 30 kHz
2. decrease	5. 3 kHz
3. 100 kHz	6. 4.42 kHz

THE SILICON CONTROLLED RECTIFIER (SCR)

OBJECTIVES

1. To observe how gate current is used to switch an SCR on and off
2. To verify that gate current levels may be used to control collector load current levels

INTRODUCTORY INFORMATION

The silicon controlled rectifier is a four-layer NPNP solid-state rectifier which has three electrodes: an anode, a cathode, and a gate, which serves as a control element. The SCR differs from the two-element-diode rectifier in that it will not pass any appreciable current, though forward-biased, until the anode voltage equals or exceeds a value called the *forward breakover voltage*, V_{BRF}. When V_{BRF} is reached, the SCR is switched on, that is, it becomes highly conductive. The value of V_{BRF} can be controlled by the level of gate current. The gate then provides a new dimension in rectifier operation in which low levels of gate current control high levels of anode or load current.

Various-capacity SCRs are manufactured today. Low-current SCRs can provide anode currents less than 1 A. High-current silicon controlled rectifiers pass anode or load currents in the hundreds of amperes. The low-current SCR resembles a transistor in physical appearance, with three leads issuing from a hermetically sealed housing. The high-current SCR looks like a stud-mounted silicon rectifier with an added terminal (Fig. 49-1). The circuit symbol for an SCR and its four-layer representation are shown in Fig. 49-2.

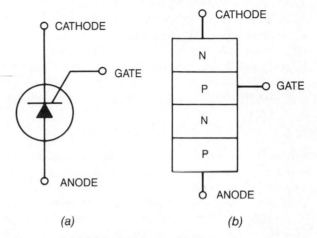

Fig. 49-2. *(a)* SCR circuit symbol; *(b)* four-layer representation.

Voltage-Current Characteristic

Figure 49-3 is the voltage-current characteristic of an SCR whose gate is not connected (open). When the anode-cathode circuit is reverse-biased, there is a slight reverse-leakage current called the *reverse blocking current*. This current remains small until the *peak reverse voltage* V_{ROM} is exceeded. At that point the reverse avalanche region begins, and the leakage current increases sharply.

When the SCR is forward-biased, there is a small forward-leakage current, called the *forward blocking current*, which remains small until the forward breakover voltage is reached.

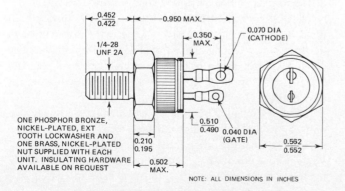

Fig. 49-1. A stud-mounted SCR, GE-type C20.

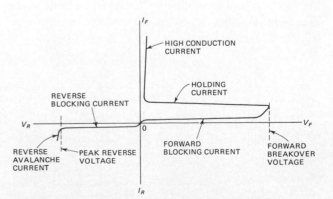

Fig. 49-3. Voltage-current characteristic of an SCR whose gate is open.

This is the forward avalanche region. At that point the current jumps rapidly to the high-conduction level. Here the anode-to-cathode resistance of the SCR becomes very small, and the SCR acts like a closed switch. In this high forward-conduction region, the voltage across the SCR drops to a very low value, and to all intents and purposes almost the entire source voltage appears across the load which is in series with the rectifier (Fig. 49-4). It is the external load resistance, then, which must limit the current through the SCR and hold it within the rated value of the SCR.

The two operating stages of the silicon controlled rectifier correspond to the two states of an on-off switch. When the applied voltage is below the breakover point, the switch is OFF. When the voltage increases to a value equal to or greater than the breakover voltage, the rectifier is turned on. The rectifier remains ON, that is, in its high-current state, as long as the current stays above a certain value called the "holding" current. When the voltage across the rectifier drops to a value too low to sustain the holding current, the rectifier turns off.

Gate Control of Forward Breakover Voltage

When the gate-cathode junction is forward-biased, the rectifier is turned on at a lower anode-voltage level than with the gate open. That is, the value of forward breakover voltage is reduced with forward bias. This is evident from the characteristic curves in Fig. 49-5. The maximum forward breakover voltage V_{BRF0} occurs when the gate is open; that is, when gate current $I_{G0} = 0$. When there is gate current at the level I_{G1}, the forward breakover voltage V_{BRF1} is lower than V_{BRF0}. Similarly V_{BRF2}, which is determined by I_{G2} which is greater than I_{G1}, is lower than V_{BRF1}, and V_{BRF3} is lower than V_{BRF2}.

As the gate current is increased, the SCR begins to act like an ordinary silicon rectifier.

After the SCR has been turned on by gate current, the gate loses control, and reducing gate current has no effect on anode current. The SCR remains on until anode forward voltage has been removed or until the anode forward-voltage level is below that required to maintain the *holding current*. When the anode source is alternating current, the rectifier turns off during the negative alternation, when the anode-cathode is reverse-biased.

Since they are controlled by forward gate bias, silicon controlled rectifiers are operated below the maximum forward breakover point.

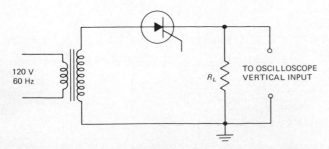

Fig. 49-4. AC source driving an SCR and load, gate open.

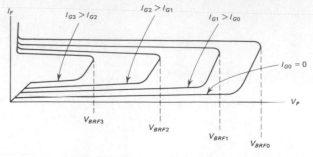

Fig. 49-5. Effect of gate current on forward voltage breakover level.

The gate circuit is rated as to its power dissipation and its forward and reverse gate-voltage capabilities.

SCR Ratings

To ensure trouble-free operation, the manufacturer's maximum ratings should not be exceeded. Permanent damage to the rectifier may result if operation beyond these ratings is attempted. SCR ratings include:

PFV	repetitive peak forward voltage, gate open
I_F	rms forward current, ON state
$I_{F_{av}}$	average forward current, ON state
V_{RO_M}	peak reverse voltage, gate open
P_{G_M}	peak gate power dissipation
$P_{G_{av}}$	average gate power dissipation
V_{GR_M}	peak reverse gate voltage
T_{stg}	storage temperature
T_j	operating junction temperature

SCR Used As a Rectifier

Silicon controlled rectifiers are very useful in ac circuits where they may serve as rectifiers whose output current can be controlled by controlling their gate current. An example of this type of application is the use of SCRs to operate and control dc motors or any dc load from an ac supply.

Refer to Fig. 49-4. In this circuit the SCR is connected as a half-wave rectifier supplying current to R_L. The gate is open. The SCR switches on when the forward breakover voltage is reached, at point V_{BRO} on the positive alternation of the applied sine wave. During the interval V_{BRO} through V_X the SCR conducts. When the anode voltage drops to V_X, which is below the holding current potential, the SCR turns off. It remains off during the negative alternation. Current through the load flows during the interval V_{BRO} through V_X. The applied voltage, anode-to-cathode voltage, load voltage, and current waveforms observed in the preceding circuit are shown in Fig. 49-6.

In Fig. 49-7 the voltage V_{BRO1} at which the SCR is turned on is controlled by the gate current, which in turn is determined by the setting of V_{GG}, the gate current source, and R_2. V_{BRO1} is lower than V_{BRO}. Therefore the period during which the load draws current V_{BRO1} through V_X is increased, and more current is delivered by the rectifier to the load. The

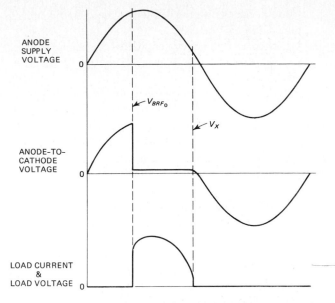

Fig. 49-6. Waveforms in SCR circuit.

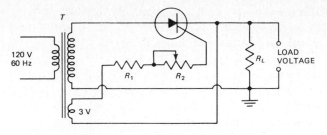

Fig. 49-8. SCR controlled by ac gate current.

resistance R_L were replaced by a light bulb, or by a bank of bulbs, the circuit of Fig. 49-8 could serve as a light dimmer.

NOTE: For a sizable load, full-wave rather than half-wave rectification would be employed. A full-wave controlled rectifier would require two or more SCRs.

CAUTION: *The peak reverse gate voltage must not be exceeded.*

conduction period can be further increased by increasing gate current beyond the level in the preceding example.

Note that gate current can be and is relatively low compared with anode current. In the circuit of Fig. 49-7, a dc milliammeter M_2 is used to measure the gate control current. The average load current through R_L may measure many amperes, depending on the amplitude of the applied voltage and on the size of R_L. However, the average anode current should not exceed the average anode-current rating of the rectifier.

Control of anode current by direct gate current was illustrated in Fig. 49-7. The circuit of Fig. 49-8 illustrates ac control, which is more frequently employed than dc control. In this circuit the anode-to-cathode voltage for the SCR is supplied by the high-voltage secondary of transformer T. The gate current is delivered by the low-voltage secondary winding. Rheostat R_2 is used to adjust the level of gate current, hence the conduction period of the SCR. If the load

SUMMARY

1. The silicon controlled rectifier (SCR) is a four-layer NPNP solid-state rectifier which contains an anode, a cathode, and a gate.
2. Unlike the diode rectifier, the SCR will not conduct, though its anode-to-cathode is forward-biased, until the anode voltage equals or exceeds a value called the *forward breakover voltage* (V_{BRF}).
3. The breakover voltage can be controlled by the value of gate current.
4. SCRs vary from low- to high-current devices.
5. The forward and reverse voltage-current characteristics of an SCR are shown in Fig. 49-3.
6. The forward breakover voltage is affected by the level of gate current. Figure 49-5 shows that as gate current is increased, V_{BRF} is decreased; that is, the forward anode-to-cathode voltage required to fire the SCR is reduced.
7. After an SCR has been turned on the gate loses control.

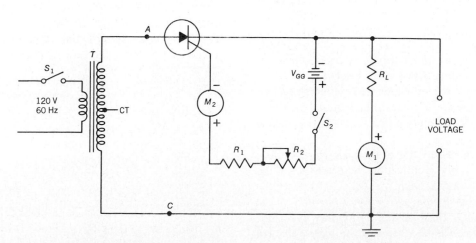

Fig. 49-7. Voltage at which SCR is turned on is controlled by dc gate current.

Increasing or decreasing gate current will then not affect anode current.

8. An SCR, therefore, is like an on-off switch. When it fires, it is ON. Otherwise it is OFF.

9. The SCR can be connected as a rectifier supplying unfiltered dc current to a load as in Fig. 49-7. Here dc gate current controls the firing point. Figure 49-6 shows the waveforms which will be seen in this circuit.

10. The firing point of an SCR can also be controlled by ac gate current, as in Fig. 49-8. If the load in this circuit were a light bulb, varying R_2, the gate current control would affect the brightness of the bulb. This circuit may therefore be used as a light-dimming circuit.

11. The SCR is used in electronics to supply rectified current to a load. The level of the rectified current is determined by the setting of gate current control and by the parameters of the circuit.

SELF-TEST

Check your understanding by answering these questions.

1. An SCR is a more versatile device than the diode silicon rectifier because anode current can be controlled by the level of _____ current.

2. The forward breakover voltage (V_{BRF}) of a particular SCR is 150 V, anode to cathode, when there is 1.5 mA of gate current. If the gate current is increased to 2.5 mA, the forward breakover voltage will become _____ (higher, lower) than 150 V.

3. An SCR may be used as an alternating-current _____ .

4. Once an SCR fires, it remains ON until the anode forward voltage has been _____ , or until the anode

forward-voltage level is below that required to maintain the holding current.

5. When the anode source of an SCR is alternating voltage, the rectifier turns _____ during the negative alternation.

6. In Fig. 49-6, the point on the anode-to-cathode voltage waveform at which the SCR fires is _____ . It remains conducting until the anode-to-cathode voltage _____ is reached, when it turns off.

7. The current waveform in Fig. 49-6 shows that load current remains constant during the time that the SCR is ON. _____ (true, false)

8. The gate current source of an SCR may be either _____ or _____ .

9. The forward-bias voltage of an SCR, anode to cathode, is dc. The level of gate current is adjusted to fire the SCR. Once it has fired the SCR may be turned off by reducing the gate current. _____ (true, false)

10. The holding current is the _____ (gate, anode) current below which the SCR turns off.

MATERIALS REQUIRED

- Power supply: Variable regulated 0–50-V dc source
- Equipment: DC oscilloscope; EVM; 0–10-mA dc milliammeter; 0–100-mA dc milliammeter or equivalent ranges on a VOM
- Resistors: 220-, 1000-, 5100-Ω ½-W; 500-Ω 5-W
- Semiconductors: 2N1596 SCR or equivalent, with heat sink; 1N5625 or equivalent
- Miscellaneous: Power transformer 120-V primary, 25-V 1-A secondary (Triad F-40X or equivalent); 2-W 5000-Ω potentiometer; two SPST switches

PROCEDURE

NOTE: The 2N1596 low-current SCR is rated at 1.6 A rms anode current, peak positive gate current 100 mA, peak gate voltage ±6 V. These ratings should not be exceeded.

The gate current required to turn on the SCR may vary widely. Moreover, the dc gate currents required to trigger the 2N1596 over the range of anode supply voltages which this SCR can handle may be grouped very closely together. The first part of this experiment will therefore be exploratory. The student will determine the dc gate current required to turn on the SCR when the anode supply is first set at 15 V dc, then at 40 V dc.

After the firing characteristics of the SCR are observed, the student will study the rectification characteristics of the 2N1596, using an ac anode supply.

The experimental techniques are carefully set forth in the following instructions. These should be followed precisely. Note that when using an anode dc supply, the only way to turn off the SCR after it has fired is to turn off the anode supply

voltage, or to reduce the anode current below the holding current level.

DC Anode Voltage, DC Gate Current

1. Connect the circuit of Fig. 49-9. Switches S_1 and S_2 are both open. V_{AA} is a variable regulated power source.

2. Set V_{AA} at 15 V as measured across the power supply with S_1 open. Do not vary V_{AA} again until instructed to do so. M_1 measures anode current after the SCR turns on; M_2 measures gate current. V_1 measures anode-to-cathode voltage. When the SCR has fired, the voltage across the SCR will drop to a very low value (about 0.1 to 3 V) and the anode current I_A will be limited by V_{AA} and the size of R_L. For with the SCR ON

$$I_A = \frac{V_{AA}}{R_L} \text{ (approx)}$$

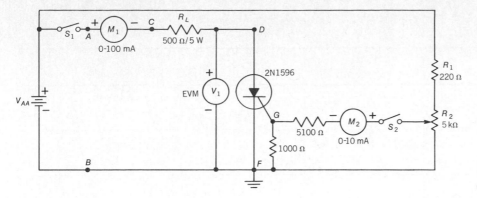

Fig. 49-9. Experimental dc circuit to determine SCR characteristics.

3. Set R_2 (gate current control) output at 0 V. *Close S_2. Close S_1.* Is the SCR on? _____
4. Adjust R_2 *very gradually,* monitoring gate current, just to the point where the SCR is on.

NOTE: If the SCR does *not* turn on, it may be necessary to reduce the value of the 5100-Ω resistor in the gate circuit, or to eliminate it altogether.

Measure and record in Table 49-1, under trial 1, the gate current required to turn on the SCR. Measure also and

record anode voltage V_{DF} across the SCR after it has turned on. Measure and record anode current I_A after the SCR is on.

5. *Open S_1.* Readjust R_2 so that gate voltage is again 0 V. Repeat step 4 and record your results under trial 2.
6. Repeat step 5 for trial 3.
7. After the SCR has fired, in step 6, *open S_2,* removing gate current. Observe and record in Table 49-1 the effect, if any, on anode current, after the SCR is on, of opening gate switch S_2.
8. *Open S_1.* Now *both* S_1 and S_2 are open.
9. Set V_{AA} at 40 V as measured across the power supply with S_1 *open.* Set R_2 gate current control at 0 V. *Close S_2 and S_1.* Determine in three trials the gate current required to trigger the SCR, following the procedure outlined in steps 4 and 5. Record your results in Table 49-1.
10. Repeat step 7 after the third trial.

DC Gate Control and AC Anode Source

11. Connect the circuit of Fig. 49-10. S_1 and S_2 are *off.* T is the power transformer used in early power-supply experiments. Set V_{GG}, the dc gate supply, at 6 V.
12. S_1 *on.* S_2 remains *off.* Set your oscilloscope on *line* trigger/sync. Observe the waveform from A to C. Adjust the oscilloscope controls until this *reference* waveform

TABLE 49-1. DC Anode Voltage, DC Gate Current

V_{AA}, V	I_G, mA	V_{DF}, V	I_A, mA	Trial Number
				1
15				2
				3
				1
40				2
				3

Effect on anode current, after SCR is on, of opening gate switch S_2:

Step 7 _____

Step 10 _____

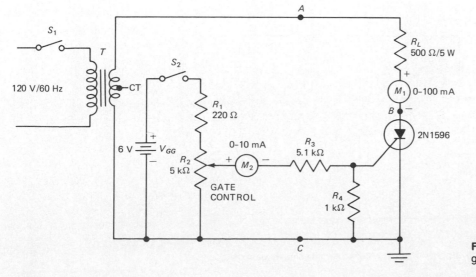

Fig. 49-10. Experimental circuit using dc gate control.

The Silicon Controlled Rectifier (SCR) **291**

TABLE 49-2. AC Anode Source, DC Gate Control

Waveform No.	Waveform	Volts, Peak Positive	I_L, mA	I_G, mA	Conduction Angle
Reference			X	X	X
1 (max.)					
2 (min.)					

appears as in Table 49-2. Measure and record in Table 49-2 its peak-to-peak voltage.

Observe the waveform across the SCR, points B to C. This should be the same as the reference waveform as long as the SCR is OFF. When the SCR is ON, the anode-to-cathode voltage waveform (Fig. 49-6) will appear on the oscilloscope.

13. *Close S_2.* Slowly vary R_2, the gate current control, over its entire range and observe the effect on the waveform from *anode to cathode* (V_{BC}).

In Table 49-2 draw in proper phase with the reference waveform V_{BC} for *maximum conduction* period (waveform 1) and *minimum conduction* period (waveform 2). Measure for each the peak positive amplitude of the waveform, the load current I_L, the gate current I_G, and the conduction angle (duration in degrees of the on interval in any one cycle). Record the data in Table 49-2.

AC Gate Control and AC Anode Source

14. **Power off.** Modify the preceding experimental circuit to conform with that in Fig. 49-11. Note that an ac gate source is used, rectified so that the *positive* alternations appear on the gate.

15. **Power on.** With an oscilloscope connected across the load resistor (V_{BC}) observe the load voltage waveform as R_2, the gate current control, is varied from minimum to maximum. Record in Table 49-3 the minimum and maximum conduction angles, and the corresponding load waveforms.

16. In Table 49-3 also record the measured values of gate current I_G and load current I_L for the minimum and maximum conduction angles.

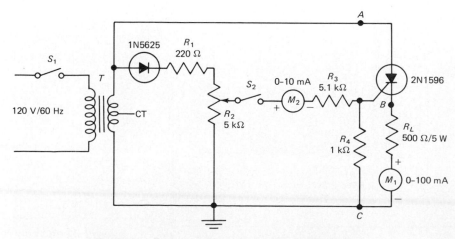

Fig. 49-11. Experimental ac circuit using ac gate control.

TABLE 49-3. AC Anode Source, AC Gate Control

Waveform No.	Waveform	Conduction Angle, Degrees	V p-p	I_L, mA	I_G, mA
Reference	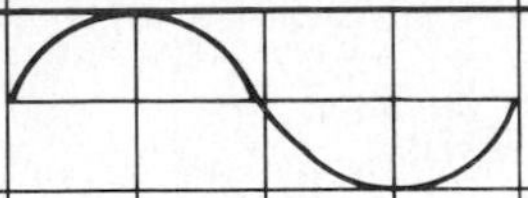	X		X	X
Maximum (1)					
Minimum (2)					

QUESTIONS

1. What are the differences between an ordinary silicon rectifier and an SCR?
2. After the rectifier turned on, what effect did the anode voltage supply have on anode current?
3. Will the size of load resistance affect the level of load current in the circuit of Fig. 49-9? How?
4. How does the forward breakover voltage vary with bias current? Refer specifically to your data to confirm your answer.
5. How does load current vary with conduction angle? Refer to your data in Table 49-2.
6. Refer to waveform 1, Table 49-2. Identify the forward breakover voltage point.

7. What triggering/synchronizing arrangement in this experiment makes it possible for you to observe the load waveforms in proper time phase with the reference waveform?
8. Compare the effectiveness of the ac versus the dc gate triggering source on control of anode current in this experiment.

Answers to Self-Test

1. gate
2. lower
3. rectifier
4. removed
5. off

6. V_{BRFO}; V_X
7. false
8. dc, ac
9. false
10. anode

The Silicon Controlled Rectifier (SCR)

THE UNIJUNCTION TRANSISTOR (UJT)

OBJECTIVES

1. To determine experimentally the emitter characteristic (V_E versus I_E) of a UJT
2. To connect the UJT as a relaxation oscillator and observe the output waveform
3. To trigger an SCR with a UJT relaxation oscillator

INTRODUCTORY INFORMATION

Many techniques are used for triggering an SCR. We have experimented with dc gate current and with a sinusoidal ac gate source. In designing a trigger source, these factors must be considered: A low-power source may cause erratic SCR triggering; a high-power source, while ensuring consistent SCR turn-on, may overheat the gate and cause it to burn out. An ideal solution would suggest triggering the SCR with sharp, high-powered pulses of short duration, whose peak and average power do not exceed the power capabilities of the SCR gate for which they are intended. The unijunction transistor (UJT) is frequently employed as a trigger source, because it can generate the required pulses.

UJT Construction and Characteristics

A cross-sectional diagram of a UJT is shown in Fig. 50-1. A ceramic disk serves as the base for this device, assuring it rigidity and ruggedness. A conductive gold film is deposited on this base on both sides of a very narrow slit in the center. An N-type silicon bar is symmetrically pressed against the gold film, forming two resistive (nonrectifying) contacts which are called the *bases*, designated as base 1, B_1, and base 2, B_2. A PN junction is formed by pressing a P-type emitter against the N-type bar, at a point closer to B_2 than to B_1. The entire device is encapsulated and enclosed in a case. Physically, it appears no different from a similarly cased transistor—but its electrical characteristics are unique.

Figure 50-2 is the circuit symbol for this three-terminal solid-state device, showing the emitter E and the two bases B_1 and B_2. A simplified equivalent circuit of a UJT appears in Fig. 50-3. The PN emitter-to-base junction is shown as a diode D_1. The interbase resistance R_{BB} of the N-type silicon bar appears as two resistors R_{B1} and R_{B2}, where R_{BB} equals the sum of R_{B1} and R_{B2}. The doping and geometry of the bar

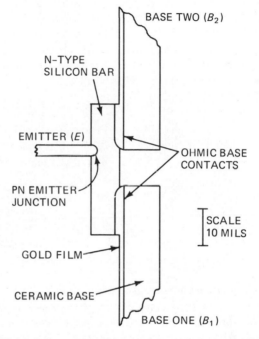

Fig. 50-1. Cross-sectional diagram of a UJT. *(General Electric)*

determine this interbase resistance, which for the 2N2160 lies in the range of 4000 to 12,000 Ω. When emitter-to-base-1 current is permitted to flow and the UJT is turned on, the resistance of R_{B1} decreases sharply. The resistance of R_{B1} varies inversely with emitter current. Since the conductivity of R_{B1} is a function of emitter current, the variation of the resistance of R_{B1} caused by changes in emitter current is called *conductivity modulation*.

When there is no emitter current I_E, the voltage V_{AB1} from point A to B_1 may be written as

$$V_{AB1} = V_{BB} \times \frac{R_{B1}}{R_{B1} + R_{B2}} = V_{BB} \times \frac{R_{B1}}{R_{BB}} \quad (50\text{-}1)$$

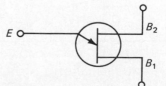

Fig. 50-2. UJT circuit symbol.

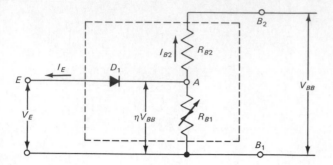

Fig. 50-3. UJT simplified equivalent circuit.

where V_{BB} is the interbase voltage. R_{B1} is the base-1 resistance, and R_{BB} is the interbase resistance. The ratio R_{B1}/R_{BB} is called the *intrinsic standoff ratio* of the unijunction transistor and is designated by the Greek letter η (eta). Equation (50-1) may therefore be written

$$V_{AB1} = \eta \times V_{BB} \qquad (50\text{-}2)$$

The ratio η is determined by the geometry of the UJT and depends on the spacing between the emitter junction and the two base contacts. The value of η lies in the range 0.51 to 0.81.

If the bias voltage V_E is less than $\eta \times V_{BB}$, the emitter-to-base-1 junction is reverse-biased and there is no emitter current I_E except for reverse emitter leakage current. When the applied voltage V_E is greater than $\eta \times V_{BB}$, the emitter-to-base-1 junction is forward-biased, and emitter current will flow.

The emitter conductivity characteristic is such that, as I_E increases, the emitter-to-base-1 voltage decreases, as can be seen from the emitter characteristic curve in Fig. 50-4. At the "peak point" V_P and the "valley point" V_V, the slope of the emitter characteristic curve is 0. At points to the left of V_P, the emitter-to-base-1 is reverse-biased, and there is no emitter current. This is called the *cutoff region*. For points to the right of V_P, the emitter-to-base-1 is forward-biased, and there is I_E. Between V_P and V_V, increases in I_E are accom-

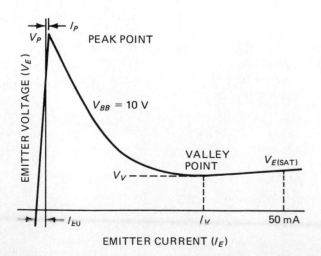

Fig. 50-4. Static emitter characteristic curve showing important parameters. *(General Electric)*

panied by a reduction in emitter voltage V_E. This is the *negative resistance region* of the UJT. Beyond the valley point V_V, an increase in I_E is accompanied by an increase in V_E. The region to the right of V_V is known as the *saturation region*.

The peak point voltage V_P is given by the equation

$$V_P = \eta V_{BB} + V_D \qquad (50\text{-}3)$$

It is evident, therefore, that V_P is dependent on the interbase voltage V_{BB} and on the forward voltage V_D across the emitter-to-base diode. V_D varies inversely with temperature. Stabilization of V_P is partly achieved by connecting a small resistor R_2 in series with base 2, as in Fig. 50-5. In this circuit, R_1, R_{BB}, and R_2 constitute a voltage divider. Since the interbase resistance R_{BB} increases with an increase in UJT temperature, the voltage V_{BB} will also increase with an increase in temperature. If the value of R_2 is properly chosen, the increase in interbase voltage will compensate for the decrease in V_D. The proper value of R_2 to achieve stabilization is given approximately by the equation

$$R_2 \cong \frac{0.7R_{BB}}{\eta V_1} + \frac{(1 - \eta)R_1}{\eta} \qquad (50\text{-}4)$$

Figure 50-5 shows a typical UJT bias and stabilization circuit. In order for the UJT to be turned on, the voltage V_E applied to the emitter must at least equal the peak-point voltage V_P, and the emitter current I_E must be greater than the peak-point current I_P.

UJT Connected As a Relaxation Oscillator

The UJT, connected as a relaxation oscillator as in Fig. 50-6a, generates a voltage waveform V_{B1} (Fig. 50-6b), which can be applied as a triggering pulse to an SCR gate to turn on the SCR. Operation of the circuit is as follows: When switch S_1 is first closed, applying power to the circuit, capacitor C_T starts charging exponentially through R_T to the applied voltage V_1. The voltage across C_T is the voltage V_E applied to the emitter of the UJT. When C_T has charged to the peak-point voltage V_P of the UJT, the UJT is turned on, decreasing greatly the effective resistance R_{B1} between the emitter and base 1. A sharp pulse of current I_E (limited only by R_1) flows from base 1 into the emitter, discharging C_T. When the

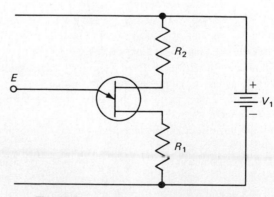

Fig. 50-5. UJT bias and stabilization circuit.

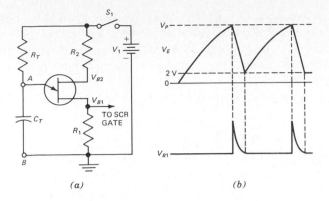

(a)   (b)

Fig. 50-6. UJT relaxation oscillator.

voltage across C_T has dropped to approximately 2 V, the UJT turns off and the cycle is repeated. The waveforms in Fig. 50-6b illustrate the sawtooth voltage V_E, generated by the charging of C_T and the output pulse V_{B1} developed across R_1. V_{B1} is the pulse which will be applied to the gate of an SCR to trigger the SCR.

The frequency f of the relaxation oscillator depends on the time constant $C_T R_T$ and on the characteristics of the UJT. For values of $R_1 \lesssim 100\ \Omega$, the period of oscillation T is given approximately by the equation

$$T = \frac{1}{f} \approx R_T C_T \ln\frac{1}{1-\eta} \qquad (50\text{-}5)$$

The value of R_T is limited to the range 3000 Ω to 3 MΩ. The supply voltage V_1 normally used lies in the range of 10 to 35 V.

SCR Triggered by UJT Relaxation Oscillator

In the experimental circuit of Fig. 50-7, the pulses developed across R_1, in base 1 of the UJT, are used to trigger the SCR, as in the circuit of Fig. 50-8. The UJT is connected as a relaxation oscillator. The frequency of the sawtooth voltage developed across C is determined by the time constant $R_4 C$. R_4 is made variable so that the timing of the trigger pulses, developed across R_1, may be adjusted to control the firing of the SCR at different points on the pulsating input anode wave. The fact that *full-wave* rectified ac (pulsating dc) is employed as a power source for the SCR, rather than a straight sinusoidal input, doubles the load current capabilities of the circuit, because this arrangement eliminates the half-cycle during the negative alternation when the SCR would normally be cut off. The pulsating dc source, however, permits the SCR to turn off when the anode voltage is reduced below the holding level of the controlled rectifier.

Of interest is the arrangement of resistor R_3 and zener diode Z_1. The zener clips the tops of the positive alternations and provides a relatively stable voltage level to which capacitor C can charge through resistor R_4. Straight dc cannot be used as the charging level for capacitor C in this relaxation oscillator, because the frequency of the pulse output of the UJT would then not be synchronized to the frequency of the pulsating dc applied to the SCR. The flat-topped voltage

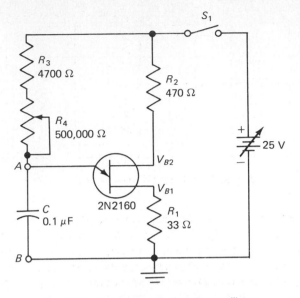

Fig. 50-7. Experimental relaxation oscillator.

alternations to which C charges effect synchronization, because the sawtooth waveforms developed across C occur in groups at a *recurrent* frequency which is the same as the frequency of the full-wave rectified ac source. Thus, though the basic frequency of the oscillator is still determined by the time constant $R_4 C$, the pulse-recurrent frequency is fixed by the power source. In this manner, the full-wave rectified signal, obtained from an appropriate rectifying source, supplies both power to the SCR and synchronization to the trigger circuit.

SUMMARY

1. The unijunction transistor (UJT) is a semiconductor which has a *single* PN junction, a P-type emitter and an N-type bar.
2. The N-type bar is symmetrically pressed against a split gold film deposited on a ceramic base (Fig. 50-1) and the bar makes resistive (nonrectifying) contacts with the film. The bottom gold film is called *base 1* (B_1); the top is called *base 2* (B_2). The emitter junction is physically closer to B_2 than to B_1.
3. The geometry and construction of this transistor give it its unique characteristics. Its circuit symbol is shown in Fig. 50-2.
4. The equivalent circuit of a UJT (Fig. 50-3) shows the diodelike PN junction, the base-2 resistance R_{B2}, and the base-1 resistance R_{B1}. The sum of R_{B1} and R_{B2} is called the *interbase resistance* R_{BB}.
5. When emitter-to-base-1 current I_E is permitted to flow, the UJT is turned on.
6. Figure 50-4 is the emitter characteristic curve for a UJT.
7. In order to dc-stabilize the UJT, a small resistor R_2 must be connected in the circuit of B_2 (Fig. 50-5).
8. The UJT, connected as in Fig. 50-6, is turned on by the rising voltage across capacitor C_T, charging toward V_1

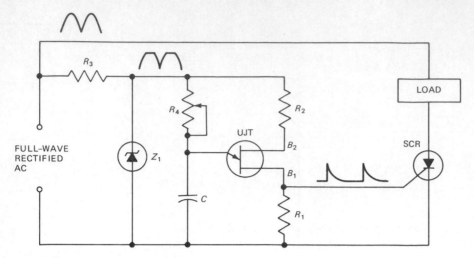

Fig. 50-8. SCR triggered by a UJT.

through R_T. When the voltage across C_T rises to the UJT turn-on level, the UJT draws current and the emitter current discharges C_T. When the voltage on the emitter drops to 2 V (approx), the UJT turns off. C_T starts charging again and the on-off cycle is repeated. This circuit is a UJT relaxation oscillator.

9. The current pulses through R_1 in Fig. 50-6 generate positive voltage spikes which are used to trigger an SCR (Fig. 50-8).

10. Resistance R_4 in Fig. 50-8 is made variable to change the frequency of the oscillator. It is thus possible to change the firing point and hence the conduction angle of the SCR.

11. Circuit operation limits R_4 to the range 3000 Ω to 3 MΩ (approx).

SELF-TEST

Check your understanding by answering these questions.

1. The UJT is a bipolar transistor with two PN junctions. _____ (*true, false*)
2. The emitter of a UJT is _____ (*P, N*)-type semiconductor material.
3. The UJT can turn on only when the emitter-to-base-1 is _____ (*forward, reverse*)-biased.
4. In the UJT equivalent circuit (Fig. 50-3) $R_{B1} = 3500$ Ω, $R_{BB} = 6000$ Ω, and $V_{BB} = 10$ V. The emitter voltage required to turn on the UJT must be greater than _____ V.
5. In the graph of Fig. 50-4, the peak point V_P is the point at which emitter current I_E is maximum. _____ (*true, false*)

6. In the graph of Fig. 50-6, the UJT will turn on when the voltage across capacitor C_T has reached the value _____ .
7. In Fig. 50-6, the load resistor is _____ , and the dc stabilization resistor is _____ .
8. In a properly operating UJT relaxation oscillator, the voltage across capacitor C_T is a _____ (*sawtooth, positive spike*).
9. In order to change the frequency of the experimental relaxation oscillator in Fig. 50-7, we must vary _____ .
10. By varying R_4 in Fig. 50-8 we can _____ the conduction time of the SCR.

MATERIALS REQUIRED

- Power supply: Variable regulated dc voltage source; variable 60-Hz source (variable autotransformer); isolation transformer
- Equipment: Oscilloscope; dc milliammeter; EVM or VOM; frequency-calibrated AF sine-wave oscillator (as a comparison source for checking frequency)
- Resistors: 33-, 100-, 220-, 470-, 1200-, 4700-Ω ½-W; 1000-Ω 1-W; 250-, 5000-Ω 5-W
- Capacitors: 0.1-μF 400-V
- Semiconductors: SCR 2N1596; UJT 2N2160; 1N4746 (18-V 1-W zener); four silicon rectifiers, type 1N5625 or equivalent
- Miscellaneous: Two SPST switches; transformer, 120-V primary, 25-V 1-A secondary; 500,000-Ω 2-W potentiometer

PROCEDURE

Emitter Characteristic (V_E versus I_E)

See note at end of step 3 before proceeding.

1. Connect the circuit of Fig. 50-9. The autotransformer is plugged into an isolation transformer. Switches S_1 and

S_2 are *open*. **Power off.** Output of the autotransformer is set at 0 V. Output of the dc supply is set at 0 V, as measured by voltmeter V.

2. The *horizontal sweep* switch of the oscilloscope is set on "Ext."

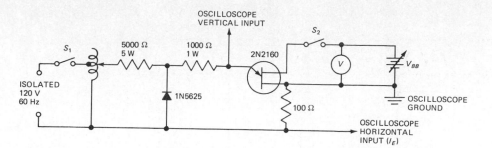

Fig. 50-9. Test setup to display V_E versus I_E characteristic of a UJT.

Calibrate the vertical amplifier of the oscilloscope at 2.5 V/cm, and the horizontal amplifier at 100 mV/cm. Position the horizontal trace of the scope on the lowest horizontal line of the graticule.

3. *Close S_2* and set V_{BB} at 5 V dc. *Close S_1* and adjust output of autotransformer until the horizontal deflection on the oscilloscope is 10 cm. (NOTE: This limits I_E to 10 mA, because each centimeter of horizontal deflection corresponds to 1 mA of emitter current.) Observe the trace on the screen and record on the graph in Table 50-1. Mark this curve $V_{BB} = 5$ V. Identify V_P, the peak voltage point.

NOTE: The curve observed on the oscilloscope may appear reversed, as compared to that in Fig. 50-4.

4. Repeat step 3 in turn for **(a)** $V_{BB} = 10$ V, **(b)** $V_{BB} = 15$ V, **(c)** $V_{BB} = 20$ V.

Draw and identify by V_{BB} all the curves in Table 50-1. Identify also the peak voltage point V_P for each curve. *Open S_1* and *S_2*. **Power off.**

NOTE: If a curve tracer is available which can check the response of a UJT, skip steps 1, 2, 3, and 4. Instead check the response of the UJT with a curve tracer for $V_{BB} = 5$ V, 10 V, and 15 V in turn. Draw the graphs in Table 50-1. Identify them and also identify the peak voltage point V_P for each curve.

TABLE 50-1. UJT Emitter Characteristic Curve

Emitter voltage—V_E, V

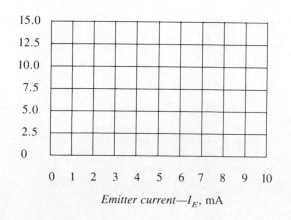

Emitter current—I_E, mA

Relaxation Oscillator

5. Connect the circuit of Fig. 50-7. Set R_4 for maximum resistance. S_1 is *open*. Adjust the output of the power supply V_1 for 25 V. Calibrate the dc vertical amplifiers of the oscilloscope for 5 V/cm. Set the trace on the lowest horizontal line of the graticule. The oscilloscope is set on triggered (or free-running) sweep. Set R_4 in the middle of its range.

6. *Close S_1*, applying power to the circuit. Connect the vertical input leads of the oscilloscope across capacitor C_T, hot lead on A, ground lead on B. *Externally* trigger/sync the oscilloscope with the voltage waveform V_{B2} at base 2. Adjust scope sweep controls for at least two or three complete waveforms. Draw and record the waveform, labeled V_E, in Table 50-2. Measure and record the peak-to-peak amplitude of the waveform. Measure and record the voltage level to which the waveform falls (still using the dc vertical amplifiers of the oscilloscope). Measure also and record the frequency of the waveform.

TABLE 50-2. UJT Relaxation Oscillator Waveforms

Step		Waveform	V p-p	Frequency, Hz
6	V_E			
7	V_{B1}			
	V_E			
8	V_{B1}			
	V_E			
9	V_{B1}			

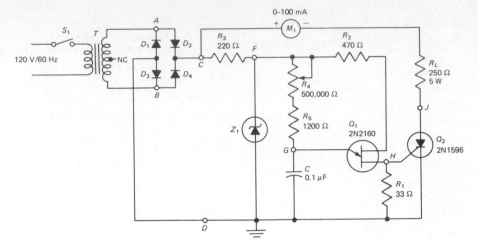

Fig. 50-10. Experimental UJT trigger controls the firing of the SCR.

7. Observe the waveform V_{B1} at base 1. Draw it in Table 50-2, in proper time phase with the waveform V_E. Measure also and record its peak-to-peak amplitude.

8. Adjust R_4 for the minimum resistance. Observe and record the waveforms V_E and V_{B1}; measure and record in Table 50-2 their peak-to-peak amplitude and frequency as in steps 6 and 7.

9. Adjust R_4 for half of its total resistance and repeat step 7. **Power off.**

SCR Triggered by UJT

10. Connect the circuit of Fig. 50-10. T is a voltage step-down transformer (120-V primary, 25-V center-tapped secondary). *Close S_1.* **Power on.**

11. Calibrate the vertical amplifiers of your oscilloscope at 10 and 3 V/cm. With the oscilloscope set on line trigger/sync or externally synchronized by the voltage from point A in the secondary of T, observe the waveform V_{AB} across the secondary of T, vertical lead of the oscilloscope on point A, ground lead on point B. Adjust the sweep, trigger/sync, and centering controls, until the reference waveform appears as in Table 50-3. Measure and record in Table 50-3 the peak-to-peak amplitude of the waveform.

12. Observe, measure, and record in Table 50-3, in proper time phase with the reference, the waveforms V_{CD} and V_{FD}.

13. With R_4 set at minimum (zero resistance) observe, measure, and record in Table 50-3 the waveform V_{JD}, in proper time phase with the reference. Measure on milliammeter M_1 and record the load current I_F.

14. Vary R_4 over its entire range. Observe the effect on load current and on the load waveform.

15. With R_4 set at maximum resistance, observe, measure, and record in Table 50-3 the waveform V_{JD} and load current.

TABLE 50-3. Measurements in UJT-Triggered SCR Ciruit

Step	Condition	Test Points	Waveform	V p-p	Load Current, mA
11	X	AB			X
12	X	CD			X
		FD			
13	$R_4 = 0$	JD			
15	$R_4 = 500,000\ \Omega$	JD			

QUESTIONS

1. What are the desirable characteristics for an SCR gate-triggering source?
2. Does a phase-shift sinusoidal gate-triggering source meet the characteristics you listed in answer to question 1? If not, why not?
3. Draw the circuit diagram of a UJT circuit which can be used as a gate trigger.
4. Explain how the circuit in answer to question 3 operates.
5. Assume the circuit of Fig. 50-6 acts as the gate-trigger source for an SCR. The anode-to-cathode circuit of the SCR is powered from a 60-Hz sinusoidal source. What problem, if any, do you see in the use of a dc voltage for the gate trigger and a 60-Hz sine wave for the anode-to-cathode of the SCR?
6. Would it be possible to achieve 180° (approx) control of an SCR using a UJT trigger circuit? If yes, explain why.
7. Explain in detail, with waveforms, the operation of the experimental circuit of Fig. 50-7.
8. What relationship, if any, is there between the resistance of R_4 in Fig. 50-7 and the frequency of the output waveform? Refer to your experimental data in Table 50-2 to confirm your answer.
9. What relationship, if any, is there between the resistance of R_4 in Fig. 50-10 and the conduction angle of the SCR? Refer to your experimental data in Table 50-3 to confirm your answer.
10. From your measurements in Table 50-3, what is the zener voltage (approximately) for Z_1?

Answers to Self-Test

1. false
2. P
3. forward
4. 5.83
5. false
6. V_P
7. R_1; R_2
8. sawtooth
9. R_4
10. change (or vary)

CHARACTERISTICS OF A CATHODE-RAY TUBE (CRT)

OBJECTIVES

1. To determine how a CRT beam may be deflected (*a*) electrostatically, (*b*) magnetically
2. To measure the deflection sensitivity of an electrostatically deflected CRT
3. To observe how a trace is produced on the screen

INTRODUCTORY INFORMATION

Cathode-ray tubes (CRTs) are specially constructed vacuum tubes which act as visual indicators of electrical voltages. There are many types of CRTs exhibiting a wide variety of physical and electrical characteristics. The range of face, or "screen," diameter varies from 1 to 30 in, the length of the bulb from 3 to 29 in. Operating (accelerating) voltages may vary from 500 to 33,000 V. Cathode-ray tubes may be electrostatically or electromagnetically focused and deflected. They may serve as a "picture" tube, an oscilloscope screen, a radar scope, or in any one of many other applications. It is important therefore that the technician understand the operation of a CRT.

Construction

An electrostatically deflected CRT consists of an outer envelope, or bulb; a glass face, or screen, whose inner surface is phosphor-coated; an electron-gun assembly rigidly supported inside the neck of the housing; deflection plates; and a tube base to which connections from the elements inside the CRT are made (Fig. 51-1). The bulb is usually all glass.

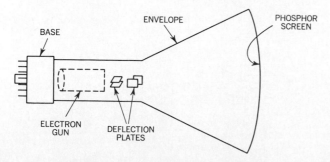

Fig. 51-1. Electrostatically deflected CRT.

Electromagnetically deflected CRTs are similarly constructed except that they do not have deflection plates. Moreover, though the envelope is usually made of glass, metal cones are sometimes used instead.

Because of its size and the fact that the envelope is highly evacuated, CRTs present special safety problems and therefore require special care in handling. The pressure on the face of a CRT is proportional to the surface area. Thus, at sea level the pressure on the screen of a 21-in (diameter) round CRT is approximately 2 tons. To withstand this tremendous pressure (there is no counteratmospheric pressure in the envelope) very heavy glass is used for the face.

The tube must be protected against imploding, since the effect of an implosion is the same as that of an explosion. Flying glass resulting from an implosion can cause severe injury or death. *For that reason the CRT must be handled carefully.* The tube must not be struck, scratched, or jarred. The use of shatterproof goggles is recommended in handling picture tubes.

The voltages required to accelerate the electron beam in a CRT are much higher than voltages found in other electronic circuitry. This then is another factor the technician must consider when working with the CRT or its associated circuits.

Electron Gun

The purpose of the electron gun is to form a well-focused electron beam and accelerate it in the direction of the phosphor screen. Under the impact of the beam, the phosphors give off light (fluoresce). A sharply focused beam will cause a small, bright spot of light to appear on the screen.

The electron gun in an electrostatically focused CRT consists of an indirectly heated cathode; a control grid; a first, or focusing, anode; and a second, or accelerating, anode (Fig. 51-2). The heated cathode emits a stream of electrons, whose density (number of electrons) is controlled by the bias voltage between control grid and cathode. The voltage on the first anode is used for focusing the beam. The voltage on the second anode, highly positive relative to the cathode, accelerates the electron beam.

The cylindrical cathode is terminated by a small plate coated with barium and strontium oxides to provide a plentiful supply of electrons. Surrounding the cathode is the con-

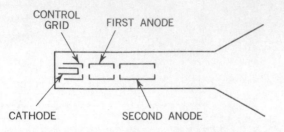

Fig. 51-2. Electron gun in electrostatically deflected CRT.

trol grid, a cylinder with a tiny aperture at its center, an aperture smaller than the cathode emitter. Another cylinder, the first, or focusing, anode, is in line with the control grid, facing it. There are also small apertures in the baffles at the center of the focusing anode. The second, or accelerating, anode is next to the first anode. It also contains two baffles, each of which has a small aperture at the center. This entire assembly, aligned so that the apertures are collinear, is rigidly mounted in the neck of the CRT. In the absence of any deflection voltages on the deflection plates, the electron gun is aligned so that the electron beam will strike closer to the center of the screen.

Electrostatic Deflection

Negatively charged electrons are attracted to a positive, and repelled by a negative, voltage source. Hence the electron beam emitted by a CRT gun can be deflected from its path by having it pass through an electrostatic field or fields. This is the deflection method used in electrostatically deflected CRTs.

Figure 51-3a shows the effect of sending an electron beam through two parallel, vertically oriented, rectangular deflection plates, D_1 and D_2. In the absence of these plates, the beam would strike the fluorescent screen in the center at 0. If plate D_1 is made more positive than D_2, the beam will be attracted toward plate 1 and will strike the screen at some point 1, to the *right* of 0. On the other hand, if D_2 is made more positive than D_1, the beam will be deflected to the *left* and will strike the screen at some point 2. D_1 is called the *right horizontal deflection plate* (RHDP), D_2 the *left horizontal deflection plate* (LHDP).

The amount that the beam is deflected on the screen depends on the *speed* with which the electrons in the beam are traveling, the voltage applied across the deflection plates,

the geometry of the plates, and the distance of the screen from the plates. A slow-moving electron remains in the field between the two plates longer than a fast-moving electron. Therefore the electrostatic field has a longer time to act on the slow-moving electron, and it is deflected more than a fast-moving electron. The velocity of the beam is proportional to the potential on the accelerating anode: the higher the accelerating potential, the greater the velocity.

The amount of deflection is also proportional to the potential across the deflection plates. The deflection will be large when the voltage across the deflection plates is high. However, the voltage across the plates must not be so high that the deflected electron beam will strike the deflection plates instead of striking the screen. To offset the possibility of having the beam strike D_1 and D_2, these plates may be flared, as in Fig. 51-4.

Even though they are vertically oriented, deflection plates D_1 and D_2 are called *horizontal plates*. There is also a pair of vertical deflection plates, D_3 and D_4 (Fig. 51-3b), further back from the screen than the horizontal. These are so oriented that a more positive D_3 will cause the beam to be deflected up to point 3 above 0; a more positive D_4 will cause the beam to be deflected *down* to some point 4. D_3 is called the *upper vertical deflection plate* (UVDP), D_4 the *lower vertical deflection plate* (LVDP). Note that by rotating the CRT through 90° the positions of the vertical and horizontal deflection plates are interchanged.

In the manufacture of a cathode-ray tube and in the design of the external CRT circuits, the accelerating potential (hence velocity of the beam), the geometry of the deflection plates, and their distances from the screen are fixed. The only

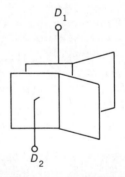

Fig. 51-4. Flared deflection plates.

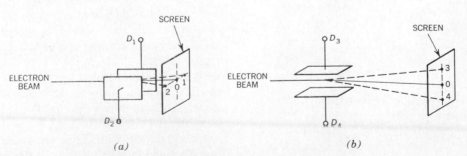

Fig. 51-3. Effect of deflection plates on CRT beam.

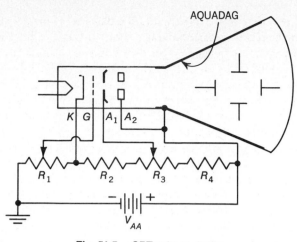

Fig. 51-5. CRT voltage divider.

variable is the deflection voltage across the plates. Therefore the position of the spot on the screen will depend on the voltages applied to the deflection plates. The distance that the beam will be moved on the screen by a potential difference of 1 V across a pair of deflection plates is called the *deflection sensitivity* of these plates. This distance is given as a rating in millimeters per volt dc, or in inches per volt dc. Another way in which a manufacturer may show the deflection sensitivity of a tube is in inches per volts, as, for example, 1 in per 100 V horizontal and 1 in per 95 V vertical.

Electrostatically deflected CRTs are usually used in oscilloscopes.

CRT Voltage Divider

Figure 51-5 shows the circuit symbol for an electrostatically deflected CRT, where K is the cathode, G the control grid, A_1 the focusing anode and A_2 the accelerating anode. V_{AA} is a high-voltage dc source, and R_1 to R_4 constitute a voltage divider across V_{AA}. The bias voltage between grid and cathode is made variable by R_1, the "intensity" potentiometer. Note that the grid is normally negative with respect to the cathode, as in an ordinary vacuum-tube triode. The voltage on the focusing anode is made variable by means of R_3, the focus control. The highest positive dc voltage is connected to the accelerating anode. Note the graphite coating labeled "aquadag" on the inside of the envelope. This conductive coating is also connected to the most positive voltage point

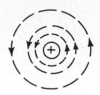

Fig. 51-7. Magnetic field about an electron beam. The beam is moving into the page.

on the voltage divider. The purpose of the aquadag coating is to attract any electrons dislodged from the phosphors on the screen as they are struck by the CRT beam. The emission of electrons from the screen due to the electron-beam bombardment is called *secondary emission*. The electrons resulting from secondary emission are therefore returned to the cathode via the aquadag coating and power supply.

Magnetic Deflection

A basic difference between a magnetically deflected and an electrostatically deflected CRT is that the former does not contain deflection plates. An external magnetic field is employed to deflect the beam. Figure 51-6 shows a deflection coil placed around the neck of the CRT past the electron-gun assembly. Current flowing in this coil produces a magnetic field which interacts with the magnetic field about the electron beam inside the CRT, and causes deflection. The amount of deflection is proportional to the strength of the magnetic fields, which in turn is proportional to the current in the deflection coil. A permanent magnet, just as well as an electromagnet, may be used to deflect an electron beam.

An electron beam produces a circular magnetic field around it just like a wire carrying current. The direction of the field depends on the direction of the current. Figure 51-7 shows the magnetic field about an electron beam in which the beam is moving into the page. When a uniform magnetic field (Fig. 51-8a) is made to interact with the field about the electron beam (Fig. 51-8b), the magnetic lines of force are distorted, and the beam is deflected, as shown. If the direction of either of the magnetic fields is reversed, the direction of deflection is reversed. Thus, by reversing the poles of the external magnetic field, the beam can be made to move *up*, rather than down.

By rotating the uniform magnetic field 90° clockwise and shooting the electron beam through the field (into the page), the beam can be deflected to the left. A deflection to the right can be achieved by rotating the external magnetic field of Fig. 51-8a 90° counterclockwise.

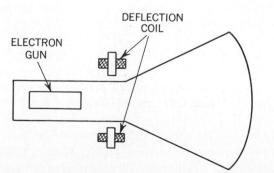

Fig. 51-6. Deflection coils around a magnetically deflected CRT.

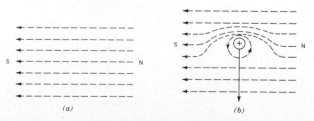
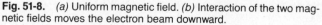

Fig. 51-8. *(a)* Uniform magnetic field. *(b)* Interaction of the two magnetic fields moves the electron beam downward.

Coils wound on an assembly called a *yoke* placed about the neck of the CRT serve as electromagnets for magnetic deflection of a CRT beam. If deflection in only one direction is required, the yoke takes the form shown in Fig. 51-9. It consists of two coils wound in series in such a manner that the fields produced by the two coils oppose each other. The resulting magnetic lines of force then take the direction shown in Fig. 51-9. The deflection of the beam is then either *up* or *down*, depending on the direction of current flow through the windings of the coil. Horizontal deflection may be achieved by rotating the yoke 90°.

If both vertical and horizontal deflection is required simultaneously, two pairs of coils, one for horizontal, the other for vertical, are wound around the core.

CRT Trace

In our discussion of beam deflection we considered only dc voltages for electrostatic deflection, dc currents for electromagnetic deflection. Here, the beam was deflected to a spot on the CRT and remained there. However, if an ac voltage is applied across the deflection plates of a CRT, the amplitude of the voltage across the plates will vary, causing the deflection of the CRT beam to vary. The position of the spot of light on the screen will therefore change in accordance with the changes of the ac voltage waveform. Suppose, for example, that a sinusoidal voltage is applied to the left horizontal deflection plate, while the right horizontal deflection plate is kept at zero reference (Fig. 51-10*a*). At the instant 1, the ac voltage is 0 and the CRT spot will be in the center, since there is zero voltage across the two deflection plates (Fig. 51-10*b*). However, during the first 90°, as the waveform increases from 0 to its peak positive voltage at 2, the electron beam is deflected from the center to the left. The maximum deflection on the left occurs at the positive peak of the voltage waveform. As the ac voltage drops from its peak at 2 to 0 at 3, the beam moves back from the left to the center of the screen. During the first half of the negative alternation, the electron beam is repelled by the negative voltage on the LHDP and moves to the right. Maximum deflection on the right occurs

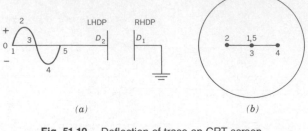

Fig. 51-10. Deflection of trace on CRT screen.

when the ac voltage peak is at maximum negative, point 4. During the interval from 4 to 5, the spot moves back from the right to the center of the screen.

If the frequency of the ac voltage applied to the LHDP were very low, say 1 Hz, the eye could follow the movement of the spot on the screen, from center to left, back to center, to right, and back to center. This motion would be repeated again and again for every cycle in the ac waveform. However, if the frequency were increased to, say, 60 Hz, the motion of the spot would be too rapid, and the eye could not follow it. Instead of a moving spot, a continuous line of light would be seen on the screen. This phenomenon is known as "persistence of vision." The light trace which results is called simply the *CRT trace*.

There is another reason the eye perceives this trace. It is that the screen lights up as the electron beam strikes it and continues to glow after the beam has passed. This ability of the phosphors to continue giving off light, after excitation, is called *phosphorescence*.

Magnetically, a trace can also be produced by having an alternating current flow in the windings of a deflection coil.

SUMMARY

1. Cathode-ray tubes (CRTs) are vacuum tubes which give a *visual* indication of electrical voltages. CRTs are used as the *screen* in oscilloscopes. Picture tubes used in TV sets are cathode-ray tubes.
2. CRT faces are round or rectangular.
3. A CRT is made up of a highly evacuated glass or metal bulb containing an *electron gun* assembly. The glass face or screen of the CRT is internally coated with phosphors, chemicals which emit light when struck by a fast-moving electron beam. The elements of the electron gun are connected to pins on the base.
4. The glass of a CRT screen is thick and strong to withstand the pressure exerted by the atmosphere.
5. A CRT may *implode* if struck or jarred, or if its screen is scratched. An *imploding* CRT may cause severe injury. For this reason CRTs should be handled very carefully. *Safety goggles should be worn when working with CRTs.*
6. The electron gun is rigidly mounted and carefully aligned inside the bulb. The gun emits a stream of electrons which strike the phosphor coating on the screen.

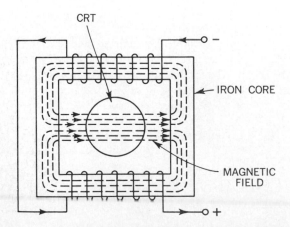

Fig. 51-9. Deflection coil and magnetic field for deflection of electron beam.

7. The electron gun consists of a cylindrical cathode which is heated by a filament inside the cathode, a control grid, a first or focusing anode, and a second or accelerating anode (Fig. 51-2). Inside the neck of the CRT is a graphite coating called the *aquadag*.
8. The heated cathode emits electrons.
9. The voltage on the control grid is made variable and acts as an *intensity* control. The voltage on the first anode is also manually variable to focus the beam (Fig. 51-5). The accelerating anode carries the highest voltage in the CRT. Accelerating voltages may be as high as 33,000 V.
10. To make the CRT useful, deflection plates or coils are provided to deflect the beam. The beam can be deflected up, down, right, or left.
11. There are two types of deflection, electrostatic and electromagnetic.
12. An electrostatically deflected CRT has two sets of deflection plates, vertical and horizontal. The two vertical plates deflect the electron beam in a vertical direction; the horizontal plates in a horizontal direction.
13. Deflection coils, wound on an assembly called a *yoke*, fit around the neck of the electromagnetically deflected CRT (Fig. 51-6).
14. The horizontal deflection plates or coils create a light *trace* on the screen. This is achieved by feeding a periodic ac voltage to the horizontal plates or coils.

SELF-TEST

Check your understanding by answering these questions.

1. A CRT is a solid-state device. _____ (*true, false*)
2. The envelope or bulb of the CRT is made of glass or _____ and is highly _____ .

3. A CRT must be handled carefully because the device can _____ if struck, jarred, or if its face is scratched.
4. _____ _____ should be worn when working with CRTs.
5. The elements of the _____ _____ assembly inside the CRT are connected to pins on the CRT base.
6. The electron gun _____ and _____ a beam of electrons.
7. If the electrons in a CRT beam are sufficiently accelerated, they will strike the _____ of the CRT and cause the phosphors to emit _____ .
8. When the voltage on the RHDP is made more _____ than the voltage on the LHDP, the electron beam will be deflected to the right.
9. A vertical light trace may be formed on the screen of the CRT if a(n) _____ _____ is applied across the _____ deflection plates of the CRT.
10. CRTs serve as the _____ in TV receivers, oscilloscopes, and _____ .

MATERIALS REQUIRED

- Power supply: Variable regulated 400-V dc with 6.3-V ac 60 Hz source; 40-V dc source
- Equipment: EVM
- Tube: 2 BP1 CRT or equivalent with associated socket
- Resistors: Two 47,000-, 470,000-Ω; 2.2-MΩ, two 3.9-MΩ; two 4.7-MΩ ½-W
- Miscellaneous: Two 10,000-, two 500,000-Ω 2-W potentiometers; two SPST switches; horseshoe magnet; ruler with marked scale

PROCEDURE

1. Connect the circuit of Fig. 51-11. The four deflection plates are short-circuited, as shown. One side of the 6.3-V filament is short-circuited to the cathode.
2. **Power on.** Set the regulated dc supply for 400- to 500-V output. *Close* S_1. Set R_1 so that the bias is 0 V as measured with an EVM between grid and cathode. A spot of light should appear on the screen. Adjust R_3 for best focus.

NOTE: Because the potential applied to the accelerating anode is substantially below the typical operating value, the danger of "burning" the phosphor in the CRT screen is reduced.

3. Vary R_1 and observe the effect on the brightness of the spot. Measure the bias voltage at which the spot is no longer visible (cutoff voltage). Record this voltage in Table 51-1. Readjust R_1 for maximum brightness. Mea-

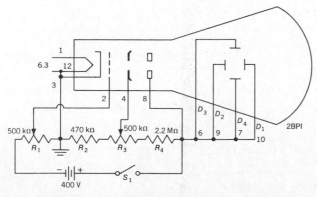

Fig. 51-11. Experimental CRT electron-gun circuit.

TABLE 51-1. CRT Voltages

	Volts	
Cut off	*Focusing*	*Accelerating*

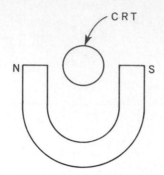

Fig. 51-12. Horseshoe magnet used to deflect CRT beam (cross-sectional top view).

sure and record the voltage of the focusing and accelerating anode with respect to the cathode.

Magnetic Deflection

NOTE: The principle of magnetic deflection can be demonstrated with a tube containing deflection plates as long as there is no voltage across these plates. For convenience, an electrostatic tube will be used also in this part of the experiment.

4. Slowly bring a horseshoe magnet, horizontally oriented, close to the CRT until the marked poles of the magnet are at the junction of the neck and bell, as in Fig. 51-12. Observe, and record in Table 51-2, the direction of deflection of the beam.
5. Rotate the magnet 180° along its long axis (see Fig. 51-13), thus reversing the poles, and repeat step 4.
6. Place the magnet, horizontally oriented as in step 4, at right angles to its original position (Fig. 51-14). Repeat step 4.

Electrostatic Deflection

7. **Power off.** Modify the CRT circuit according to Fig. 51-15. V_{PP} is a second dc source to provide voltage for the deflection plates. Note that deflection plates 6 and 9 are connected through 3.9-MΩ isolating resistors to point P, the midpoint of the 40-V supply, and that

TABLE 51-2. Magnetic Deflection

Show direction of deflection of the beam with an arrow, and label each arrow with its corresponding step number.

Beam

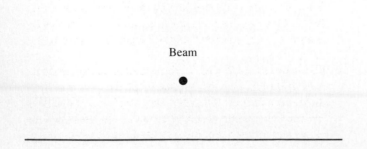

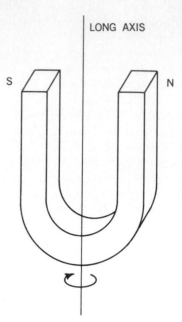

Fig. 51-13. Magnet rotated 180° along its long axis.

deflection plates 7 and 10 are connected, respectively, to the center arms of potentiometers R_5 and R_6 through 2.2-MΩ isolating resistors. When R_5 is set in the middle of its range, the net voltage across the vertical deflection plates is 0. As R_5 is varied to the right or left of center, the LVDP can be made more positive or negative, respectively, than the UVDP. Similarly, as R_6 is varied to the right or left of center, the RHDP can be made more positive or negative, respectively, than the LHDP.

8. **Power on.** *Close S_1 and S_2.* Adjust R_5 so that the voltage measured across the vertical deflection plates is 0. Also adjust R_6 so that the voltage measured across the horizontal deflection plates is 0. The light spot should be in the center of the CRT (approximately). Readjust R_5 and R_6, if necessary, to center the spot.
9. With R_6 set as in step 8, vary R_5 on either side of center and observe the effect on the deflection of the spot. Orient the CRT (and its socket assembly), if necessary, so that the deflection of the spot is vertical. Readjust R_5 so that the spot is again centered vertically. Mark the position of the spot on the CRT.

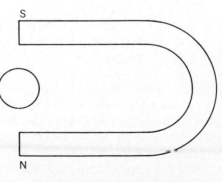

Fig. 51-14. Magnet rotated 90°.

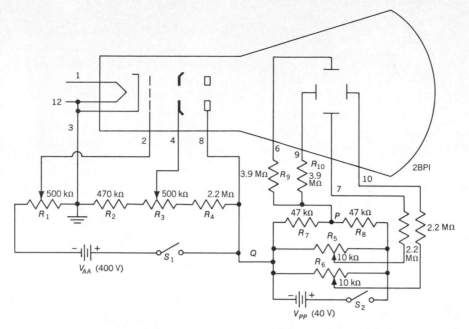

Fig. 51-15. CRT circuit electrostatic deflection.

TABLE 51-3. Electrostatic Deflection

Position of Spot	Voltage across Vertical Deflection Plates	Polarity of Voltage	
		UVDP	LVDP
¼ in above center			
¼ in below center			

Position of Spot	Volts across Horizontal Deflection Plates	Polarity of Voltage	
		RHDP	LHDP
⅛ in to right of center			
⅛ in to left of center			

10. Vary R_5 until the spot is ¼ in above center. Measure the voltage across the vertical deflection plates. Record this voltage in Table 51-3. Show the polarity of voltage of the UVDP with respect to the LVDP.

11. Readjust R_5 until the spot is ¼ in below center, and repeat the measurements as in step 10. Record the results in Table 51-3. Return the spot to its *marked-center* position.

12. Leave R_5 set as above. Vary R_6 until the spot is ⅛ in to the right of center. Measure, and record in Table 51-3, the voltage across the horizontal deflection plates. Show the polarity of the RHDP with respect to the LHDP.

13. Readjust R_6 until the spot is ⅛ in to the left of center. Repeat the measurements as in step 12. Record the results in Table 51-3.

Producing a CRT Trace

14. **Power off.** *Open* S_1 and S_2. Change the connection for UVDP pin 6 from P to Q, Fig. 51-15. Otherwise, leave the dc circuit connected as in Fig. 51-15. Connect an AF sine-wave generator to the horizontal deflection plates through two 0.1-µF dc blocking capacitors, as in Fig. 51-16. Observe that the capacitors are in *series* with the leads of the AF generator, and isolate the *generator* from the dc circuits.

15. **Power on.** *Close* S_1 and S_2. Turn on AF generator and set it as 1000 Hz. Set the attenuator of the generator for maximum output. Observe the effect on the CRT. Reduce the output of the generator, and observe the effect on the trace. Set the output of the generator until

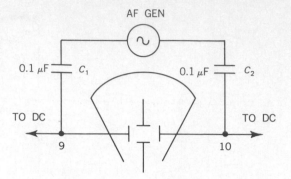

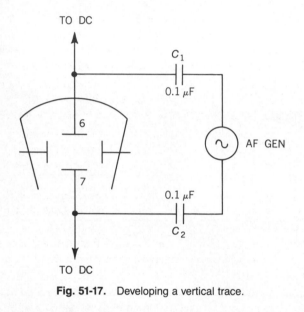

Fig. 51-16. Developing a horizontal trace.

Fig. 51-17. Developing a vertical trace.

there is a ¼-in trace on the screen. Measure, and record in Table 51-4, the peak-to-peak amplitude of the ac voltage across the horizontal deflection plates. Vary R_5 and R_6 and observe the effect on the trace. Recenter the trace.

16. **Power off.** *Open* S_1 and S_2. Again modify the circuit by interchanging the connections from pin 6 and 9. Pin 6 is now connected to P, and pin 9 to Q, Fig. 51-15. Connect the dc isolated output of the AF sine-wave generator set at 1000 Hz to the vertical deflection plates, as in Fig. 51-17.

17. **Power on.** *Close* S_1 and S_2. Repeat step 15, measuring the peak-to-peak amplitude of *ac* voltage across the vertical deflection plates required for ¼ in of trace. Record the result in Table 51-4.

Extra Credit

18. Explain in detail the procedure you would follow to obtain a figure-8 Lissajous pattern on the CRT. Show the ac circuit diagram.

19. After your circuit is approved, connect it. Observe and draw the resulting pattern. Indicate all frequencies.

TABLE 51-4.   CRT Trace

Length of Trace	Voltage, Peak-to-peak across Deflection Plates	
	Horizontal	Vertical
¼ in		

QUESTIONS

1. In what respects is a CRT like an ordinary vacuum tube?
2. In what respects is a CRT different from an ordinary vacuum tube?
3. What is the purpose of the (*a*) electron gun? (*b*) horizontal deflection plates? (*c*) vertical deflection plates? (*d*) aquadag?
4. What are the differences between an electrostatically and a magnetically deflected CRT?
5. Under what conditions will the electron beam in an electrostatically deflected CRT be centered (*a*) vertically, (*b*) horizontally?
6. What is meant by deflection sensitivity of an electrostatic CRT?
7. What factors determine the deflection sensitivity of an electrostatic CRT?
8. Why is there a difference between the vertical and horizontal deflection sensitivity in an electrostatic CRT?
9. From your data determine in two ways the vertical deflection sensitivity of the CRT. Show your computations.

10. From your data determine in two ways the horizontal deflection sensitivity of the CRT. Show your computations.
11. Explain how a (*a*) vertical trace and (*b*) horizontal trace may be produced on an electrostatic CRT.
12. Explain how a (*a*) vertical trace and ((*b*) horizontal trace may be produced on a magnetically deflected CRT.

Extra Credit

13. Explain how you would produce (*a*) a straight-line trace inclined 45° to the right and (*b*) a circular trace in an electrostatically deflected CRT.

Answers to Self-Test

1. false
2. metal; evacuated
3. implode
4. safety goggles
5. electron gun
6. forms; emits
7. phosphor screen; light
8. positive
9. ac voltage; vertical
10. screen; oscilloscopes

DIGITAL INTEGRATED CIRCUITS:
AND, OR GATES

OBJECTIVES

1. To become familiar with the characteristics and symbols of an AND gate and an OR gate
2. To determine experimentally the truth table of a combined AND gate and OR gate
3. To determine experimentally the truth table of a combined AND/OR gate

INTRODUCTORY INFORMATION

In preceding experiments you worked with linear ICs. In the remaining experiments you will study digital ICs. Digital ICs are *logic circuits*, the building blocks of digital computers and calculators. The basic digital circuits are rather simple and will serve as an introduction to digital ICs.

Logic Circuits

In Experiment 1 you observed the unidirectional characteristics of a solid-state semiconductor diode. When forward-biased, this diode acted like a closed switch; that is, it permitted current to flow in a complete circuit. When reverse-biased, it acted like an open switch and permitted little, if any, current to flow. This unique characteristic of a diode is employed in the design of basic logic circuits.

In digital electronics a *gate* is a logic circuit with one output and one or more inputs; an output signal occurs for certain combinations of input signals. In this experiment we examine the AND gate and the OR gate.

Modern computers do many jobs, from scientific computation to business accounting. Computers are also used in the automatic control of manufacturing processes. Though these devices perform highly complex jobs, they consist of arrangements of simple logic circuits.

To solve problems a computer has to make decisions as it progresses through the steps in the problem solving. Computer circuits that make decisions and comparisons are called *logic circuits*.

Computer decisions are of the yes/no variety, also known as *two-state* logic. Logic circuits can be in one of two states, such as on or off, high or low, magnetized or unmagnetized, etc. A toggle switch is a simple example of a two-state device.

AND Gate

Figure 52-1 shows a diode circuit with a switch input and a load resistor of 100 kΩ. The supply voltage is +5 V. When the switch is in the ground position, the diode is forward-biased and approximately 0.7 V appears across the diode. Therefore, the output voltage is low when the input is low.

On the other hand, when the switch is at +5 V, the net voltage across the diode-resistor combination is 0. As a result, the diode is nonconducting. Since there is no current through the load resistor, the output is pulled up to the supply voltage. In other words, the output is high (+5 V) when the input is high.

Now look at the two-input AND gate of Fig. 52-2a. When both switches are in the ground position, both diodes are conducting and the output is low. If S_1 is switched to +5 V and S_2 is left in the ground position, then the output is still low because D_2 still conducts. Conversely, if S_1 is in the ground position and S_2 is at +5 V, diode D_1 is conducting and the output is still low.

The only way to get a high output with an AND gate is to have all inputs high. If S_1 and S_2 are both at +5 V, both diodes are nonconducting. In this case, the output is pulled up to the supply voltage because there is no current through the load resistor.

By adding more diodes and switches, we can get three-input AND gates, four-input AND gates, and so on. Regardless of how many inputs an AND gate has, the operation is the same because it is an all-or-nothing gate. That is, all inputs must be high to get a high output. If any input is low, the output is low.

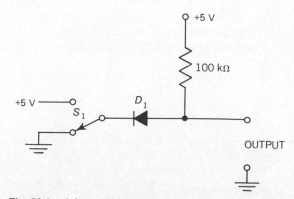

Fig. 52-1. A forward-biased diode acts like a closed switch.

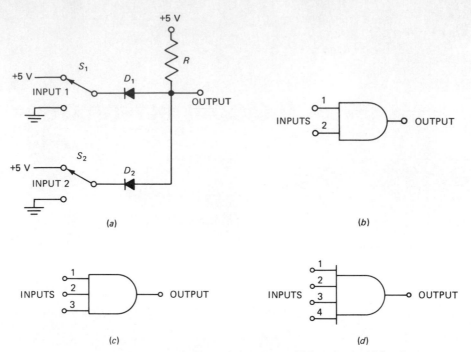

Fig. 52-2. AND gate: *(a)* Diode circuit; *(b)* two-input; *(c)* three-input; *(d)* four-input.

Transistors, MOSFETS, and other devices can also be used in the construction of AND gates. Figure 52-2b shows the schematic symbol for a two-input AND gate of any design. Figures 52-2c shows the symbol for a three-input AND gate, while Fig. 52-2d is the four-input AND gate. For these AND gates the action can be summarized like this: All inputs must be high to get a high output.

Truth Table for Two-Input AND Gate

The action of logic circuit is usually summarized in the form of *truth tables*. These are tables that show the output for all combinations of the input signals. Table 52-1 shows the truth table for a two-input AND gate.

Binary means "two." Computers use the binary number system. Rather than having digits 0 to 9, a binary number system has only digits 0 and 1. This is better suited to digital electronics where the signals are low or high, switches are open or closed, lights are off or on, etc. In our experiments we will use positive logic; this means binary 0 represents the low state and binary 1 the high state. With this is mind, Table 52-2 is the truth table of a two-input AND gate as it is usually shown. This gives the same information as Table 52-1, except it uses a binary code where 0 is low and 1 is high.

TABLE 52-2. Two-Input AND Gate

Inputs		
A	B	Output
0	0	0
0	1	0
1	0	0
1	1	1

OR Gate and Truth Table

Figure 52-3a shows a two-input OR gate. When both switches are in the ground position, the diodes are nonconducting, and the output is low. If either switch is set to +5 V, then its diode conducts and the output is approximately +4.3 V. In fact, both switches can be at +5 V and the output will be around +4.3 V. (The diodes are in parallel.)

Therefore, if either input is high or if both are high, the output is high. Table 52-3 summarizes the operation of a two-input OR gate in terms of binary 0s and 1s. As you see, if both inputs are low, the output is low. If either input is high, the output is high. If both inputs are high, the output is high.

TABLE 52-1. Two-Input AND Gate

Inputs		
A	B	Output
Low	Low	Low
Low	High	Low
High	Low	Low
High	High	High

TABLE 52-3. Two-Input OR Gate

Inputs		
A	B	Output
0	0	0
0	1	1
1	0	1
1	1	1

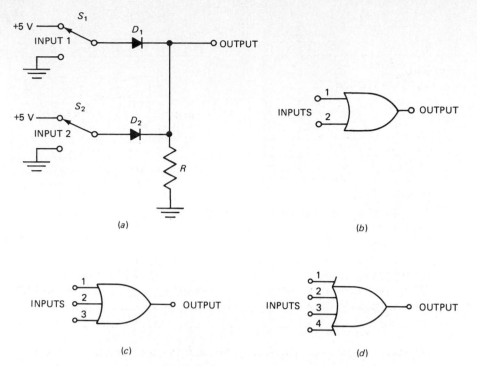

Fig. 52-3. OR gate: *(a)* Diode circuit; *(b)* two-input; *(c)* three-input; *(d)* four-input.

Unlike the AND gate where all inputs must be high to get a high output, the OR gate has a high output if any input is high.

Figure 52-3*b* shows the symbol for a two-input OR gate. By adding more diodes to the gate, we can produce three-input OR gates, four-input OR gates, and so on. Figure 52-3*c* and *d* show the schematic symbols for three- and four-input OR gates of any design.

Combined AND-OR Gates

Combinations of AND and OR gates may be used to perform complex logic operation in computers. Figure 52-4 is an example of combining AND and OR gates. To analyze this circuit, consider what happens for all possible inputs starting with all low, one low, and so on. For instance, if all inputs are low, the AND gate has a low output; therefore, both inputs to the OR gate are low and the final output is low. This is the first entry shown in Table 52-4.

Next, consider *A* low, *B* low, and *C* high. The OR gate has a high input; therefore, its final output is high. This is the second entry in Table 52-4. By analyzing the remaining input combinations, you can get the other entries shown in the truth table. (You should analyze the remaining entries.)

TABLE 52-4. AND-OR Circuit

Inputs			
A	*B*	*C*	*Output*
0	0	0	0
0	0	1	1
0	1	0	0
0	1	1	1
1	0	0	0
1	0	1	1
1	1	0	1
1	1	1	1

IC Gates

Nowadays, most logic circuits are available as ICs. The earlier digital ICs used only resistors and transistors. Known as *resistor-transistor logic* (RTL), this type of IC is now obsolete.

Next came *diode-transistor logic* (DTL). This used diodes and transistors in various designs for OR gates, AND gates, and other logic circuits. Although still occasionally used, DTL is rapidly heading toward obsolescence.

Transistor-transistor logic (TTL) became commercially available in 1964. Since then, it has become the most popular family of digital ICs. In this experiment you will work with TTL gates.

Figure 52-5*a* shows a 7408, one of the many available ICs in the TTL family. As you see, this dual-in-line package contains four AND gates. For this reason, it is called *quad two-input* AND *gate*. Notice that pin 14 is the supply pin. For

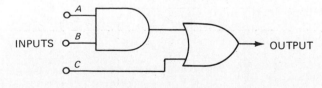

Fig. 52-4. AND-OR circuit.

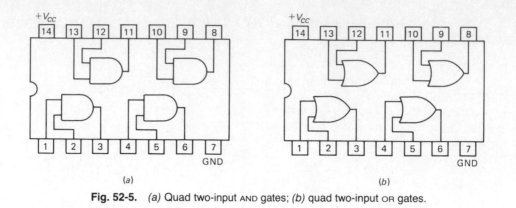

Fig. 52-5. *(a)* Quad two-input AND gates; *(b)* quad two-input OR gates.

TTL devices to work properly, the supply voltage must be between +4.75 and +5.25 V. This is why +5 V is the nominal supply voltage specified for all TTL devices. Notice also pin 7, the common ground for the chip. The other pins are for inputs and outputs.

The four AND gates are independent of each other. In other words, they can be connected to each other or to other TTL devices such as the quad two-input OR gate shown in Fig. 52-5b. Again, notice pin 14 connects to the supply voltage and pin 7 to ground.

Boolean Equations

Boolean algebra is a special algebra used with logic circuits. In boolean algebra, the variables can have only one of two values: 0 or 1. Another thing that is different about boolean algebra is the meaning of the plus and times signs. In boolean algebra, the + sign stands for the OR operation. For instance, if the inputs to an OR gate are A and B, then the output Y is given by

$$Y = A + B$$

Read this equation as Y equals A OR B. Similarly, the • sign is used for the AND operation. Therefore, the output of a two-input AND gate is written as

$$Y = A \cdot B$$

or simply as

$$Y = AB$$

Read this as Y equals A AND B.

SUMMARY

1. Digital electronics deals with voltages that are in one of two states, either high or low.
2. Digital circuits are called *logic circuits* because certain combinations of inputs determine the output.
3. In positive logic, a binary 0 represents low voltage and a binary 1 is high voltage.

4. The simplest logic circuits are two-input OR gates and two-input AND gates.
5. All inputs must be high to get a high output from an AND gate.
6. An OR gate has a high output if any input is high.
7. A *truth table* is a concise summary of all input/output combinations.
8. TTL is the most popular family of digital ICs.

SELF-TEST

Check your understanding by answering these questions.

1. Are digital circuits the same as linear circuits?
2. In a three-input AND gate all inputs must be _____ to get a _____ output.
3. In a four-input OR gate at least _____ input must be high to get a _____ output.
4. With positive logic, a binary 0 represents the _____ state and a binary 1 the _____ state.
5. _____ is the most popular family of digital ICs, two examples being the 7408 and the 7432. The first is a _____ two-input AND gate and the second is a quad two-input OR gate.
6. The nominal supply voltage for TTL is _____ .

MATERIALS REQUIRED

- Power supply: +5-V
- Equipment: Voltmeter
- ICs: 7408, 7432
- Miscellaneous: logic breadboard if available; otherwise, three SPDT switches

PROCEDURE

AND Gate

1. Connect the circuit of Fig. 52-6. (Remember to connect pin 14 to +5V and pin 7 to ground.)
2. Use the voltmeter to measure the output voltage. Set the switches as needed to get the different input combinations shown in Table 52-5. Record the state of the output as a 0 or 1 for each input possibility.

OR Gate

3. Connect the circuit of Fig. 52-7.
4. Measure the output voltage for each input combination of Table 52-6. Record the results as 0s or 1s.

Combined AND-OR Gate

5. Connect the circuit of Fig. 52-8.
6. Set the switches for each input shown in Table 52-7. Record the output states as 0s or 1s.

Extra Credit

7. Design a three-input circuit with any combination of gates to get a high output only when all inputs are high. Draw the circuit.
8. Verify the circuit experimentally. Record your results in a truth table.

Extra, Extra Credit

9. Design a four-input OR gate using any combination of gates. Draw the circuit.
10. Verify the circuit experimentally and record the results in a truth table.

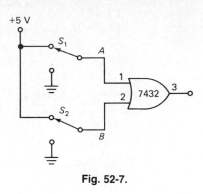

Fig. 52-7.

TABLE 52-6. Two-Input OR Gate

Inputs		
A	B	Output
0	0	
0	1	
1	0	
1	1	

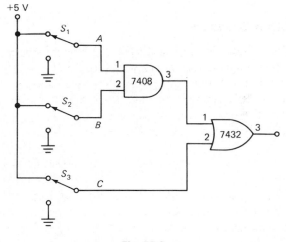

Fig. 52-8.

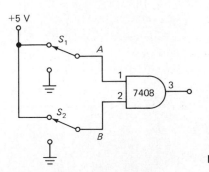

Fig. 52-6.

TABLE 52-5. Two-Input AND Gate

Inputs		
A	B	Output
0	0	
0	1	
1	0	
1	1	

TABLE 52-7. AND-OR Circuit

Inputs			
A	B	C	Output
0	0	0	
0	0	1	
0	1	0	
0	1	1	
1	0	0	
1	0	1	
1	1	0	
1	1	1	

QUESTIONS

1. What is a logic circuit?
2. Identify the two states of a logic circuit.
3. From the data in Table 52-5, what can you conclude about an AND gate?
4. From the data in Table 52-6, what can you conclude about an OR gate?
5. Refer to Fig. 52-9 and fill in the blanks of Table 52-8.

INPUTS

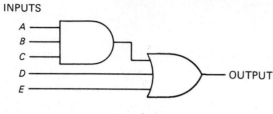

Fig. 52-9.

TABLE 52-8.

Inputs					
A	B	C	D	E	Output
1	0	0	0	0	
1	1		0	0	1
1	0	1		0	1
0	0	0	0		1
	1	1	0	0	0
0	0	0	1	1	

DIGITAL ICs: THE INVERTER, THE NOR GATE, THE NAND GATE

OBJECTIVES

1. To determine experimentally the truth table for a NOR gate
2. To use NOR logic to construct a logic inverter
3. To use NOR logic to construct a NAND gate and determine a truth table for this gate

INTRODUCTORY INFORMATION

Logic Building Blocks

Experiment 52 introduced two basic logic building blocks, the AND gate and the OR gate. These gated circuits have two or more inputs, A, B, . . ., N, and deliver an output for specific input conditions. They can evaluate input-signal levels and respond predictably when certain input conditions are met. It seems almost as though these gates were making logical decisions, hence the term "logic circuits."

There are other logic gates. In this experiment you will study the inverter, or NOT circuit, and the NOR and NAND gates.

NOT Logic

We have defined the logic operations AND and OR. There is another logic operation termed "NOT." A NOT circuit is simply an *inverter*, as in Fig. 53-1a—an amplifier, biased to cut off—whose output is 180° out of phase with its input. In the grounded-emitter configuration in Fig. 53-1a, the output is taken from the collector. In the *absence* of a signal in the input, the transistor is cut off and the output is at $+V_{CC}$; that is, it is high. When a positive pulse is applied to the base, the transistor conducts, and the voltage at the collector drops. Thus when a positive input is applied, the output goes low. So in a NOT circuit an output (high level) is present when an input is *not* applied; an output signal is *not* present (low) when an input signal is applied. The schematic symbol for a NOT circuit is shown in Fig. 53-1b.

A mathematical statement of the characteristics of an inverter is given in Eq. (53-1).

If $$V_{\text{in}} = A, \quad V_{\text{out}} = \overline{A} \qquad (53\text{-}1)$$

The *bar* over the A represents NOT. Thus, if the letter A represents a high level (1), $\overline{A}$ represents low (0), and if $A = 0$, $\overline{A} = 1$. We may also state that $\overline{0} = 1$ and that $\overline{1} = 0$.

In this experiment you will use part of a 7404, a TTL gate with six inverters. As with the 7408 and the 7432, pin 14 is the supply and pin 7 the ground.

NOR Gate and Truth Table

The three building-block circuits, AND, OR, and NOT, serve as the basis for other logic circuits. The NOR gate combines NOT and OR logic. What characterizes a NOR circuit is that an output is produced when a signal is *not* applied to input A,

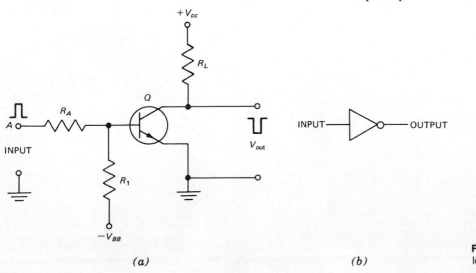

(a)   *(b)*

Fig. 53-1. *(a)* NOT or inverter circuit; *(b)* logic symbol.

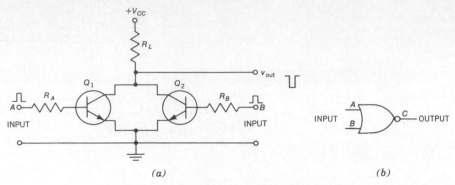

Fig. 53-2. *(a)* Two-input NOR gate; *(b)* logic symbol.

nor to input B . . . , *nor* to input N, *nor* to any combination of inputs. Figure 53-2 is a NOR gate with two inputs. Consider the base circuits of Q_1 and Q_2. In the absence of an input signal, both bases are returned to ground through the input circuits (not shown). Hence, Q_1 and Q_2 are cut off for lack of forward bias. The output, taken from the collector resistor R_L, common to both Q_1 and Q_2, is therefore equal to $+V_{CC}$ (high) when an input signal is *not* present on A *or* on B. When a positive pulse appears on either A or B or both, Q_1 or Q_2 or both, respectively, conduct, and a negative pulse appears in the output; that is, the output is low. We therefore have the effect of an inverter (NOT) and an OR circuit in the operation of this gate. The schematic symbol in Fig. 53-2*b* will be used for the NOR gate in this book. Mathematically, the operation of a NOR gate is given by

$$\overline{A + B} = V_{out} \tag{53-2}$$

The characteristics of a two-input NOR gate are given in Table 53-1.

NAND Gate and Truth Table

A circuit which combines the NOT and AND functions is called a NAND gate. A two-input NAND gate is shown in Fig. 53-3*a*. In this circuit Q_1 and Q_2 are both forward-biased and conducting heavily in the absence of an input. The output voltage taken from collector to ground is therefore close to 0, or high (1), when there is *not* a signal on A *or* on B. If a positive pulse is applied at input A but not at B, Q_1 cuts off but Q_2 still conducts and the output is still high. Similarly if a

TABLE 53-1. Two-Input NOR Gate

A	B	C	
0	0	1	$\overline{A + B} = C$
0	1	0	or
1	0	0	$\overline{A} \cdot \overline{B} = C$
1	1	0	

positive pulse is applied at input B but not at A, the output is high. However, if positive pulses are applied simultaneously at A *and* B, both Q_1 and Q_2 are cut off and the voltage at the collector drops to $-V_{CC}$; that is, the output is low or 0. This result therefore is like that which would be produced by a NOT AND circuit; hence the term "NAND." The NAND gate is therefore an AND gate with its output inverted.

A mathematical statement of the characteristics of a NAND gate is

or
$$\overline{A \cdot B} = C$$
$$\overline{A} + \overline{B} = C \tag{53-3}$$

A truth table for a two-input NAND gate is given in Table 53-2.

TTL Logic Block

Present state of the art employs integrated-circuit TTL logic in the manufacture of NOT, NOR, and NAND gates. In this experiment you will use the 7427, a TTL positive-logic IC.

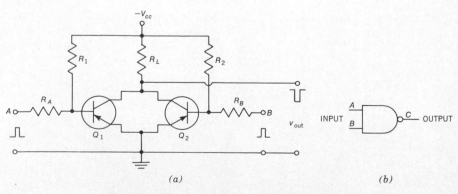

Fig. 53-3. Two-input NAND gate; *(b)* logic symbol.

TABLE 53-2. Two-Input NAND Gate

A	B	C
0	0	1
0	1	1
1	0	1
1	1	0

$$\overline{A \cdot B} = C$$
or
$$\overline{A} + \overline{B} = C$$

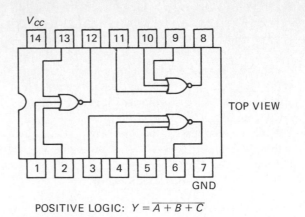

TOP VIEW

POSITIVE LOGIC: $Y = \overline{A + B + C}$

Fig. 53-5. Top view and block diagram of a 7427.

This device is a triple three-input NOR gate. The circuit of one of the three identical NOR gates (Fig. 53-4) is somewhat more complex than the NOR gate circuit (Fig. 53-2). However, since we cannot break *into* the IC circuit or change any of its components, we will work with the gates in this IC, utilizing block symbols (Fig. 53-5) rather than circuit diagrams.

Figure 53-5 is a top view of the 7427 showing the inputs and outputs of each of the three gates. Figure 53-5 shows also the connection for $+V_{CC}$, terminal 14, and the connection for ground terminal 7. The 7427 operates with a supply of +5 V.

It is possible to convert a NOR gate into an *inverter* and a combination of NOR gates into a NAND gate. Using NOR and NOT logic, we will develop the truth tables for each device.

SUMMARY

1. Logic circuits are made up of building blocks called AND, OR, NOT, NOR, and NAND circuits.
2. A NOT circuit is a logic inverter, converting a binary 1 into a 0 or a 0 into a 1.
3. A NOR gate is an OR circuit whose output is inverted. It is a NOT OR gate (or simply a NOR).
4. Table 53-1 is the truth table of a two-input NOR gate. NOR gates may have two, three, or more inputs.

5. A NAND gate is an AND gate whose output has been inverted.
6. The truth table of a NAND gate is that of an AND gate, with the output inverted.
7. NAND gates may have two, three, or more inputs.
8. A bar ($-$) placed over a letter or letters representing a logic state means that the logic level has been inverted. The bar represents NOT.
9. Mathematically a NOR gate is characterized by the expression $\overline{A + B} = C$.
10. A NAND gate is characterized mathematically by the expression $\overline{A \cdot B} = C$.

SELF-TEST

Check your understanding by answering these questions.

1. If *each* of the inputs of a three-input NAND gate is high, the output is _____ .

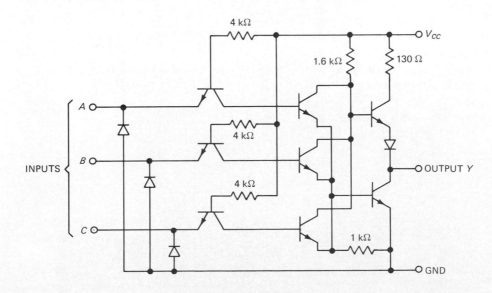

4 kΩ

V_{CC}

1.6 kΩ   130 Ω

A

4 kΩ

B

INPUTS

4 kΩ

OUTPUT Y

C

1 kΩ

GND

COMPONENT VALUES SHOWN ARE TYPICAL.

Fig. 53-4. Circuit diagram of one of the triple three-input NOR gates in a 7427.

2. The expression 1 + 0 represents a _____ gate, one of whose inputs is _____, the other _____ .

3. A binary 1 is changed into a binary 0 by a circuit called a(n) _____ or _____ circuit.

4. A circuit whose logic is the inverse of AND logic is called a(n) _____ gate.

5. A circuit whose logic is the inverse of NOR logic is called a(n) _____ gate.

6. The symbol for an inverter is _____ .

7. The symbol for a two-input NAND gate is

_____ .

8. The symbol for a three-input NOR gate is

_____ .

MATERIALS REQUIRED

- Power supply: Regulated low-voltage dc source
- Equipment: EVM or VOM
- ICs: 7427 (three-input NOR gate), 7404
- Resistor: Two 1000-, 1800-, and 4700-Ω ½-W
- Miscellaneous: Three SPDT switches

PROCEDURE

NOTE: Use the 7427 shown in Fig. 53-5.

NOR Gate Connected as Inverter

1. Connect one of the NOR gates in the 7427, as in Fig. 53-6. The voltage divider provides the proper dc input levels for the gate. In this IC, a binary 1 is represented by +2.4 to +5.0 V (approx) and binary 0 is represented by 0 to +0.4 V (approx). S_1 delivers either 0 or +3.2 V to the input of the gate. In this and subsequent steps pin 14 of the IC is connected to +5 V, and pin 7 is connected to ground.

2. Complete the truth table of Fig. 53-6.

NOR Gate Logic

3. Connect the circuit shown in Fig. 53-7 and complete the truth table.

4. Connect the circuit shown in Fig. 53-8 and complete the truth table.

5. Connect the circuit shown in Fig. 53-9 and complete the truth table.

6. Connect the circuit in Fig. 53-10 and complete the truth table.

Extra Credit

7. Design a three-input AND gate utilizing the 7427 and the 7404. Draw the circuit showing all connections, numbering all terminals.

8. Connect the circuit and complete a truth table.

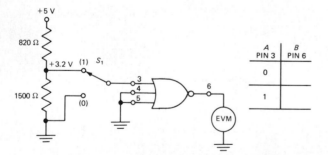

Fig. 53-6. Experimental circuit 1 and truth table 1.

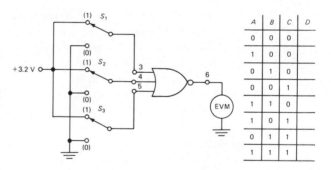

Fig. 53-7. Experimental circuit 2 and truth table 2.

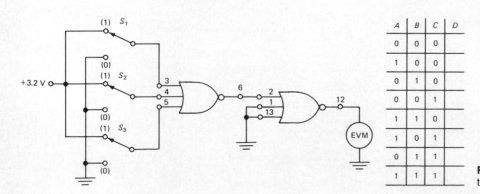

Fig. 53-8. Experimental circuit 3 and truth table 3.

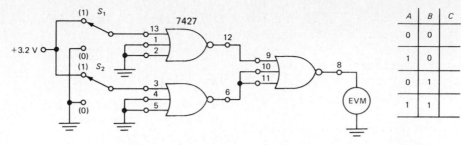

A	B	C
0	0	
1	0	
0	1	
1	1	

Fig. 53-9. Experimental circuit 4 and truth table 4.

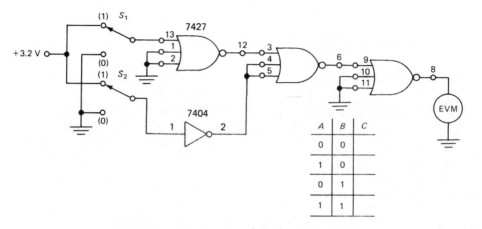

A	B	C
0	0	
1	0	
0	1	
1	1	

Fig. 53-10. Experimental circuit 5 and truth table 5.

QUESTIONS

1. What are the characteristics of an inverter?
2. What are the characteristics of a NOR gate?
3. What are the characteristics of a NAND gate?
4. Identify each of the experimental circuits by its logic name (inverter, three-input NOR gate, etc).
   (a) Circuit 1 (Fig. 53-6) _____
   (b) Circuit 2 (Fig. 53-7) _____
   (c) Circuit 3 (Fig. 53-8) _____
   (d) Circuit 4 (Fig. 53-9) _____
   (e) Circuit 5 (Fig. 53-10) _____

Answers to Self-Test

1. low
2. NOR; UP; high
3. inverter; NOT
4. NAND
5. NOR
6.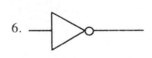
7.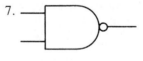
8.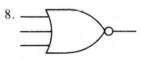

DIGITAL ICs: BINARY ADDITION AND THE FULL ADDER

OBJECTIVES

1. To learn the rules of binary addition
2. To convert a decimal into a binary number, and a binary into a decimal number
3. To construct a full adder using IC logic blocks

INTRODUCTORY INFORMATION

Two-State Nature of Computer Components

As explained in a previous experiment, computer components generally have two operating states. A hole is either present or absent in a given location in a card; an electromagnetic relay maintains its contacts either opened or closed; a piece of magnetic material may be magnetized in one direction to represent information or magnetized in the opposite direction to represent lack of information. Presence or absence of information can also be indicated by a dc level on a signal line.

The binary system of arithmetic uses only two symbols (0 and 1) to represent all quantities. This system finds wide use in computers because the 0 and 1 are easily represented by the two-state digital circuits.

A correlation between the binary digits 0 and 1 and computer components may be made. For example, a hole in a card can represent a binary 1; no hole can represent a binary 0. In the case of a magnetic material, magnetic flux in one direction can represent 1; flux in the opposite direction can represent 0.

In transmitting digits from one part of a computer to another, a binary 1 can be represented by a positive level on a line; a binary 0 can be represented by a negative level, or vice versa.

A computer must do more than store digits and transmit them from one place to another; it must calculate by proper manipulation of the digits. In general, the calculating operations can be reduced to the four functions of adding, subtracting, multiplying, and dividing. The purpose of this experiment is to study the operation of a binary adder; but first, a brief study of binary addition will be necessary.

Binary Arithmetic

Counting is started in the binary system in the same way as in the decimal system with 0 for zero and 1 for one. But at two in the binary system there are no more symbols. Therefore, the same move must be taken at two in the binary system that is taken at ten in the decimal system: Thus it is necessary to place a 1 in the position to the left and start again with a 0 in the original position. Table 54-1 is a list of numbers shown in both decimal and binary form.

The order of a binary number is not designated unit, tens, hundreds, thousands, etc., as in the decimal system. Instead, the order is 1, 2, 4, 8, 16, 32, 64, 128, and so on, reading from right to left with the position farthest to the right being 1. Table 54-2 shows more decimal quantities and their equivalents in binary form. Note how the positions are numbered right to left.

To understand how a decimal quantity is converted to a binary quantity, consider the quantity 75. The first step is to find the largest number in the binary order that can be subtracted from 75. In this case the number is 64, so a binary 1 is placed in the column under 64 in the table. The first step has left a remainder of 11. The next step is to find the largest number in the binary order that can be subtracted from 11. Because 8 is the largest number meeting the requirement, a 1 is placed under 8 in the table. The remainder is now 3. The largest number in the binary order that can be subtracted is 2, so a 1 is placed under position 2 in the table. Next, 1 is subtracted from the remainder of 1, leaving 0, and a 1 is

TABLE 54-1. Decimal and Binary Numbers

Decimal	Binary	Decimal	Binary
0	0	6	110
1	1	7	111
2	10	8	1000
3	11	9	1001
4	100	10	1010
5	101	11	1011

TABLE 54-2. Decimal Numbers and Their Binary Equivalents

				Binary					
Decimal	256	128	64	32	16	8	4	2	1
34				1	0	0	0	1	0
15						1	1	1	1
225		1	1	1	0	0	0	0	1
75			1	0	0	1	0	1	1

placed under the binary order 1 in the table. The quantity 75 can therefore be written in binary form as 1001011. This actually represents the quantity $64 + 8 + 2 + 1$.

Addition of binary quantities is very simple and is based on the following three rules:

1. $0 + 0 = 0$
2. $0 + 1 = 1$
3. $1 + 1 = 0$ with a 1 carry to the left

Table 54-3 is an example of binary addition using the rules stated.

The factors to be added are 75 and 225. Starting at the right, we have $1 + 1 = 0$ with a 1 carry (rule 3).

The next position to the left is added: $0 + 1 = 1$. However when we add the 1 carry, the sum becomes 0 with 1 carried to the third position. The third position consists of $0 + 0 = 0 + 1$ (carry) $= 1$. This procedure is followed until all positions are added. The sum is given in binary form as 100101100, which is equal to $256 + 32 + 8 + 4 = 300$. This sum is exactly what we would expect to get by adding the decimal quantities 225 and 75.

Binary quantities can also be subtracted, multiplied, and divided, using rules similar to those for addition.

Binary Half Adder and Truth Table

The simplest binary adder is called a *half adder* and is capable of combining two binary numbers and providing an output and a carry when necessary. The first step in understanding the operation of a half adder is to investigate the input combinations and the resulting outputs based on the rules of binary addition. Table 54-4 is a truth table showing these combinations.

Table 54-4 shows that a binary 1 on either input results in a binary 1 sum and binary 0 carry. A binary 1 on both inputs results in a binary 0 sum and a binary 1 carry. A binary 0 on both inputs results in a binary 0 sum and binary 0 carry.

TABLE 54-3. Adding Binary Numbers

Decimal Value	Binary Value								
	256	128	64	32	16	8	4	2	1
		1	1					1	1
225	0	1	1	1	0	0	0	0	1
75	0	0	1	0	0	1	0	1	1
	1	0	0	1	0	1	1	0	0

TABLE 54-4. Truth Table for Half Adder

Input		Sum	Carry
A	B		
0	0	0	0
0	1	1	0
1	0	1	0
1	1	0	1

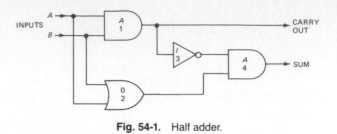

Fig. 54-1. Half adder.

Figure 54-1 is a block diagram of AND-OR logic circuits and a dc inverter that will function according to the rules of binary addition.

Using a positive level for a binary 1 and a negative level for a binary 0, the circuit operates as follows:

A binary 1 on either input A or input B produces an up level in the output of OR block 2. This up level is combined with the output of the inverter block 3 to produce an up level (binary 1 sum) at the output of AND block 4. Note that the input to the inverter must be down to produce an up level at its output. The input to the inverter is down, except where there is an up level on inputs A and B at the same time. When both inputs are at a down level, there is a down level on the sum and carry lines. A simultaneous up level on inputs A and B results in an up level (binary 1) on the carry line. At this time, however, the output of the inverter is down, deconditioning AND block 4; the deconditioned AND block produces a down level on the sum line.

The half adder has only limited use because there are no provisions for a carry input from a previous adder.

Binary Full Adder and Truth Table

When a carry and the two quantities to be added are considered as inputs, the input combinations increase to eight as shown in Table 54-5. An adder capable of producing the required outputs for the eight input combinations is called a *full adder*. The full adder is shown in the block diagram of Fig. 54-2.

Using the rules of AND-OR logic, it can be shown that any input combination to the full adder of Fig. 54-2 produces an output according to the rules of binary addition.

TABLE 54-5. Truth Table for a Full Adder

Inputs			Outputs	
A	B	C	Sum	Carry
0	0	0	0	0
1	0	0	1	0
0	1	0	1	0
0	0	1	1	0
1	1	0	0	1
1	0	1	0	1
0	1	1	0	1
1	1	1	1	1

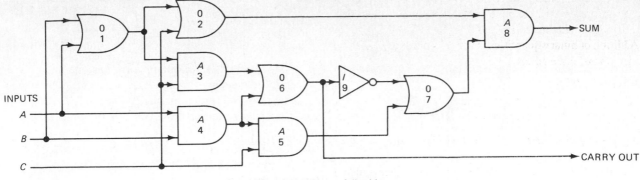

Fig. 54-2. Binary full adder.

Consider the condition where a binary 1 is present on each of the three inputs *A*, *B*, and *C*. From the truth table we find that the sum and carry output lines should each contain a binary 1. With the input conditions stated, AND block 4 has an up level on its output. This up level is combined with the up level on the input carry line in AND block 5 to produce an up level on the lower input to OR block 7. The output of OR block 7 is now up, producing an up level on the lower leg of AND block 8.

Going back to the input, we find that OR block 1 has an up level on its output, which in turn produces an up level on the output of OR block 2. The output of OR block 2 and OR block 7 (both outputs being up at this time) produces an up level (binary 1) on the sum line at the output of AND block 8. At this point, one of the conditions in the truth table has been satisfied.

We must now obtain an up level on the output carry line. Because the output of AND block 4 is up at this time, the output of OR block 6 also has an up level, which also appears on the output carry line. Following the same analysis procedure for any of the other input combinations shown in the truth table, the indicated output can be obtained.

The full adder shown represents a single position in a binary-adder system. Because many such adders are combined in a large computer, each full adder is represented as a block in the computer logic diagram. An example of a five-position binary adder is shown in Fig. 54-3. The actual number of positions in such an adder depends on the size of the computer and the type of calculations the computer is designed for.

Experimental Full Adder

The AND and OR gates of the binary full adder in Fig. 54-2 may consist of diodes or transistors, or they may be incorporated in a single IC block consisting of three inputs and two outputs—the sum and carry lines. In this experiment, we will build a full adder using the AND gates included in a quad two-input AND-gate IC and the OR gates in a quad two-input OR-gate IC. We used these two ICs in Experiment 52. Figure 54-2 shows that a binary full adder requires four AND and four OR gates and an inverter. The ICs contain the required number of gates. The inverter, block *I*-9 in Fig. 54-2, will be constructed using a transistor with discrete components, as in the preceding experiment.

SUMMARY

1. The binary number system for digital computers uses only two symbols, 1 and 0. These have the same meaning as 1 and 0 in the decimal number system with which you are so familiar.

2. In the decimal or base-10 system the value of each digit in a number is some power of 10 and depends on its position in the number. For example, in the number 527, the 7 is in the units (10^0) column and counts for 1×7, or 7; the 2 is in the tens (10^1) column and counts for 2×10, or 20. The 5 is in the hundreds (10^2) column and counts for 5×10^2, or 500.

3. Numbers in the binary system are formed exactly as they are in the decimal, except the value of a column is a power

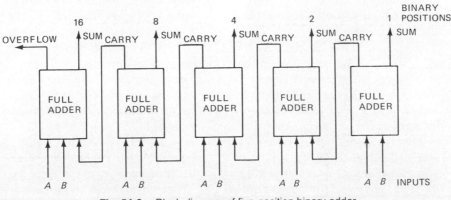

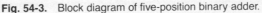

Fig. 54-3. Block diagram of five-position binary adder.

of 2 rather than of 10, with the extreme right-hand column having the value 2^0 or 1. The next column on the left has the value 2^1 or 2; the next 2^2 or 4; the next 2^3 or 8, etc. The value of the first seven binary columns, reading from right to left, are:

... 64	32	16	8	4	2	1
		1	0	1	1	0

4. To convert a binary number into its base-10 (decimal) equivalent, place each of the binary digits in its proper binary column, just as the number 10110 is placed above. Now multiply each binary digit (1 or 0) by its value in the column and add the result. Thus the binary number 10110 $= 1 \times 16 + 0 \times 8 + 1 \times 4 + 1 \times 2 + 0 \times 1 = 22$, the base-10 equivalent.

5. Addition of binary numbers is based on the following rules:

(1) $0 + 0 = 0$
(2) $1 + 0 = 1$
(3) $1 + 1 = 0$ with a carry to the left

6. Table 54-2 shows how a decimal number is converted into its equivalent binary number.

7. The reason digital computers use the binary number system is that digital circuits are two-state circuits—they are either ON (1) or OFF (0).

8. A *half adder* (Fig. 54-2) is a binary adder which combines two binary digits and provides an output and a carry. A half adder has four possible input combinations (Table 54-4).

9. A *full adder* (Fig. 54-2) is a binary adder which combines three binary digits and provides an output and a carry. One of the inputs may be a carry from a previous arithmetic operation. A full adder has eight possible input combinations (Table 54-5).

SELF-TEST

Check your understanding by answering these questions.

1. A number written in binary form has one and only one equivalent decimal value. _____ (*true, false*)
2. The number 7 written in binary form is _____ .
3. The result of adding these two binary numbers is:

   1011011
   1101001
   _____

4. The value of the number 1011011 in decimal form is _____ .
5. The value of the number 1101001 in decimal form is _____ .
6. The number 196 written in binary form is _____ .
7. Complete the truth table for the half adder if the inverter input is shorted to the output:

A	B	Carry	Sum
0	0		
1	0		
0	1		
1	1		

MATERIALS REQUIRED

- Power supply: Variable regulated low-voltage dc
- Equipment: EVM or VOM
- Resistors: 4700-, 10,000-Ω ½-W
- Integrated circuits: 7408, 7432, 7404
- Miscellaneous: Three SPDT toggle switches

PROCEDURE

1. Connect the full-adder circuit (Fig. 54-4). The AND gates are contained in the 7408 and the OR gates in the 7432. The inverter is a 7404. Connect pin 14 of each of the ICs to +5 V of the supply and pin 7 to ground. The other pins are interconnected as shown.

2. The level on each of the three input lines is controlled by the respective SPDT switch, S_A, S_B, or S_C. When the switch is closed on the +5-V side, a binary 1 is present on the input line. When the switch is closed on the 0-V side, a down level representing a binary 0 is present.

   Using the notations high and low, set the input switches for the combinations shown in Table 54-6 and record the conditions for the sum and carry output lines. When a positive level exists on an output line, record high in the proper position in the table. For a negative output level, record low.

   Do the output results recorded in the chart agree with the outputs in Table 54-5?

3. Refer to Fig. 54-4 and assume the following conditions: Input A = binary 1; input B = binary 1; input C = binary 0; and terminal 13 in AND block 4 is open.

   What is the level on the sum line under these conditions? Is this level correct for the input conditions given? Check your answer by operating your laboratory circuit with the terminal open and with the input switches set for the conditions shown. Reconnect the terminal when finished.

4. Refer to Fig. 54-4 and assume the following conditions: Input A = binary 1; input B = binary 1; input C = binary 0; and terminal 10 in OR block 3 is open.

   Determine the levels of the output lines by analyzing the circuit operation under the conditions stated. Is output

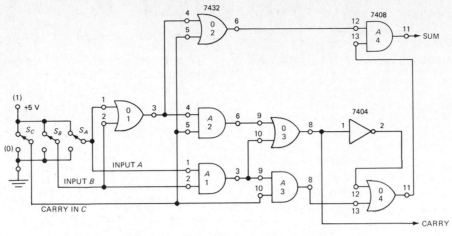

Fig. 54-4. Full-adder experimental circuit.

correct for the input conditions used? Check your answer by operating your laboratory circuit with the terminal open. Reconnect the terminal when the test is completed.

5. Assume that there is no inverter in Fig. 54-4, and that the output of OR block 3 (terminal 8) is connected directly to terminal 12 or OR block 4. Fill in the blanks of Table 54-7

for the input combinations shown. Check your answers by operating your laboratory circuit with the inverter removed and terminal 8 of OR block 3 shorted to terminal 12 of OR block 4. Observe the outputs for the various input combinations and compare the results with your answers in Table 54-7.

TABLE 54-6. Logic of a Full Adder

Inputs			Outputs	
A	B	C	Sum	Carry
Low	Low	Low		
High	Low	Low		
Low	High	Low		
Low	Low	High		
High	High	Low		
High	Low	High		
Low	High	High		
High	High	High		

TABLE 54-7. Defective Full Adder

Inputs			Outputs	
A	B	C	Sum	Carry
0	0	0		
1	0	0		
0	1	0		
0	0	1		
1	1	0		
1	0	1		
0	1	1		
1	1	1		

QUESTIONS

1. Why is the binary number system preferred to the decimal system for use in computers?
2. What is the main difference between a half adder and a full adder?
3. Write the quantity 896 in binary form.
4. Convert the binary quantity 1001000111 to its decimal equivalent.

Answers to Self-Test

1. true
2. 111
3. 11000100
4. 91
5. 105
6. 11000100
7.

Carry	Sum
0	0
0	0
0	0
1	1

OBJECTIVES

1. To construct an RS flip-flop using NOR gates
2. To observe the action of a D flip-flop
3. To observe the action of a JK flip-flop

INTRODUCTORY INFORMATION

Bistable multivibrators, called *flip-flops*, find extensive use in computers. Because a flip-flop is bistable, it can remain in either of two stable states until an external trigger forces it into the other state. As a result, a flip-flop is a memory element that can store binary data.

RS Flip-Flop

Experiment 46 discussed a type of flip-flop called the *RS flip-flop*, symbolized in Fig. 55-1. Table 55-1 summarizes the operation. When both control inputs are low, no change can occur in the output and the circuit remains latched in its last

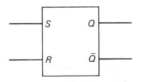

Fig. 55-1. Symbol for RS flip-flop.

state. This condition is called the *inactive* state because nothing changes.

When R is low and S is high, the circuit sets the Q output to a high. On the other hand, if R is high and S is low, the Q output resets to a low.

Look at the final entry of Table 55-1. R and S are high simultaneously. This is called a *race condition*; it is never used because it leads to unpredictable operation. It means you are trying to set and reset the flip-flop at the same time, which is a contradiction. From now on, an asterisk in a truth table indicates a race condition, sometimes called a *forbidden* or *invalid* state.

NOR Latches

A discrete circuit with separate transistors, resistors, etc., is rarely used because we are in the age of integrated circuits. Nowadays, we build RS flip-flops with cross-coupled NOR or NAND gates.

Figure 55-2a is a NOR *latch*, equivalent to an RS flip-flop. As shown in Table 55-2, a low R and a low S produce the inactive state; in this state, the circuit stores or remembers. A low R and a high S represent the set state, while a high R and a low S give the reset state. Finally, a high R and a high S produce a race condition, where the output is uncertain; therefore, we must avoid $R = 1$ and $S = 1$ when using a NOR latch.

TABLE 55-1.   RS Latch

R	S	Q	Comment
0	0	NC	No change
0	1	1	Set
1	0	0	Reset
1	1	*	Race

TABLE 55-2.   NOR Latch

R	S	Q	Comment
0	0	NC	No change
0	1	1	Set
1	0	0	Reset
1	1	*	Race

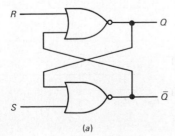

(a)

(b)

Fig. 55-2. *(a)* NOR latch; *(b)* timing diagram.

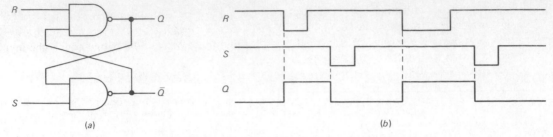

Fig. 55-3. *(a)* NAND latch; *(b)* timing diagram.

Figure 55-2*b* is a timing diagram; it shows how the input signals interact to produce the output signal. As you see, the *Q* output goes high when *S* goes high. *Q* remains high after *S* goes low. *Q* returns to low when *R* goes high, and stays low after *R* returns low.

NAND Latches

Figure 55-3*a* shows an RS latch built with cross-coupled NAND gates. Because of the NAND-gate inversion, the inactive and race conditions are reversed as shown in Table 55-3. Therefore, whenever you use a NAND latch, you must avoid having both inputs low at the same time.

Figure 55-3*b* shows the timing diagram of a NAND latch. *R* and *S* are normally high to avoid the race condition. Only one of them goes low at any time. As you see, the *Q* output goes high whenever *R* goes low; the *Q* output goes low whenever *S* goes low.

Clocking

Computers use thousands of flip-flops. To coordinate the overall action, a square-wave signal called the *clock* is sent to each flip-flop. This signal prevents the flip-flops from changing states until the right time.

Figure 55-4*a* shows a clocked RS flip-flop. The idea is simple. When the clock is low, the AND gates are disabled,

and the *S* and *R* signals cannot reach the flip-flop. But when the clock goes high, the *S* and *R* signals can drive the flip-flop, which then sets, resets, or does nothing depending on the values of *S* and *R*. The point is the clock controls the timing of the flip-flop action.

Figure 55-4*b* shows the timing diagram. *Q* goes high when *S* is high and *CLK* goes high. *Q* returns to the low state when *R* is high and *CLK* goes high. Using a common clock signal to drive many flip-flops allows us to synchronize the operation of the different sections of a computer.

Table 55-4 summarizes the operation of the clocked RS flip-flop. When the clock is low, the output is latched in its last state. When the clock is high, the circuit will set if *S* is high or reset if *R* is high. *CLK, R,* and *S* all high simultaneously is a race condition, which is never used deliberately.

D Latches

Since the RS flip-flop is susceptible to a race condition, we can modify the design to eliminate the possibility of a race condition. The result is a new kind of flip-flop called the *D latch*.

TABLE 55-4. Clocked NAND Latch

CLK	*R*	*S*	*Q*
0	0	0	NC
0	0	1	NC
0	1	0	NC
0	1	1	NC
1	0	0	NC
1	0	1	1
1	1	0	0
1	1	1	*

TABLE 55-3. NAND Latch

R	*S*	*Q*	*Comment*
0	0	*	Race
0	1	1	Set
1	0	0	Reset
1	1	NC	No change

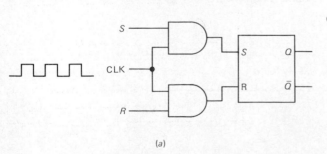

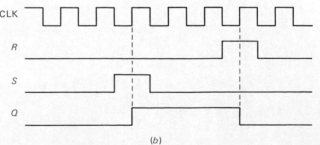

Fig. 55-4. *(a)* Clocked RS flip-flop; *(b)* timing diagram.

Figure 55-5 shows one way to build a D latch. Because of the inverter, data bit D drives the S input and the complement $\bar{D}$ drives the R input. Therefore, a high D sets the latch, and a low D resets it. Table 55-5 summarizes the operation of the D latch. Especially important, there is no race condition in this truth table. The inverter guarantees that S and R are always in opposite states; therefore, it is impossible to set up a race condition.

Usually, a D flip-flop is clocked as shown in Fig. 55-6. When CLK is low, the AND gates are disabled and the RS latch remains inactive. When CLK is high, D and $\bar{D}$ can pass through the AND gates and set or reset the latch.

Table 55-6 summarizes the operation. X represents a

"don't-care" condition; it stands for either 0 or 1. While CLK is low, the output cannot change, no matter what D is. When CLK is high, however, the output equals the input.

Edge-Triggered D Flip-Flops

Look at Fig. 55-7a. By deliberate design, the time constant of the input RC circuit is much smaller than the clock pulse width. Because of this, the capacitor can charge fully when CLK goes high; this exponential charging produces a narrow positive voltage spike across the resistor. Later, the trailing edge of the clock pulse results in a narrow negative spike.

The narrow positive spike enables the AND gates for an instant; the narrow negative spike does nothing. The effect is to activate the input gates during the positive spike, equivalent to sampling the value of D for an instant. At this unique point in time, D and its complement hit the latch inputs, forcing Q to set or reset.

This kind of operation is called *edge triggering* because the flip-flop responds only when the clock is changing states. The triggering of Fig. 55-7a occurs on the positive-going edge of the clock; this is why it is referred to as *positive-edge triggering*.

Figure 55-7b is the timing diagram. The crucial idea is this: The output can change only on the rising edge of the clock. Put another way, data is stored only on the positive-going edge.

Table 55-7 summarizes the operation of the positive-edge-triggered D flip-flop. The up and down arrows represent the rising and falling edges of the clock. The first three entries indicate there is no output change when the clock is low, high, or on its negative edge. The last two entries indicate an output change on the positive edge of the clock.

PRESET and CLEAR

When power is first applied, flip-flops come up in random states. To get some computers started, an operator has to

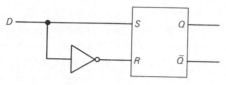

Fig. 55-5. D latch.

TABLE 55-5. Unclocked D Latch

D	Q
0	0
1	1

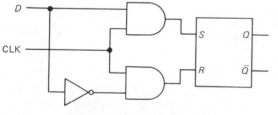

Fig. 55-6. Clocked D latch.

TABLE 55-6. Clocked D Latch

CLK	D	Q
0	X	NC
1	0	0
1	1	1

TABLE 55-7. Edge-Triggered D Flip-Flop

CLK	D	Q
0	X	NC
1	X	NC
$\downarrow$	X	NC
$\uparrow$	0	0
$\uparrow$	1	1

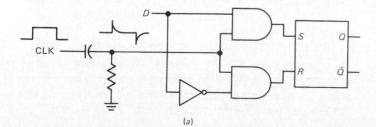

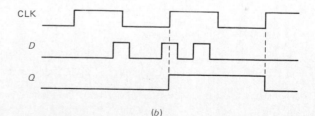

Fig. 55-7. (a) Edge-triggered D flip-flop; (b) timing diagram.

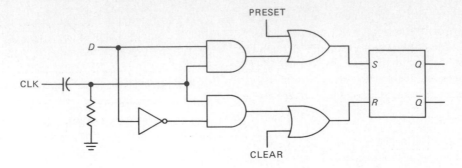

Fig. 55-8. Edge-triggered D flip-flop with PRESET and CLEAR.

push a master reset button. This is a *clear* (reset) signal to all flip-flops. Also, it is necessary in some computers to *preset* (synonymous with set) certain flip-flops before a computer run.

Figure 55-8 shows how to include both functions in a D flip-flop. The edge triggering is the same as previously described. In addition, the OR gates allow us to slip in a high PRESET or high CLEAR when desired. A high PRESET sets the latch; a high CLEAR resets it.

PRESET is sometimes called *direct set,* and CLEAR is sometimes called *direct reset.* The word "direct" means unclocked. For instance, a clear signal may come from a push button; regardless of what the clock is doing, the output will reset when the operator pushes the clear button.

Logic Symbol

Figure 55-9a is the logic symbol of a positive-edge-triggered D flip-flop. The *CLK* input has a small triangle, a reminder of the edge triggering. When you see this symbol, remember what it means: The *D* input is sampled and stored on the rising edge of the clock. Also included are the preset and clear inputs; if either of these goes high, the output is set or reset.

Figure 55-9b is another logic symbol for the positive-edge-triggered D flip-flop. The positive-edge triggering is identical to before. In some applications, it is preferable to have active-low PRESET and CLEAR. This means a low PRESET will set the flip-flop; a low CLEAR will reset it. As a reminder of the phase reversal, inversion bubbles are shown on the preset and clear inputs.

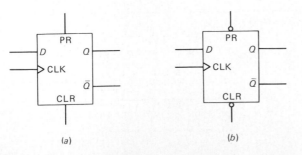

Fig. 55-9. Symbols for edge-triggered D flip-flop: *(a)* Active-high PRESET and CLEAR; *(b)* active-low PRESET and CLEAR.

Edge-Triggered JK Flip-Flops

Figure 55-10a shows one way to build a JK flip-flop. As before, an *RC* circuit with a short time constant converts the rectangular *CLK* pulse to narrow spikes. The *J* and *K* inputs are control inputs; they determine what the circuit will do on the positive clock edge. When *J* and *K* are low, both inputs are disabled and the circuit is inactive.

When *J* is low and *K* is high, the upper gate is disabled; there is no way to set the flip-flop. The only possibility is reset. When *Q* is high, the lower gate passes a reset trigger as soon as the positive clock edge arrives. This forces *Q* to become low. Therefore, $J = 0$ and $K = 1$ means a rising clock edge resets the flip-flop.

On the other hand, when *J* is high and *K* is low, the lower gate is disabled and it is impossible to reset the flip-flop. But it can be set as follows. If *Q* is low, $\overline{Q}$ is high; therefore, the upper gate can pass a set trigger on the next positive clock edge. This drives the flip-flop into the set state.

The final possibility is both *J* and *K* high. If *Q* is high, the lower gate passes a reset trigger on the next positive clock edge. On the other hand, if *Q* is low, the upper gate passes a set trigger on the next positive clock edge. Either way, *Q* changes to the complement of the last state. Therefore, $J = 1$ and $K = 1$ means the flip-flop will *toggle* on the next positive clock edge. (Toggle means switch to the opposite state.)

Figure 55-10b is a visual summary of the action. When *J* is high and *K* is low, the rising clock edge sets *Q* to high. When *J* is low and *K* is high, the rising clock edge resets *Q* to low. Finally, if both *J* and *K* are high, the output toggles once each rising clock edge.

Table 55-8 summarizes the action. The circuit is inactive when the clock is low, high, or on its negative edge. Likewise, the circuit is inactive when *J* and *K* are both low. Output changes occur only on the rising edge of the clock as indicated by the last three entries of the table. The output either resets, sets, or toggles.

A variety of JK flip-flops are available in IC form. Figure 55-11a is the symbol for one type. It uses positive-edge triggering, and responds to high PRESET and CLEAR. Figure 55-11b is a positive-edge-triggered JK flip-flop that responds to low preset and clear signals. If the IC design includes an internal inverter on the clock input, we get negative-edge triggering which is preferred in some applications. As a

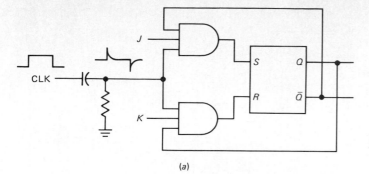

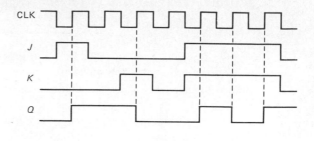

(a)

(b)

Fig. 55-10. (a) Edge-triggered JK flip-flop; (b) timing diagram.

TABLE 55-8. Positive-edge-triggered JK Flip-flop

CLK	J	K	Q
0	X	X	NC
1	X	X	NC
↓	X	X	NC
X	0	0	NC
↑	0	1	0
↑	1	0	1
↑	1	1	Toggle

reminder of this negative-edge triggering, Fig. 55-11c has a bubble at the clock input; it also has active-low PRESET and CLEAR.

SUMMARY

1. A *flip-flop* can remain in its last state until an external trigger forces it into the other state. Because of this, it is a memory element.
2. In the inactive state, a flip-flop stores or remembers because it remains in its last state.
3. A *race condition* exists when both R and S are high in an RS flip-flop. This undesirable state is forbidden because it represents a contradiction.
4. One way to build an RS flip-flop is with cross-coupled NOR gates. Alternatively, NAND gates can be used.
5. Usually, a signal called the *clock* determines when a flip-flop can change states.
6. By including an inverter, we can convert an RS flip-flop into a D flip-flop. The big advantage of the D flip-flop is the lack of a race condition.

7. A *positive-edge-triggered* D flip-flop stores the data bit only on the rising edge of the clock.
8. PRESET and CLEAR allow a direct set or a direct reset of a flip-flop, regardless of what the clock is doing.
9. Depending on the values of J and K, a JK flip-flop will either do nothing, set, reset, or toggle.

SELF-TEST

Check your understanding by answering these questions.

1. A _____ condition is sometimes called a *forbidden* or *invalid* state.
2. RS flip-flops can be built with cross-coupled _____ or _____ gates.
3. A square-wave signal called the _____ can synchronize the operation of many flip-flops.
4. The D flip-flop has the advantage of no _____ condition.
5. A flip-flop that responds only on the rising _____ of the clock is called a _____ edge-triggered flip-flop.
6. For a JK flip-flop to toggle, J must be _____ and K must be _____ .

MATERIALS REQUIRED

- Power supply: +5-V
- Equipment: Voltmeter; for extra credit a square-wave generator and oscilloscope
- ICs: 7402, 7404, 7474, 7476
- Miscellaneous: Logic breadboard if available; otherwise, three SPDT switches

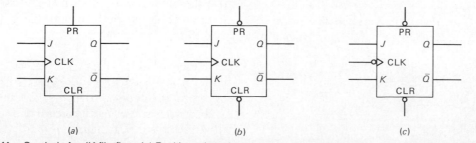

(a)   (b)   (c)

Fig. 55-11. Symbols for JK flip-flop: (a) Positive edge-triggering with active-high PRESET and CLEAR; (b) positive-edge triggering with active-low PRESET and CLEAR; (c) negative-edge triggering with active-low PRESET and CLEAR.

PROCEDURE

RS Latch

1. Connect the NOR latch of Fig. 55-12a. (Remember pin 14 goes to +5 V and pin 7 to ground.)
2. Use the voltmeter to measure the output voltages from pins 1 and 4.
3. Set the R and S switches to the input combinations of Table 55-9. Follow the order shown; record the Q and $\overline{Q}$ outputs for each input.

D Latch

4. Connect the D latch of Fig. 55-12b.
5. Set the D switch to the low input. Measure and record Q and $\overline{Q}$ in Table 55-10.
6. Repeat the preceding step for the D switch at the high input.

Edge-Triggered D Flip-Flop

7. When TTL inputs float, they are automatically high because they are internally returned to a high voltage through resistors. Connect the circuit of Fig. 55-13 and notice that input pins 1 and 4 are floating.
8. Put switches S_1 and S_2 in the ground position. Measure the Q output (pin 5) with a voltmeter. Ground pin 1

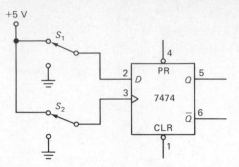

Fig. 55-13. Experimental edge-triggered D flip-flop.

(CLEAR) and notice the output is low. Float pin 1 and the output should stay low.
9. Ground pin 4 (PRESET) and the output should go high. Float pin 4 and it should stay high.
10. Set the D switch to the low position. Simulate a positive clock edge by throwing the CLK switch from ground to +5 V. Record the Q output in Table 55-11.
11. Set the D switch to the high position. Throw the CLK switch from low to high; then measure and record the Q output.

JK Flip-Flop

12. Connect the circuit of Fig. 55-14. With a 7476, pin 5 connects to +5 V and pin 13 to ground.
13. Measure the Q output (pin 15) with a voltmeter. Alternately ground pin 2 and 3. Notice how this presets and clears the Q output. Leave pins 2 and 3 floating.
14. Set J and K to low (first entry in Table 55-12). Notice whether Q is low or high. Apply a positive clock edge by throwing S_2 from low to high. Q should not change. If this is what happens, write "NC" in Table 55-12.

TABLE 55-9.   RS Latch

R	S	Q	$\overline{Q}$
0	1		
0	0		
1	0		
0	0		

TABLE 55-10.   D Latch

D	CLK	$\overline{Q}$
0		
1		

TABLE 55-11.   Edge-Triggered D Flip-Flop

D	CLK	Q
0	↑	
1	↑	

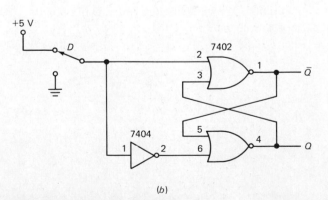

Fig. 55-12. (a) Experimental RS latch; (b) experimental D latch.

334   *Basic Electronics*

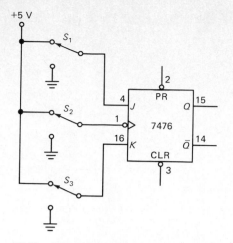

Fig. 55-14. Experimental edge-triggered JK flip-flop.

15. Set up the other J and K inputs listed in Table 55-12. Apply positive clock edges by throwing S_2 from low to high. Record the Q outputs. (Record "Toggle" for the last entry if it is working correctly.)

TABLE 55-12. JK Flip-Flop

J	K	CLK	Q
0	0	↑	
1	0	↑	
0	1	↑	
1	1	↑	

Extra Credit

16. The switch S_2 of Fig. 55-14 is replaced by a square-wave generator with a frequency of 1 kHz. Calculate the frequency of the Q output.
17. Replace switch S_2 of Fig. 55-14 by a square-wave generator with a frequency of 1 kHz. Adjust the signal level as needed to get a rectangular Q output (use oscilloscope). Measure the frequency of the Q output and record here:

$$f = \underline{\hspace{3cm}}$$

QUESTIONS

1. How does a flip-flop differ from an astable multivibrator?
2. With the data of Table 55-9, describe what the Q output did when you changed the R and S switches.
3. Using data of Table 55-10, describe what a D latch does.
4. Is the D flip-flop of Fig. 55-13 positive- or negative-edge-triggered?
5. With Table 55-11, describe what the D flip-flop of Fig. 55-13 did.
6. In Fig. 55-14 are the PRESET and CLEAR active-low or active-high?
7. Use the data of Table 55-12 to describe the action of a JK flip-flop.

Extra Credit

8. What was the frequency of the Q output you calculated and measured in steps 16 and 17 of the Procedure section? Explain why this change in frequency occurred.

Answers to Self-Test

1. race
2. NOR, NAND
3. clock
4. race
5. edge; positive
6. high; high

Appendix

PARTS REQUIREMENTS

Resistors, ½-W

Resistance	Quantity	Resistance	Quantity	Resistance	Quantity
33 Ω	1	1.8 kΩ	1	33 kΩ	2
47 Ω	1	2.2 kΩ	1	47 kΩ	2
68 Ω	1	2.7 kΩ	1	56 kΩ	1
100 Ω	2	3.3 kΩ	2	68 kΩ	1
150 Ω	1	4.7 kΩ	1	100 kΩ	4
180 Ω	1	5.1 kΩ	1	120 kΩ	1
220 Ω	2	5.6 kΩ	1	200 kΩ	2
330 Ω	1	6.8 kΩ	2	220 kΩ	1
390 Ω	1	8.2 kΩ	2	270 kΩ	2
470 Ω	1	10 kΩ	4	470 kΩ	1
560 Ω	1	12 kΩ	1	560 kΩ	1
680 Ω	1	15 kΩ	1	1 MΩ	3
820 Ω	2	18 kΩ	2	2.2 MΩ	3
1 kΩ	3	20 kΩ	1	3.9 MΩ	2
1.2 kΩ	2	22 kΩ	1	4.7 MΩ	2

Capacitors (50-V Rating or Better unless Otherwise Noted)

Capacitance	Quantity	Capacitance	Quantity	Capacitance	Quantity
15 pF	1	0.01 μF	3	1 μF (16 V)	2
47 pF	1	0.015 μF	1	10 μF (16 V)	1
250 pF	1	0.022 μF	2	25 μF	2
0.001 μF	3	0.047 μF	3	40 μF	1
0.0022 μF	1	0.1 μF	4	100 μF	4
0.0047 μF	1	0.47 μF	1		

Resistance	Wattage, W	Quantity
5 Ω	5	1
100 Ω	1	2
220 Ω	1	1
250 Ω	2	1
250 Ω	5	1
270 Ω	1	2
500 Ω	5	1
1 kΩ	1	1
5 kΩ	5	1
25 kΩ	5	1

Potentiometers (2-W)

One each unless otherwise specified.
500-Ω, 1-kΩ, 2.5-kΩ, 5-kΩ, two 10-kΩ, 50-kΩ, two 100-kΩ, two 500-kΩ, 2.5-MΩ

Cathode-Ray Tube

2BP1 together with tube-socket assembly

Diodes

One each unless otherwise specified. A † indicates diodes that are new to this (fifth) edition of *Basic Electronics*.
1N34A, †1N753 (zener), †1N914, †1N3020 (zener), 1N4154, †1N4746, four 1N5625

Transistors

One each unless other specified. A † indicates transistors that are new to this (fifth) edition of *Basic Electronics*.
†2N1596, 2N2160 (UJT), two 2N2102 with heat sinks, two 2N3440, †2N3904, †2N3906, 2N4036, 2N5484, two 2N6004, two 2N6005, 3N187 (MOSFET)

Integrated Circuits

A † indicates ICs that are new to this (fifth) edition of *Basic Electronics*.
Linear: 741C, †CA3020, †NE555, †NE565
Digital: †7402, †7404, †7408, 7427, †7432, †7474, 7476

Optoelectronic Devices

LEDs: TIL221, TIL222, TIL312

Transformers, Chokes, and Coils

A † indicates a transformer that is new to this (fifth) edition of *Basic Electronics*.
- Filament transformer: 120-V primary, 25-V center-tapped secondary @ 1 A (Triad F40X or equivalent)
- Audio output transformer: †Kelvin 149-18 or equivalent
- Audio input transformer: Push-pull, primary 10-kΩ, secondary 2-kΩ 100-mW; dc resistance 500-Ω primary, 50-Ω secondary, †Kelvin 155-08 or equivalent
- Audio output transformer: Push-pull, primary 500 CT, secondary 3.2, 100-mW; dc resistance 20-Ω primary, 0.3-Ω secondary, †Kelvin 175-45 or equivalent
- RF coil: 30-mH
- Oscillator coil: Hartley, broadcast band, Miller 2021 or equivalent

Miscellaneous

- Breadboarding device, with terminals and connectors
- Dry cells: one 1.5-V
- Switches: SPDT (three); SPST (two); DPDT (one)
- Fused line cord
- Pilot light 49 and socket (one)
- 25-W light bulb and socket (one)
- PM speaker: 3.2-Ω
- Defective components: For troubleshooting, Experiments 8, 16, 19
- Horseshoe magnet

NOTE: In one experiment five (5) SPST switches are required. SPDT switches may be used instead.